Drone and UAV Forensics

Hudan Studiawan • Kim-Kwang Raymond Choo

Drone and UAV Forensics

A Hands-On Approach

Hudan Studiawan
Department of Informatics
Institut Teknologi Sepuluh Nopember
East Java, Indonesia

Kim-Kwang Raymond Choo
Department of Information Systems
and Cyber Security
University of Texas at San Antonio
San Antonio, Texas, USA

ISBN 978-3-031-93510-7   ISBN 978-3-031-93511-4   (eBook)
https://doi.org/10.1007/978-3-031-93511-4

© The Editor(s) (if applicable) and The Author(s), under exclusive license to Springer Nature Switzerland AG 2026

This work is subject to copyright. All rights are solely and exclusively licensed by the Publisher, whether the whole or part of the material is concerned, specifically the rights of translation, reprinting, reuse of illustrations, recitation, broadcasting, reproduction on microfilms or in any other physical way, and transmission or information storage and retrieval, electronic adaptation, computer software, or by similar or dissimilar methodology now known or hereafter developed.

The use of general descriptive names, registered names, trademarks, service marks, etc. in this publication does not imply, even in the absence of a specific statement, that such names are exempt from the relevant protective laws and regulations and therefore free for general use.

The publisher, the authors and the editors are safe to assume that the advice and information in this book are believed to be true and accurate at the date of publication. Neither the publisher nor the authors or the editors give a warranty, expressed or implied, with respect to the material contained herein or for any errors or omissions that may have been made. The publisher remains neutral with regard to jurisdictional claims in published maps and institutional affiliations.

This Springer imprint is published by the registered company Springer Nature Switzerland AG
The registered company address is: Gewerbestrasse 11, 6330 Cham, Switzerland

If disposing of this product, please recycle the paper.

Foreword

Drones are becoming very common today, and we're using them for all sorts of things, from delivering packages to helping the military. But, like any technology, drones can be used for wrong reasons. That's why it's important to understand how to investigate when drones are involved in incidents. This is where drone and UAV forensics comes in.

Drone and UAV Forensics: A Hands-On Approach is a helpful guide that shows you how to examine drones and figure out what happened in different situations. Kim-Kwang Raymond Choo and Hudan Studiawan, who are experts in this field, wrote this book to give you the knowledge and skills you need. Whether you're just starting out or already have some experience, this book will teach you the basics and give you practical advice.

Inside, you'll learn about all the parts of a drone, how to get forensic images, how to analyze flight data, and even how to use computer programs to understand drone information. The book gives you step-by-step instructions on setting up a drone forensics lab, reading flight logs, and spotting problems.

This book is important for anyone working with digital evidence, law enforcement, cybersecurity, or even the drone industry itself. It will help you deal with the special challenges that come with drone technology and make sure drones are used safely and responsibly.

University of Amsterdam / Netherlands   Zeno Geradts
Forensic Institute
Amsterdam, Netherlands
February 14, 2025

Acknowledgement

We are extremely grateful to Springer and their staff for their support in this book, as well as our family and our loved ones for their constant support and encouragement.

In addition, we extend our gratitude to the digital forensics community for providing open-source tools, which have greatly supported this book.

Declarations

Competing Interests The authors have no competing interests to declare that are relevant to the content of this manuscript.

Contents

1	**Introduction to Drone and UAV Technology**	1
	1.1 Introduction	1
	1.2 Definition of UAV and Drones: Are They Different or the Same?.	3
	1.3 Components of UAVs	3
	1.4 Application of UAVs	5
	1.5 Drone Delivery of Goods	6
	1.6 UAV and Drone Technology for Military Applications	8
	1.7 Swarm Drones	10
	1.8 UAV and Drone Forensics	12
	1.9 Organization of This Book	13
	1.10 Summary	16
	1.11 Exercises	16
	References	17
2	**UAV Components and Ecosystem**	21
	2.1 Introduction	21
	2.2 UAV Components	22
	2.3 UAV Ecosystem	23
	2.4 Integration of Cloud Services	24
	2.5 UAVs and Drones Based on Types of Wings and Rotors	25
	2.6 Understanding Throttle, Yaw, Pitch, and Roll in Drones	26
	2.7 Drone Specification Fields	27
	2.8 Drone Features Explained	30
	2.9 Hands on Examples: DJI Mavic Air 2	32
	2.10 Hands on Examples: DJI FPV	38
	2.11 Summary	47
	2.12 Exercises	48
	References	48
3	**Survey on Drone and UAV Forensics**	51
	3.1 Introduction	51
	3.2 UAV Forensic Artifacts	53

		3.3	UAV Flight Logs Analysis	58

	3.3	UAV Flight Logs Analysis	58
	3.4	UAV Identification	59
	3.5	UAV Forensic Tools and Datasets	59
		3.5.1 Tools	60
		3.5.2 Datasets	61
	3.6	UAV Forensic Readiness	62
	3.7	Conceptual Drone Forensics Framework (CDFF)	62
	3.8	Challenges and Future Research Directions	65
		3.8.1 Forensic Investigation of UAV Ecosystems	65
		3.8.2 UAV Forensics Support by Artificial Intelligence and Machine Learning	67
		3.8.3 Quality of Data for UAV Forensics Investigations	67
	3.9	Summary	68
	3.10	Exercises	68
	References		69
4	**Setting Up a Drone Forensics Laboratory**		**73**
	4.1	Introduction	73
	4.2	Preparing the Forensic Workstation	74
	4.3	Forensic Workstation Used in this Book	76
	4.4	Installation of Anaconda	77
	4.5	Installation of FTK Imager	82
	4.6	Installation of Autopsy	86
	4.7	Downloading Datasets for Experiments	90
		4.7.1 VTO Labs dataset	91
		4.7.2 ALFA dataset	92
	4.8	Summary	92
	4.9	Exercises	93
	References		93
5	**Data Acquisition from Drone and UAV**		**95**
	5.1	Introduction	95
	5.2	Acquisition of iPhone Drone Controller	96
		5.2.1 Using macOS Features	96
		5.2.2 Using Open-source Tools	99
		5.2.3 Verifying iOS Controller Acquisition	103
	5.3	Acquisition of Android Drone Controller	113
	5.4	Acquisition of Drone SD Card using FTK Imager	119
	5.5	Acquisition of Drone SD Card with Command Line	127
	5.6	Summary	135
	5.7	Exercises	136
	References		136
6	**Understanding Drone and UAV Forensic Images and Artifacts**		**137**
	6.1	Introduction	137
	6.2	Related Work	138

	6.3	Type of Drone and UAV Forensic Images	140
	6.4	Autopsy	141
	6.5	Creating a New Case and Importing a Forensic Image to Autopsy	142
		6.5.1 Dataset Preparation	142
		6.5.2 Case Analysis in Autopsy	143
	6.6	Autopsy DJI Drone Analyzer	149
	6.7	Understanding Drone and UAV Artifacts	152
		6.7.1 Flight Logs	152
		6.7.2 Media: Pictures and Videos	153
		6.7.3 Controller Artifacts	154
	6.8	Summary	163
	6.9	Exercises	163
	References	164
7	**Forensic Analysis of Drone and UAV Flight Data**	165
	7.1	Introduction	165
	7.2	Related Work	166
	7.3	Format of Drone Flight Logs	168
	7.4	DatCon: A Tool for Drone Flight Log Analysis	172
	7.5	CsvView: Viewer of DatCon Results	177
	7.6	DROP: DRone Open source Parser	181
	7.7	Visualization of Flight Logs with Maraudrone's Map	186
	7.8	DRDP: An Alternative for DatCon and DROP	191
	7.9	Digital Drone Forensics Software	192
		7.9.1 Environment Settings for Digital Drone Forensics Software	193
		7.9.2 Dataset Requirements for DJI Phantom 4	198
		7.9.3 Running Digital Drone Forensics Software	201
	7.10	Summary	210
	7.11	Exercises	210
	References	210
8	**Forensic Investigation of UAV Faults and Anomalies**	213
	8.1	Introduction	213
	8.2	Related Work	214
	8.3	Obtaining Faults and Anomalies Dataset	216
	8.4	Classification of Faults and Anomalies	219
	8.5	Run the model to unseen data	227
	8.6	Summary	228
	8.7	Exercises	228
	References	229
9	**Forensic Analysis of Drone and UAV Telemetry Logs**	231
	9.1	Introduction	231
	9.2	Related Work	232

	9.3	ArduPilot and Telemetry Data Format	233
	9.4	Installation and Usage of GRYPHON	235
	9.5	Telemetry Parsing Results	237
	9.6	Other Telemetry Logs	240
	9.7	Summary	242
	9.8	Exercises	242
	References		243
10	**Forensic Timeline Analysis of Drones and UAVs**		**245**
	10.1	Introduction	245
	10.2	Related Work	246
	10.3	Creating Forensic Timeline with log2timeline Plaso	248
		10.3.1 Installation of Docker Desktop	249
		10.3.2 Installation of Plaso	252
	10.4	Drone Dataset for Forensic Timeline Analysis	253
	10.5	Building a Timeline from a iOS Drone Controller	254
	10.6	Building a Timeline from a Android Drone Controller	257
	10.7	Timeline Analysis	258
	10.8	Summary	260
	10.9	Exercises	260
	References		261
11	**Bringing Natural Language Processing to Drone and UAV Forensics**		**263**
	11.1	Introduction	263
	11.2	Related Work	264
	11.3	Drone Flight Log Messages	267
	11.4	Named Entity Recognition for Drone Log Forensics	268
		11.4.1 DroNER Dataset	269
		11.4.2 Running NER on Drone Logs	269
	11.5	DFLER: Drone Flight Log Entity Recognizer	275
	11.6	Sentiment Analysis for Drone Log Forensics	278
		11.6.1 Dataset for Sentiment Analysis for Drone Logs	279
		11.6.2 Running Sentiment Analysis on Drone Logs	280
	11.7	Summary	282
	11.8	Exercises	283
	References		283

About the Authors

Hudan Studiawan received his Bachelor's and Master's degrees from the Institut Teknologi Sepuluh Nopember, Indonesia, in 2009 and 2011, respectively, and Ph.D. degree from Murdoch University, Australia, in 2021. He is currently an Associate Professor at the Department of Informatics, Institut Teknologi Sepuluh Nopember, Indonesia. He is a member of the Advisory Board of *Forensic Science International: Digital Investigation* journal since 2024. His current research interests include digital forensics and natural language processing.

Kim-Kwang Raymond Choo received his Ph.D. in Information Security in 2006 from Queensland University of Technology, Australia. He currently holds the Cloud Technology Endowed Professorship at The University of Texas at San Antonio. He is the founding co-Editor-in-Chief of *ACM Distributed Ledger Technologies: Research and Practice*, and the founding Chair of IEEE TEMS Technical Committee on Blockchain and Distributed Ledger Technologies.

Chapter 1
Introduction to Drone and UAV Technology

Abstract Unmanned Aerial Vehicles (UAVs), commonly referred to as drones, have evolved from niche military tools to versatile instruments with a wide range of applications. This chapter provides an introduction to UAV technology by exploring its historical development, key components, and the diverse spectrum of industries and sectors in which UAVs are utilized today. From early military experiments in the twentieth century to modern-day innovations driven by miniaturization and control systems, UAVs have undergone continuous transformation. They revolutionize areas such as surveillance, agriculture, and logistics.

1.1 Introduction

The concept of UAVs dates back to the early twentieth century, with initial developments primarily focused on military applications. The earliest recorded use of a UAV was in 1916, when the United States Navy experimented with the "Aerial Target", an early drone prototype. The progress was slow, but by the mid-twentieth century, UAVs began to see more practical applications, especially during the Vietnam War, where they were used for reconnaissance.

The concept of UAVs can be traced back to the early twentieth century, with developments beginning during World War I. The Kettering Bug, an early example of a UAV, was a rudimentary flying bomb developed by the U.S. Army Signal Corps in 1918. Despite its limited use, it laid the groundwork for future UAV innovations [1]. During World War II, both the Allied and Axis powers experimented with UAV technology. The development of the German V-1 flying bomb marked another advancement, showing the potential of UAVs in warfare. In the same era, the U.S. developed the TDR-1, an assault drone used in the Pacific theater [2].

The Cold War period saw an accelerated development of UAVs, driven by the need for intelligence and reconnaissance during the standoff between the United States and the Soviet Union. The Ryan Firebee, developed in the 1950s, became one of the first successful jet-powered drones used extensively for reconnaissance and target practice [3]. In the 1970s and 1980s, advancements in electronics and control systems led to the development of more sophisticated UAVs. The Israeli-made Scout and Pioneer UAVs were notable for their use in real-time battlefield surveillance. These two vehicles influenced the design and deployment of future UAVs [4].

The 1990s marked a turning point for UAV technology with the development of the RQ-1 Predator by General Atomics. Initially used for reconnaissance, the Predator was later armed, becoming a critical tool in the U.S. military arsenal for targeted strikes. This period also saw the integration of GPS technology and advances in satellite communications, enhancing the operational capabilities of UAVs [5]. The twenty-first century has witnessed the expansion of UAV applications beyond military use. Advances in miniaturization, battery technology, and sensor integration have supported the development of smaller, more affordable UAVs suitable for a wide range of civilian applications. These include aerial photography, agriculture, infrastructure inspection, and environmental monitoring [6]. These innovations opened the door to civilian and commercial applications and they revolutionize various industries. The UAV history timeline is shown in Fig. 1.1.

The evolution of UAV technology has been influenced by the intersection between military needs, technological innovations, and economic considerations. The collaboration between military and private sectors has driven innovation and the commercialization of UAV technology. This synergy has resulted in the use of UAVs in various sectors, including healthcare, logistics, and public safety [7].

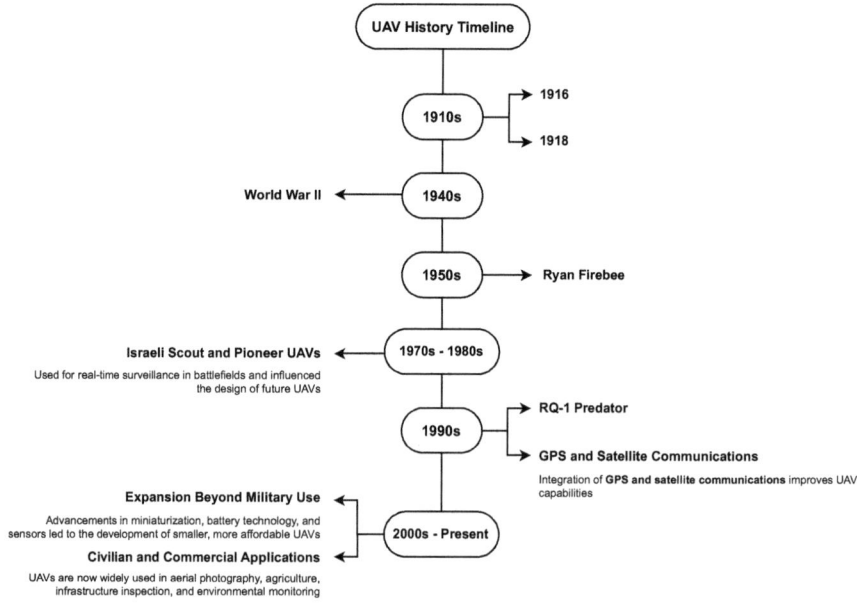

Fig. 1.1 UAV history timeline

1.2 Definition of UAV and Drones: Are They Different or the Same?

Unmanned Aerial Vehicles (UAVs) and drones are terms that are often used interchangeably in both technical and common language. However, there are differences in their definitions and connotations that are worth noting. A UAV is an aircraft that operates without a human pilot onboard. It can be controlled remotely by a human operator or autonomously by onboard computers. UAVs are part of a broader system known as Unmanned Aircraft Systems (UAS), which includes the UAV itself, a ground-based controller, and a communication system connecting the two [8].

The term "drone" is commonly used to describe UAVs but can also refer to any unmanned system, including those used in terrestrial or maritime environments. The term has become popularized due to its simplicity and broader use in public and media contexts [9]. While both UAVs and drones refer to unmanned aerial technology, their usage and connotations can differ based on context.

UAV is often used in technical, regulatory, and military contexts, focusing on the capability and operational framework of the aircraft, including its integration with ground control systems and its potential for autonomous operation [10]. On the other hand, the term "drone" is more commonly used in public and media contexts. It serves as a catch-all term that covers a wide range of unmanned vehicles, including those used for recreational, commercial, and hobbyist purposes [11].

UAV carries a more formal and technical connotation, often associated with precision, military applications, and sophisticated technology. In contrast, the term "drone" has a more informal and broad connotation. It is frequently perceived as a gadget or toy in the consumer market, yet it is also increasingly recognized for its utility in commercial and industrial applications [12].

UAVs are widely used in military operations, where they perform tasks such as surveillance, reconnaissance, target acquisition, and precision strikes. In addition, they are employed in scientific research, environmental monitoring, and disaster response efforts [10]. On the other hand, drones are commonly used in commercial applications such as aerial photography, agriculture, infrastructure inspection, and delivery services. They are also popular among hobbyists and recreational users [9].

1.3 Components of UAVs

UAVs are complex systems composed of several key components that work together to ensure functionality, stability, and performance. Understanding these components is important for designing, operating, and improving UAVs for various applications. Here, we dive into the primary components of UAVs and their specific roles as illustrated in Fig. 1.2.

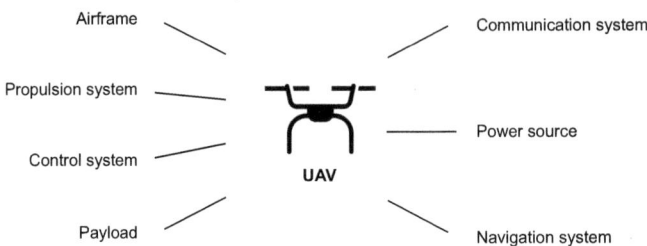

Fig. 1.2 An illustration of UAV components

Airframe The airframe is the structural framework of a UAV, designed to support all other components. It must be lightweight yet strong enough to withstand the stresses of flight. Common materials used for airframes include carbon fiber, aluminum, and composite materials. UAV airframes are designed considering aerodynamic efficiency and durability. For instance, carbon fiber composites are often used for their high strength-to-weight ratio and resistance to environmental factors [13].

Propulsion System The propulsion system includes motors, propellers, and batteries, providing the necessary thrust for the UAV to fly. The type and configuration of the propulsion system depend on the UAV's design and intended use. Effective propulsion requires precise motor and propeller design. Research shows that optimizing the design of these components can improve UAV performance and reliability [13].

Control System The control system is the brain of the UAV, comprising flight controllers, GPS modules, and various sensors. It ensures the UAV maintains stability, navigates accurately, and responds to commands from the operator or autonomous control algorithms. Open-source flight controllers such as those surveyed by [14] are important for autonomous control and navigation. They integrate hardware and software to manage the UAV's flight dynamics.

Payload The payload refers to the equipment carried by the UAV for specific tasks, such as cameras, sensors, or other instruments. The payload varies based on the UAV's application, from aerial photography to environmental monitoring. Modern UAVs integrate various sensors, including cameras, LiDAR (light detection and ranging), and infrared sensors, to collect data and perform specific missions [15].

Communication System The communication system enables data transmission and reception between the UAV and the ground control station (GCS). It includes radio links, satellite communications, and increasingly, cellular networks. In addition, ensuring secure communication is essential for UAV operations. Frameworks such as the one proposed by [16] improve the security, reliability of UAV communications, and address potential vulnerabilities.

Power Source UAVs typically use batteries, but alternative power sources like fuel cells or solar panels can also be employed, depending on the UAV's design and mission requirements. Advances in battery technology are crucial for extending the flight time and operational range of UAVs. Research into more efficient power management systems continues to evolve [17].

Navigation System The navigation system includes GPS, inertial measurement units (IMUs), and other navigational aids that help the UAV determine its position and orientation in space. Using lightweight consumer-grade sensors, UAVs can achieve precise georeferencing of data. For instance, it is important for applications such as mapping and surveying [18].

1.4 Application of UAVs

The diversity of UAVs has led to their adoption across various industries. It transforms traditional methods and enables new capabilities. This section explores the broad spectrum of UAV applications, supported by research findings.

Aerial Photography and Videography UAVs equipped with high-resolution cameras provide unique perspectives for filmmaking, journalism, and real estate marketing. Their ability to capture stunning aerial footage has revolutionized visual media. UAVs are widely used in the film industry for capturing dynamic aerial shots, which were previously difficult or expensive to achieve [19].

Agriculture Drones assist in precision agriculture by monitoring crop health, assessing irrigation needs, and applying pesticides or fertilizers with pinpoint accuracy. They provide farmers with vital data to make informed decisions. UAVs enable precise monitoring of crop conditions and lead to better resource management and higher yields [20].

Environmental Monitoring UAVs facilitate the monitoring of wildlife, forests, and water bodies. They aid in conservation efforts and environmental research. They are important tools for collecting data in remote or difficult-to-access areas. UAVs are used to inventory resources, map diseases, and monitor the effects of fires in forests [21].

Infrastructure Inspection Drones are used to inspect bridges, power lines, and pipelines. They offer a safer and more efficient alternative to traditional inspection methods. They provide detailed imagery and data, helping in maintenance and risk assessment. UAVs reduce the risks and costs associated with inspecting infrastructure such as bridges and power lines [20].

Disaster Response In emergency situations, UAVs provide real-time data and imagery to aid in search and rescue operations, damage assessment, and resource allocation. They are critical in areas where access is restricted due to disasters. UAVs can be deployed quickly in disaster-hit areas to locate survivors and assess damage [22].

Delivery Services Companies are exploring the use of drones for delivering goods, especially in remote or congested areas as it promises faster and more efficient delivery solutions. UAVs are being tested for delivering medical supplies, groceries, and other goods. Therefore, they accelerate delivery speed and reducing human intervention [23].

Law Enforcement and Security UAVs support police and security forces by providing aerial surveillance, monitoring large crowds, and assisting in crime scene investigations. They offer a discreet and efficient way to gather information and enhance public safety. UAVs are utilized for monitoring public events, tracking suspects, and providing situational awareness to law enforcement agencies [24].

Telecommunications UAVs can act as aerial base stations to enhance coverage, capacity, and reliability of wireless networks, particularly in remote areas or during large events. UAVs provide temporary network coverage and boost signal strength in areas with poor connectivity or high demand [25].

Scientific Research UAVs are invaluable tools for scientific research as it allows for the collection of high-resolution data over large areas. They are used in meteorology, geology, and other fields to gather critical information. UAVs enable detailed mapping and data collection for various scientific studies and improve the accuracy and scope of research projects [26].

1.5 Drone Delivery of Goods

Drone delivery is an emerging technology that employs unmanned aerial vehicles (UAVs) to transport goods efficiently and swiftly. This section explores the concept, applications, benefits, and challenges of drone delivery, supported by recent research and developments in the field. An illustration of drone delivery systems is shown in Fig. 1.3 [27]. Figure 1.3 outlines the operational flow of a drone delivery system using truck-based distribution with drone-enabled last-mile delivery. Initially, a truck transports packages to a local cluster center in the neighborhood, which serves as a hub for further distribution. From there, drones pick up individual packages and deliver them directly to customers, ensuring swift and efficient service. After completing their delivery, the drones return to the cluster center to collect additional packages or head back to their base as required. This hybrid system optimizes logistics by minimizing the distance traveled by drones. It improves delivery speed, reduces transportation costs, and contributes to a more sustainable delivery process.

Drone delivery involves the use of UAVs to transport packages from a central hub or warehouse to the consumer's location. This method offers a rapid and flexible solution for delivering a variety of goods, including medical supplies, food, and retail products. The integration of GPS and advanced navigation systems ensures

1.5 Drone Delivery of Goods

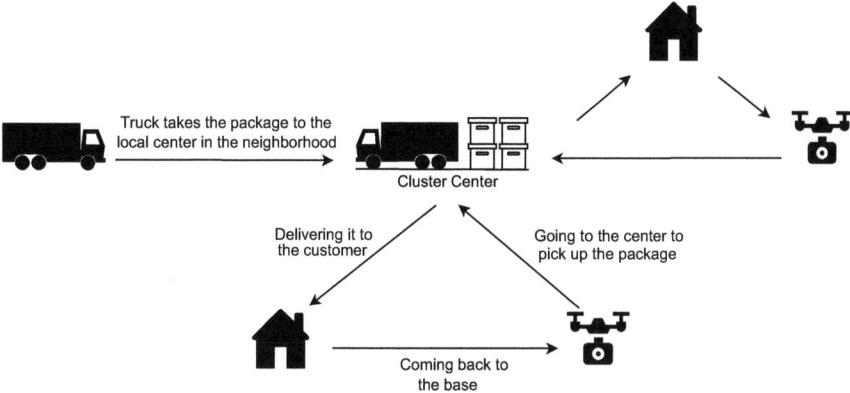

Fig. 1.3 An illustration of drone delivery systems [27]

precise delivery, enhancing customer satisfaction. A typical drone delivery system includes components like the drone itself, equipped with GPS, sensors for obstacle detection, and a payload mechanism for carrying packages. The system also features software for route planning and real-time tracking [28].

Applications of Drone Delivery Drones are used to deliver critical medical supplies, such as blood products, vaccines, and pharmaceuticals, especially to remote or inaccessible areas. This capability is crucial during emergencies when timely delivery can save lives [29]. The use of drones in food delivery is on the rise, with companies experimenting with delivering meals and beverages. Retail giants like Amazon and Walmart are exploring drone delivery for a variety of goods to enhance delivery speed and efficiency [30].

E-commerce companies are leveraging drones for last-mile delivery, which is the most expensive and time-consuming part of the delivery process. Drones help reduce delivery times and lower operational costs, making them a viable option for urban logistics [31]. Drones are increasingly used for same-day delivery services, providing customers with rapid delivery options. This is particularly effective in urban areas where traffic congestion can delay traditional delivery methods [30].

Drones can bypass ground traffic and follow direct flight paths. Therefore, it can significantly reduce delivery times. This is particularly beneficial for urgent deliveries, such as medical supplies [28]. Drones have the potential to reduce carbon emissions compared to traditional delivery vehicles, especially when powered by renewable energy sources. This contributes to more sustainable logistics practices [32]. By automating deliveries, companies can reduce labor costs and improve operational efficiency. Drones can perform multiple deliveries per day without the need for breaks, enhancing productivity [33].

Challenges of Drone Delivery One of the major hurdles in the widespread adoption of drone delivery is the regulation. Safety concerns related to UAVs flying in urban areas and the potential for accidents need to be addressed. Countries are gradually developing frameworks to regulate drone operations [34].

Drones face limitations in terms of flight endurance, payload capacity, and weather conditions. Ensuring reliable performance in various environmental conditions is crucial for the success of drone delivery systems [31]. Implementing drone delivery requires a big investment in infrastructure, such as droneports and charging stations. Efficiently managing these resources is essential for maintaining the efficiency and reliability of the delivery network [35].

1.6 UAV and Drone Technology for Military Applications

UAVs have transformed modern military operations. Their ability to perform a variety of tasks with precision, endurance, and reduced risk to human life makes them invaluable assets in military strategy and tactics. This section explores the various military applications of UAVs, their advantages, and the challenges associated with their use. Figure 1.4 illustrates of the application of this technology to military [36]. Figure 1.4 represents the operational framework of a military drone system, its key components, and processes. It begins with the Command and Control Center, which oversees mission planning, drone deployment, and real-time communication. From this center, drones are launched to execute various missions such as reconnaissance, surveillance, and tactical operations. These drones rely on

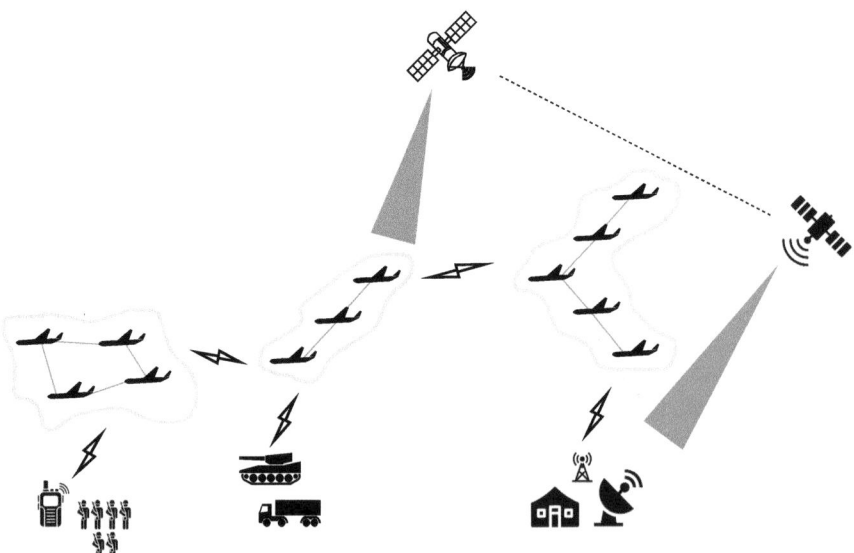

Fig. 1.4 An illustration of UAVs and drones for military [36]

1.6 UAV and Drone Technology for Military Applications

intelligence data inputs and utilize advanced technologies like GPS for navigation and secure communication links for coordination. After completing their assigned tasks, drones transmit collected data back to the command center for analysis. They may then return to base for maintenance or redeployment. This system shows strategic efficiency, enabling precision operations, real-time decision-making, and situational awareness in military contexts.

One of the primary uses of UAVs in the military is for intelligence, surveillance, and reconnaissance (ISR) missions. UAVs equipped with high-resolution cameras, sensors, and communication equipment can gather critical data and provide real-time situational awareness. UAVs are deployed to monitor enemy movements, gather intelligence, and conduct battlefield surveillance. They can cover large areas and provide continuous observation, which is crucial for strategic planning [10].

UAVs are increasingly used for target acquisition and precision strikes. Armed drones like the MQ-9 Reaper can carry out targeted attacks on high-value targets with minimal collateral damage. UAVs can carry missiles and bombs to engage targets precisely. They are used to eliminate enemy combatants, destroy infrastructure, and disrupt terrorist networks while minimizing the risk to ground troops [37]. UAVs are employed in electronic warfare to disrupt enemy communications, jam radar systems, and gather electronic intelligence. This capability is important for gaining an advantage in the electromagnetic spectrum. UAVs can carry electronic warfare payloads to interfere with enemy radar and communications and provide strategic advantages on the battlefield [38].

UAVs are also used in search and rescue operations, especially in hostile or inaccessible environments. Their ability to operate in various terrains and conditions makes them well-suited for locating and rescuing personnel. UAVs can be deployed quickly to search for missing personnel, assess the situation, and deliver supplies or communication equipment to isolated individuals [39]. UAVs are used to deliver supplies to front-line troops, providing logistic support in areas that are difficult to reach by traditional means. This capability ensures that soldiers have the necessary resources without the risks associated with ground transportation. UAVs can transport ammunition, medical supplies, food, and other essential items to remote or heavily contested areas to maintain continuous support to combat operations [40].

Advantages of Military UAVs UAVs can perform dangerous missions without putting human pilots at risk, thus saving lives and preserving valuable personnel. Compared to manned aircraft, UAVs are generally less expensive to produce, operate, and maintain. Their expendability makes them suitable for high-risk missions [41]. UAVs can loiter over targets for extended periods and provide persistent surveillance and reconnaissance capabilities that are not feasible with manned aircraft. Many UAVs are designed to be stealthy so it makes them harder to detect and allowing them to operate in enemy airspace with a lower risk of interception [42].

Challenges and Limitations UAVs are limited by battery life, payload capacity, and weather conditions. Ensuring reliable performance in various environments remains a big challenge [43]. The use of UAVs in military operations raises legal and ethical questions, particularly concerning the sovereignty of other nations and the potential for civilian casualties during drone strikes [44]. As UAV technology advances, so do counter-UAV measures. Effective countermeasures, such as jamming and shooting down drones, are being developed to mitigate the threat posed by UAVs [38]. UAVs can be vulnerable to hacking and other forms of cyber attacks, which could compromise missions and pose security risks to military operations [45].

1.7 Swarm Drones

Swarm drones represent an advanced and promising field within UAV technology. A swarm of drones involves multiple UAVs operating together in a coordinated manner to perform complex tasks that would be challenging or impossible for a single drone. This section explores the concept, applications, and challenges associated with swarm drones. Swarm drones operate based on principles inspired by natural swarms, such as those observed in bees, ants, and birds. These principles include decentralized control, local interactions, and collective behavior. They enable a group of drones to work together efficiently without a single point of failure.

The illustration of swarm drones is given in Fig. 1.5 [46]. Figure 1.5 illustrates the operational framework of a drone swarm system, its phases, and components. At the core of the system is the Mission Control Center, which defines the objectives and oversees the mission. The operation is divided into three key phases: Pre-Flight Phase, Flight Phase, and Post-Flight Phase. In the pre-flight phase, the Ground Flight Management System is utilized for planning, coordination, and setting mission parameters for the drone swarm. During the flight phase, the swarm of drones operates collaboratively, executing tasks such as surveillance, reconnaissance, or other mission-specific objectives. Each drone in the swarm communicates with others and synchronizes movements, and shared data collection. Finally, in the post-flight phase, mission logs and performance data are analyzed to assess the success

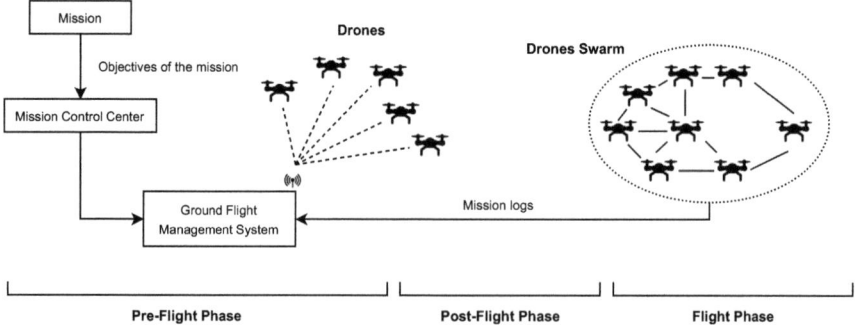

Fig. 1.5 An illustration of swarm drones [46]

1.7 Swarm Drones

of the operation. This structure shows coordinated teamwork among drones and efficient execution of complex missions with adaptability and precision.

Each drone in the swarm operates independently while following simple rules based on local information. This decentralized approach allows the swarm to adapt to changes in the environment and continue functioning even if some drones fail [47]. Drones communicate and coordinate with their immediate neighbors, facilitating real-time adjustments and collaboration without the need for a central controller [48].

Applications of Swarm Drones Swarm drones have numerous applications across various fields due to their ability to perform coordinated and distributed tasks efficiently. Swarm drones can quickly cover large areas to locate survivors in disaster-stricken zones. Their ability to navigate challenging terrains and provide real-time data improves the effectiveness of search and rescue missions [49].

Swarm drones are used for environmental monitoring tasks such as forest fire detection, wildlife tracking, and pollution assessment. Their collective sensing capabilities allow for comprehensive data collection over extensive areas [50]. In agriculture, swarm drones support precision farming by monitoring crop health, optimizing irrigation, and applying fertilizers or pesticides efficiently. This leads to improved crop yields and resource management [51].

Swarm drones provide improved surveillance capabilities for security applications. They can monitor large areas, track suspects, and provide real-time situational awareness to law enforcement agencies [52]. Swarm drones are used for inspecting infrastructure such as bridges, pipelines, and power lines. Their coordinated approach allows for thorough and efficient inspections. Thus, it reduces the need for manual labor and increasing safety [53].

Challenges in Swarm Drones Despite their potential, several challenges need to be addressed to realize the full capabilities of swarm drones. Ensuring reliable communication and coordination among drones in a swarm is challenging due to the dynamic and often harsh environments in which they operate. Advances in wireless communication technologies are needed to overcome these challenges. Preventing collisions among drones and with obstacles in the environment is essential for the safe operation of swarm drones. Effective algorithms and control systems are needed to manage collision avoidance in real-time [54].

It is essential to develop autonomous systems capable of decision-making using both local and global information. This includes navigating complex environments, adapting to changes, and executing tasks without human intervention [55]. Swarm drones need to operate efficiently to maximize their operational time and effectiveness. Energy management and optimization are essential to guarantee that drones successfully accomplish their missions without depleting their power reserves.

1.8 UAV and Drone Forensics

As the use of UAVs (Unmanned Aerial Vehicles) continues to grow, so does the need for forensic investigations into their misuse. UAV and drone forensics is an emerging field that focuses on recovering and analyzing data from drones to aid in criminal investigations and security assessments. Chapter 3 discusses the scope, challenges, and methodologies of drone forensics.

Drone forensics involves the systematic process of examining and analyzing the digital evidence from UAVs to uncover information about their activities, ownership, and potential misuse. Identifying the owner or operator of a drone, analyzing flight data to understand its routes and activities, extracting photos, videos, and other media captured by the drone, and examining potential cybersecurity breaches involving drones are all critical aspects of drone forensic analysis.

Drone forensics presents several unique challenges. Each drone model and manufacturer employs different hardware and software, complicating the standardization of forensic tools and methods [56]. Additionally, some manufacturers may incorporate anti-forensic features into their drones to protect user privacy, which further complicates forensic analysis [57]. Encrypted data storage on both drones and their controllers can also hinder investigations. Moreover, handling sensitive data and ensuring compliance with privacy laws and regulations add another layer of complexity to the forensic process.

Methodologies in Drone Forensics The first step in drone forensics is data acquisition from the drone and its associated devices. This process involves retrieving logs from the drone's onboard systems and ground control stations to reconstruct flight paths [58], extracting images and videos from the drone's internal storage or attached SD cards [59], and analyzing mobile apps used to control the drone for additional data and user information.

Forensic analysis involves examining the acquired data to uncover relevant information. This process includes using flight logs to map the drone's movements and activities [60], reviewing photos and videos for timestamps, geolocation data, and content that can provide context to the drone's operations, and analyzing metadata from files to gain insights into the drone's usage and ownership [61].

The final step is to compile the findings into a coherent report that can be used in legal proceedings or further investigations. This involves detailing the methods and tools used, the data acquired, and the results of the analysis. Additionally, maps and 3D models can be employed to visually represent flight paths and locations, providing a clearer understanding of the drone's movements [60]. Chapter 3 will discuss methodologies and framework for UAV and drone forensics in details.

Case Study Papers in Drone Forensics Research A forensic investigation on the DJI Phantom 4 revealed amounts of data stored on the drone and its associated mobile device. This included flight logs, media files, and metadata, to establish the drone's activities and ownership [59]. In addition, research on the Parrot AR Drone 2.0 demonstrated the ability to extract flight data and media files, despite the challenges posed by the drone's security features. This case highlighted the relevance of using specialized forensic tools to bypass anti-forensic measures [58].

Thornton and Zadeh [62] investigate the field of drone forensics and address the growing use of unmanned aerial systems (UAS). To set the stage, the paper explores the foundational need for standardized forensic methodologies in the context of drones. It highlights the need for standardized methodologies to guarantee the integrity of evidence collected from drones. The authors review existing frameworks and propose a new approach that comprises various stages of forensic analysis, from preparation to evidence presentation. The case study was demonstrated to DJI Phantom 4, DJI Mavic Pro, DJI Mavic Mini, and Yuneec Mantis Q.

Building on the need for flight path reconstruction, the paper by [63] focuses on the analysis of GPS data from various drone models, namely DJI Phantom 4 Pro, Parrot Bebop 2, and Yuneec Typhoon H. The research discusses about how to extract and interpret flight logs and they found out that we need tools to convert complex flight data into understandable formats. By examining the data from those three distinct drone brands, the authors illustrates the differences in data storage and log creation philosophies among manufacturers.

As another case study, Renduchintala et al. [60] presents a comprehensive forensic framework for investigating micro unmanned aerial vehicles (UAVs). To address the broader forensic needs, the study outlines a holistic framework that integrates hardware and digital forensic approaches. The framework includes both hardware/physical and digital forensic methods to analyze drones. The authors propose a model for examining drone components at crime scenes and a robust digital forensic application for analyzing flight logs. The application, developed using JavaFX, allows users to extract and visualize flight data and provides a 3D representation of the drone's flight path. The research provides case study in Yuneec Typhon H and DJI Phantom 4.

1.9 Organization of This Book

This book is structured to provide a comprehensive exploration of UAV technology and its applications in forensic investigations. Organized into distinct chapters, the book progresses from foundational concepts to forensic methodologies. Each chapter is designed to be read individually, offering readers a systematic understanding of UAV technology, its components, and the processes involved in investigating UAV-related incidents. By combining theoretical insights with practical tutorials, the book equips practitioners, researchers, students, and enthusiasts with the tools and knowledge necessary to address the challenges of modern drone forensics. Below is an overview of the topics covered in each chapter and a roadmap for navigating the book.

Chapter 1 introduces UAV technology by examining its historical evolution, essential components, and the diverse range of fields where they are now indispensable. From their origins in early twentieth-century military experiments to the cutting-edge innovations enabled by miniaturization and control systems, UAVs have undergone remarkable development. The chapter discusses that they are revolutionizing domains such as surveillance, agriculture, and logistics.

Chapter 2 explores the essential components that form the foundation of Unmanned Aerial Vehicles (UAVs) and the expansive ecosystem that underpins their development and usage. It begins by examining the core elements of UAVs, including airframe structures, propulsion systems, sensors, cameras, navigation and control systems, communication technologies, and power supplies. The discussion then broadens to the surrounding ecosystem, the contributions of manufacturers, regulatory frameworks, and the diverse applications of UAVs across various industries. Furthermore, the chapter explores into advanced topics such as the integration of cloud services for UAV management, data processing, mission planning, and the Internet of Drones (IoD).

Chapter 3 presents a systematic review of the current literature on UAV forensics. It identifies and documents emerging research trends in the digital forensic analysis of UAV-related incidents, accidents, and crimes. It begins by introducing a taxonomy that categorizes prior studies on UAV forensic artifacts, frameworks, models, forensic readiness, and investigative tools. The chapter concludes by highlighting three critical research themes that remain underexplored and provides a roadmap for future advancements in UAV forensic investigations. The content of this chapter is adapted from [64]. This chapter also provides an overview of the Conceptual Drone Forensic Framework (CDFF). It is developed to address the limitations in existing UAV forensic models. The CDFF is structured to improve the forensic investigation process by offering clearly defined phases and guiding investigators from forensic readiness to evidence disposal. These phases include preparation, preservation, and acquisition; examination and analysis; presentation and post-investigation; and the secure disposal of evidence. Furthermore, this chapter discusses how the CDFF addresses the shortcomings of earlier models and highlights key future research directions, including event correlation and reconstruction, tool development, and standardization. By applying this framework, practitioners can conduct more effective and reliable forensic investigations.

Chapter 4 serves as a guide for establishing a forensic laboratory dedicated to drone investigations. It provides detailed insights into the essential physical infrastructure, equipment, and software required for effective data examination. The chapter outlines the complete workflow of a forensic investigation, from the installation of necessary software to the analysis and reporting stages. By addressing both technical and procedural aspects, this chapter equips readers with the foundational knowledge needed to create a drone forensics laboratory capable of meeting the demands of modern forensic science. By the end of the chapter, readers will have a clear understanding of how to set up and operate a laboratory designed to effectively analyze and resolve cases involving drones and UAVs.

Chapter 5 explores the critical processes involved in collecting and securing data from UAVs and drones for forensic analysis. It reviews methods specifically tailored for macOS and Android operating systems, such as Finder, Android Debug Bridge (ADB), and forensic write blockers. Detailed step-by-step guidance is provided for

1.9 Organization of This Book

acquiring forensic images from iPhones, Android devices, and drone SD cards, using both open-source and native tools to ensure data integrity. The chapter covers both logical and physical imaging techniques. Forensic practitioners will have the practical skills needed to extract and analyze data from drones while maintaining the integrity of digital evidence throughout the process.

Chapter 6 serves as a guide to analyzing forensic images and artifacts using the Autopsy forensic tool. It begins by introducing the concept of forensic images in the context of drones and UAVs. It explains their importance as digital replicas of physical storage devices in forensic investigations. The chapter then provides a detailed overview of Autopsy, a leading open-source digital forensics platform, including step-by-step instructions for installation and configuration to prepare readers for practical analysis.

Chapter 7 describes the process of examining flight data from drones and UAVs to support forensic investigations. It provides a detailed overview of open-source forensic tools tailored for analyzing drone flight data, such as DatCon, CsvView, and DROP (DRone Open-source Parser). DatCon is explored for its ability to convert proprietary drone data logs into accessible formats, enabling in-depth analysis. CsvView is highlighted for its intuitive interface, which allows investigators to visually explore flight data captured in CSV files. The chapter also introduces DROP, a powerful open-source tool for parsing drone flight logs. Through hands-on examples and case studies, readers are guided in the practical use of these tools to uncover insights into drone flight paths, behaviors, and potential malfunctions.

Chapter 8 focuses on diagnosing, analyzing malfunctions, security breaches within UAV systems, and methodologies for fault and anomaly detection. It presents practical approaches, including case studies that utilize public datasets and hands-on Python scripting, to illustrate the application of these techniques. The chapter concludes by highlighting how these methods can assist the forensic analysis of faults and anomalies in drones and UAVs and provide tools for investigators and researchers in the field.

Chapter 9 focuses on the forensic analysis of telemetry logs from drones or UAVs. As drones become increasingly used by civilian, commercial, and potentially malicious domains, the demand for robust forensic methodologies has grown significantly. The chapter outlines techniques for analyzing flight data logs, especially on logs generated by drones operating on the ArduPilot platform. Tools such as GRYPHON, which supports the extraction and analysis of telemetry and Dataflash logs, are explored in detail. These tools enable forensic investigators to reconstruct flight paths, validate command executions, and detect anomalies such as unexpected altitude variations or hardware malfunctions.

Chapter 10 presents a systematic approach to conducting forensic timeline investigations using drone image data. It looks into the process of extracting and constructing a forensic timeline from collected drone images and how to utilize the log2timeline Plaso tool. Through a detailed, hands-on tutorial, the chapter guides readers in applying Plaso to analyze drone image data effectively. By the end of the chapter, readers will learn how to examine the constructed timeline to gain deeper insights into the security incidents under investigation.

Chapter 11 explores the integration of natural language processing (NLP) into drone forensics to support the analysis of human-readable log files, which serve as a valuable source of evidence in UAVs. By employing NLP techniques such as named-entity recognition and sentiment analysis, the process of extracting and interpreting data from these logs can be automated. It improves both the speed and accuracy of forensic investigations. This fusion of NLP and drone forensics not only optimizes the analytical workflow but also unlocks new opportunities for advancing UAV security and forensic methodologies.

1.10 Summary

The chapter begins by tracing the historical origins of UAV technology, starting from the early twentieth century, when UAVs were primarily used for military purposes. The chapter discusses landmark developments such as the first UAV, the "Aerial Target" in 1916, and the Kettering Bug during World War I, which laid the foundation for future innovations. Progress was slow initially, but by the mid-twentieth century, UAVs were deployed in practical applications like reconnaissance during the Vietnam War.

With the technological developments of the late twentieth and early twenty-first centuries, UAVs began to diversify into civilian and commercial sectors. Innovations such as sensor miniaturization, battery improvements, and sophisticated control systems enabled the development of smaller, more efficient drones. These innovations made UAVs indispensable tools in various fields, including agriculture, logistics, disaster response, and environmental monitoring. The chapter highlights the role that drones play in modern industries and discusses their impact on improving efficiency, reducing costs, and providing new capabilities across different domains. It also explores the forensic methodology and several case studies in drone forensics research. Finally, we provide the organization of this book to make it easier to the reader to navigate this book.

1.11 Exercises

1. What are Unmanned Aerial Vehicles (UAVs) commonly used for in modern applications?
2. In which century did the early military experiments with UAVs take place?
3. Name three industries or sectors where UAVs are commonly used today.
4. What technological innovations have driven the evolution of UAVs?
5. How have drones transitioned from military tools to versatile instruments?
6. Explain how miniaturization and advancements in control systems have contributed to the transformation of UAVs.

7. Discuss the significance of military experiments in the twentieth century for the development of modern UAV technology.
8. Analyze how UAVs have revolutionized agriculture or logistics, providing specific examples.
9. Compare and contrast the use of UAVs in surveillance versus agriculture, focusing on their roles and technological requirements.
10. Based on the trends mentioned in this chapter, discuss a new potential application for UAV technology.

References

1. J.M. Sullivan, Evolution or revolution? The rise of UAVs. IEEE Technol. Soc. Mag. **25**(3), 43–49 (2006)
2. K.H. Kindervater, The emergence of lethal surveillance: watching and killing in the history of drone technology. Secur. Dialogue **47**(3), 223–238 (2016)
3. T.P. Ehrhard, *Air Force UAVs: The secret history* (Mitchell Institute Press Washington, 2010)
4. V. Prisacariu, A brief history of UAVs in the 1970s. Sci. Res. Edu. Task Force, 200–210 (2022)
5. J.D. Boys, Predator's progress: the bureaucratic challenges to the Clinton administration's development and deployment of Unmanned Aerial Vehicles (1993–2001). Intell. Natl. Secur. **38**(4), 538–557 (2023)
6. K.P. Valavanis, M. Kontitsis, A historical perspective on unmanned aerial vehicles, in *Advances in Unmanned Aerial Vehicles: State of the Art and the Road to Autonomy* (Dordrecht, Springer Netherlands, 2007), pp. 15–46
7. A.R. Hall, C.J. Coyne, The political economy of drones. Defence Peace Econ. **25**(5), 445–460 (2014)
8. K. Saelam, Editorial note on unmanned aerial vehicle. J. Aeronautics Aerospace Eng. **10**(3), 1–1 (2021)
9. N. McKelvey, C. Diver, K. Curran, Drones and privacy. Int. J. Handheld Comput. Res. (IJHCR) **6**(1), 44–57 (2015)
10. S.A.H. Mohsan, et al., Towards the unmanned aerial vehicles (UAVs): a comprehensive review. Drones **6**(6), 147 (2022)
11. P.K. Garg, Characterisation of fixed-wing versus multirotors UAVs/drones. J. Geom. **16**(2), 152–159 (2022)
12. G. Martinic, 'Drones' or 'smart' unmanned aerial vehicles? Aust. Defence Force J. **189**, 46–54 (2012)
13. Y.-T. Wu, et al., Numerical investigation of the mechanical component design of a hexacopter drone for real-time fine dust monitoring. J. Mech. Sci. Technol. **35**, 3101–3111 (2021)
14. E. Ebeid, M. Skriver, J. Jin, A survey on open-source flight control platforms of unmanned aerial vehicle, in *2017 Euromicro Conference on Digital System Design (DSD)* (2017), pp. 396–402
15. P.J. Dziuban, et al., Solid state sensors: practical implementation in unmanned aerial vehicles (UAVs). Proc. Eng. **47**, 1386–1389 (2012)
16. H. Lee, et al., A robot operating system framework for secure UAV communications. Sensors **21**(4), 1369 (2021)
17. C.T. Aksland, et al., Graph-based electro-mechanical modeling of a hybrid unmanned aerial vehicle for real-time applications, in *2019 American Control Conference (ACC)* (2019), pp. 4253–4259
18. N. Pfeifer, P. Glira, C. Briese, Direct georeferencing with on board navigation components of light weight UAV platforms. Int. Arch. Photogrammetry Remote Sensing Spatial Inf. Sci. **39**, 487–492 (2012)

19. F. Nex, F. Remondino, UAV for 3D mapping applications: a review. Appl. Geom. **6**, 1–15 (2014)
20. H. Shakhatreh, et al., Unmanned aerial vehicles (UAVs): a survey on civil applications and key research challenges. IEEE Access **7**, 48572–48634 (2018)
21. C. Torresan, et al., Forestry applications of UAVs in Europe: a review. Int. J. Remote Sensing **38**(8–10), 2427–2447 (2017)
22. M.S.A. Rushdi, et al., Development of a small-scale autonomous UAV for research and development, in *2016 IEEE International Conference on Information and Automation for Sustainability (ICIAfS)* (2016), pp. 1–6
23. A. Otto, et al., Optimization approaches for civil applications of unmanned aerial vehicles (UAVs) or aerial drones: a survey. Networks **72**(4), 411–458 (2018)
24. L.-Y. Lo, et al., Dynamic object tracking on autonomous UAV system for surveillance applications. Sensors **21**(23), 7888 (2021)
25. M. Mozaffari, et al., A tutorial on UAVs for wireless networks: applications, challenges, and open problems. IEEE Commun. Surv. Tutor. **21**(3), 2334–2360 (2019)
26. G. Pajares, Overview and current status of remote sensing applications based on unmanned aerial vehicles (UAVs). Photogram. Eng. Remote Sensing **81**(4), 281–329 (2015)
27. M. Behroozi, D. Ma, Last mile delivery with drones and sharing economy. preprint (2023). arXiv:2308.16408
28. S. Rajalingam, et al., Optimizing drone delivery: an efficient design for shipper applications. J. Autonom. Intell. **7**(1) (2024)
29. V. Gatteschi, et al., New frontiers of delivery services using drones: a prototype system exploiting a quadcopter for autonomous drug shipments, in *2015 IEEE 39th Annual Computer Software and Applications Conference*, vol. 2 (2015), pp. 920–927
30. M.W. Ulmer, B.W. Thomas, Same-day delivery with heterogeneous fleets of drones and vehicles, Networks **72**(4), 475–505 (2018)
31. L. Di Puglia Pugliese, F. Guerriero, G. Macrina, Using drones for parcels delivery process. Proc. Manuf. **42**, 488–497 (2020)
32. J.K. Stolaroff, et al., Energy use and life cycle greenhouse gas emissions of drones for commercial package delivery. Nature Commun. **9**(1), 409 (2018)
33. E. Frachtenberg, Practical drone delivery. Computer **52**(12), 53–57 (2019)
34. D. Wanner, et al., UAV avionics safety, certification, accidents, redundancy, integrity, and reliability: a comprehensive review and future trends. Drone Syst. Appl. **12**, 1–23 (2024)
35. L.M. Bine, A. Boukerche, L.B. Ruiz, A.A. Loureiro, Drone Delivery: Why, Where, and When, in *Proceedings of the International ACM Symposium on Performance Evaluation of Wireless Ad Hoc, Sensor, & Ubiquitous Networks* (2023, October), pp. 35–43
36. E. Yaacoub, Synergy between 6G and AI: open future horizons and impending security risks. Preprint (2022). arXiv:2203.10534
37. U.E. Franke, Drones, drone strikes, and US policy: the politics of unmanned aerial vehicles. Parameters **44**(1), 121 (2014)
38. K. Dobija, Countering unmanned aerial systems (UAS) in military operations. Safety Defense **9**(1), 74–82 (2023)
39. C. van Tilburg, First report of using portable unmanned aircraft systems (drones) for search and rescue. Wilderness Environ. Med. **28**(2), 116–118 (2017)
40. J. Braun, et al., The promising future of drones in prehospital medical care and its application to battlefield medicine. J. Trauma Acute Care Surgery **87**(1S), S28–S34 (2019)
41. P. Mahadevan, The military utility of drones. CSS Anal. Secur. Policy **78** (2010)
42. B. Siddappaji, Pinosh Kumar Hajoary, and KB Akhilesh. UAVs/Drones-Based IoT Services. Smart Technol. Scope Appl. 159–167 (2020)
43. S. Koulali, et al., A green strategic activity scheduling for UAV networks: a sub-modular game perspective. IEEE Commun. Mag. **54**(5), 58–64 (2016)
44. B. Doyle, The autonomous debate: Drone warfare. Eng. Technol. **13**(11), 40–44 (2018)
45. W.R. Dufrene, Mobile military security with concentration on unmanned aerial vehicles, in *24th Digital Avionics Systems Conference*, vol. 2 (2005), 8 pp.

46. R.N. Akram, et al., Security, privacy and safety evaluation of dynamic and static fleets of drones, in 2017 IEEE/AIAA 36th Digital Avionics Systems Conference (DASC) (2017), pp. 1–12
47. W. Chen, et al., Toward robust and intelligent drone swarm: challenges and future directions. J. Ind. Inf. Integr. **18**, 100131 (2020)
48. C. Beck, J. Avila, M. Frye, Guidance and navigation controls for drone swarm applications, in *2022 IEEE/AIAA 41st Digital Avionics Systems Conference (DASC)* (2022), pp. 1–5
49. L. Giacomossi, F. Souza, R.G. Cortes, H.M.M. Cortez, C. Ferreira, C.A. Marcondes, V.V. Curtis, et al., Autonomous and collective intelligence for UAV swarm in target search scenario, in *2021 Latin American Robotics Symposium (LARS), 2021 Brazilian Symposium on Robotics (SBR), and 2021 Workshop on Robotics in Education (WRE)* (2021), pp. 72–77
50. S. Hauert, et al., Communication-based swarming for flying robots, in *Proceedings of the Workshop on Network Science and Systems Issues in Multi-Robot Autonomy, IEEE International Conference on Robotics and Automation* (2010)
51. S. Qamar, et al., Autonomous drone swarm navigation and multitarget tracking with island policy-based optimization framework. IEEE Access **10**, 91073–91091 (2022)
52. F. Venturini, F. Mason, F. Pase, F. Chiariotti, A. Testolin, A. Zanella, M. Zorzi, Distributed reinforcement learning for flexible and efficient UAV swarm control. IEEE Trans. Cognitive Commun. Netw. **7**(3), 955–969 (2021)
53. Z. Darush, M. Martynov, A. Fedoseev, A. Shcherbak, D. Tsetserukou, SwarmGear: Heterogeneous swarm of drones with morphogenetic leader drone and virtual impedance links for multi-agent inspection, in *2023 International Conference on Unmanned Aircraft Systems (ICUAS)* (2023), pp. 557–563
54. H.-J. Kim, H.-S. Ahn, Realization of swarm formation flying and optimal trajectory generation for multi-drone performance show, in *2016 IEEE/SICE International Symposium on System Integration (SII)* (2016), pp. 850–855
55. F. Schilling, et al., Learning vision-based flight in drone swarms by imitation. IEEE Robot. Autom. Lett. **4**(4), 4523–4530 (2019)
56. T.E.A. Barton, M.A.H.B. Azhar, Forensic analysis of popular UAV systems, in *2017 Seventh International Conference on Emerging Security Technologies (EST)* (2017), pp. 91–96
57. S. Atkinson, et al., Drone forensics: the impact and challenges, in *Digital Forensic Investigation of Internet of Things (IoT) Devices* (2021), pp. 65–124
58. H. Bouafif, F. Kamoun, F. Iqbal, A. Marrington, Drone forensics: challenges and new insights, in *2018 9th IFIP International Conference on New Technologies, Mobility and Security (NTMS)* (2018), pp. 1–6
59. D.A.A. Hamdi, F. Iqbal, S. Alam, A. Kazim, Á. MacDermott, Drone forensics: A case study on DJI phantom 4, in *2019 IEEE/ACS 16th International conference on computer systems and applications (AICCSA)* (2019), pp. 1–6
60. A. Renduchintala, et al., A comprehensive micro unmanned aerial vehicle (UAV/Drone) forensic framework. Digital Investigation **30**, 52–72 (2019)
61. F. Iqbal, et al., Drone forensics: examination and analysis. Int. J. Electron. Secur. Digital Forensics **11**(3), 245–264 (2019)
62. G. Thornton, P.B. Zadeh, An investigation into Unmanned Aerial System (UAS) forensics: data extraction & analysis. Forensic Sci. Int. Digital Investigation **41**, 301379 (2022)
63. R. Kumar, A.K. Agrawal, Drone GPS data analysis for flight path reconstruction: a study on DJI, Parrot & Yuneec make drones. Forensic Sci. Int. Digital Investigation **38**, 301182 (2021)
64. H. Studiawan, G. Grispos, K.-K.R. Choo, Unmanned Aerial Vehicle (UAV) forensics: the good, the bad, and the unaddressed. Comput. Secur. 103340 (2023)

Chapter 2
UAV Components and Ecosystem

Abstract This chapter provides an overview of the key elements that constitute Unmanned Aerial Vehicles (UAVs) and the broader ecosystem that supports their development and application. It begins by detailing the basic components of UAVs, including airframe structures, propulsion systems, sensors, cameras, navigation, control systems, communication systems, and power supplies. The discussion extends to the ecosystem surrounding UAVs. For example, the roles of manufacturers, regulatory bodies, and the diverse applications of drones across industries. Furthermore, the chapter addresses the integration of cloud services for UAV management, data processing, mission planning, and the concept of the Internet of Drones (IoD). This chapter provides insights into the interconnected components and ecosystem that enable the effective operation of UAV technology.

2.1 Introduction

Unmanned Aerial Vehicles (UAVs), commonly referred to as drones, have emerged as game-changing tools across a multitude of industries. From precision agriculture and infrastructure inspections to filmmaking and public safety, their application continues to redefine the possibilities of aerial technology. The integration of cutting-edge components such as high-resolution cameras, advanced navigation systems, and obstacle avoidance mechanisms has made UAVs necessary in both commercial and recreational contexts.

The evolution of UAV technology has been driven by improvements in sensor systems, communication protocols, and battery performance, so it can support higher efficiency and broader application scopes. Modern drones are no longer limited to basic flight operations; they are equipped with intelligent features such as GPS-based navigation, automated Return to Home (RTH) functionality, and follow-me modes that provide unparalleled ease of use and operational safety.

This chapter discusses into the key elements and features of UAVs and provides readers with an understanding of their design and capabilities. The discussion covers various hardware and software aspects, including propulsion systems, cameras, communication systems, and safety features. Furthermore, the chapter describes the integration of UAVs with cloud-based services and the emerging concept of the Internet of Drones (IoD), which is the growing ecosystem around UAV technology.

To bridge theoretical knowledge with practical insights, this chapter uses the DJI Mavic Air 2 and DJI FPV drone as case studies. Detailed examples and illustrations of its components, such as the camera system, battery specifications, and controller design, offer a hands-on perspective on how modern drones are built and operate. By the end of this chapter, readers will gain a holistic view of the UAV ecosystem, from basic elements to complex features that determine their functionality and possibilities.

2.2 UAV Components

The functionality and performance of Unmanned Aerial Vehicles (UAVs) are determined by their components where each component contributes to maintain seamless operations. From the structural foundation provided by the airframe to the precision offered by navigation systems, these components operate harmoniously to empower UAVs to perform a wide array of tasks. Modern technologies such as sensors, propulsion systems, and communication technologies have expanded UAV capabilities. This section looks into the key components of UAVs.

Airframe and Propulsion System The airframe forms the structural body of the UAV, which can vary from fixed-wing designs for long-distance flights to multi-rotor configurations for vertical takeoff and landing (VTOL). The propulsion system typically includes electric motors powered by batteries or, in some cases, internal combustion engines for larger drones [1].

Sensors and Cameras Drones are equipped with various sensors and cameras to collect data. High-resolution visible, multispectral, and thermal cameras are used for different applications such as ecological monitoring, agriculture, and infrastructure inspection [2]. LIDAR sensors and infrared cameras also play important roles in mapping and surveillance tasks [3].

Navigation and Control Systems UAVs rely on sophisticated navigation systems, including GPS for positioning and inertial measurement units (IMUs) for orientation. Advanced control systems allow for autonomous flight and precision landing and reduce the need for human intervention [4].

Communication Systems Reliable communication systems are needed by UAV operations to support command and control (C2) links, data transmission, and integration with broader networks such as the Internet of Things (IoT). Emerging technologies such as 5G are improving network capacities by providing high-throughput links for real-time data streaming and remote control [5].

Power Supply Most UAVs are powered by lithium-polymer (LiPo) batteries due to their high energy density and lightweight properties. However, limitations in battery life remain a challenge. Therefore, research into alternative power sources and energy-efficient designs are keep progressing [6].

2.3 UAV Ecosystem

The UAV ecosystem represents a domain that integrates technology, regulatory frameworks, and diverse applications to address modern challenges and opportunities. From the design and manufacture of cutting-edge drones to the integration of emerging technologies such as AI and blockchain, this ecosystem is redefining industries and services globally. Regulatory bodies and standardization efforts ensure the safe and efficient deployment of drones, while applications in agriculture, logistics, and environmental monitoring highlight their potential. As the ecosystem grows, sustainability and innovation remain a concern as UAVs also contribute to environmental progress.

Manufacturers and Integrators The UAV ecosystem includes manufacturers of drones and their components, software developers, and system integrators who ensure that different parts work together seamlessly. Leading companies like DJI, SenseFly, and Parrot are at the forefront of this industry [7].

Regulatory Bodies and Standards We need regulatory frameworks for the safe integration of drones into airspace. Organizations such as the Federal Aviation Administration (FAA) and European Union Aviation Safety Agency (EASA) provide guidelines and standards for drone operations. The development of "detect and avoid" technologies and secure communication protocols are key to meeting these regulatory requirements [8].

Applications and Services Drones are used in diverse applications including agriculture for crop monitoring, logistics for parcel delivery, and environmental monitoring for assessing ecosystem health. Each application demands specific technological adaptations and services, such as precision agriculture tools or delivery route optimization algorithms [9].

Emerging Technologies Innovations such as blockchain for secure drone operations, edge computing for real-time data processing, and AI for autonomous decision-making are shaping the future of the UAV ecosystem. These technologies boost the performance of drones and open up new possibilities for their use [10].

Environmental Impact and Sustainability UAVs contribute to environmental sustainability by enabling precise monitoring and management of natural resources. They help reduce carbon footprints in logistics and offer new methods for ecological conservation and restoration [11].

2.4 Integration of Cloud Services

Cloud services provide platforms for real-time UAV monitoring and management. These systems can handle multiple UAVs, track their positions, and ensure safe flight operations through collision avoidance algorithms [12]. Utilizing cloud infrastructure allows for near real-time processing of the vast amounts of data captured by UAVs. This approach supports seamless data sharing and collaboration among various stakeholders. In addition, it reduces the need for high-specification local hardware [13].

Cloud-based platforms enable sophisticated mission planning and management. They allow users to design, execute, and monitor UAV missions remotely. These platforms can integrate various UAVs into a cohesive network and support complex operations such as emergency response and environmental monitoring [14]. The IoD concept involves connecting drones to the internet, so they can interact with other IoT devices. This connectivity provides more advanced applications such as smart city management and disaster response [15].

Cloud services also provide robust security frameworks for the safe operation of UAVs. These frameworks assure that data transmission and storage comply with security standards and protect against cyber threats and unauthorized access [16]. As an example, Fig. 2.1 shows FlytBase,[1] a drone autonomy software platform specifically designed for system integrators. The platform offers features for the automation of repeatable and routine drone operations through the use of drone docks. It provides several tasks such as site security, asset inspection, and emergency

Fig. 2.1 FlytBase: A drone autonomy software platform

---

[1] https://www.flytbase.com/.

response. Its focus is on providing seamless operations from remote locations. FlytBase delivers its ease of use and integration capabilities. It is a robust choice for professionals aiming to enhance their drone management systems with advanced software solutions.

2.5 UAVs and Drones Based on Types of Wings and Rotors

UAVs come in various configurations, each designed for specific purposes and operational requirements. This classification primarily focuses on the types of wings and rotor systems used in UAVs, which dictate their aerodynamic properties, flight capabilities, and suitability for different applications. Fixed-wing UAVs are ideal for long-range missions, rotary-wing UAVs offer versatility and maneuverability, hybrid UAVs provide a balance of both, and flapping-wing UAVs are used for specialized tasks requiring high agility. Each type's unique features accommodate different mission requirements and make UAV technology versatile and widely applicable.

Fixed-wing UAVs Fixed-wing UAVs have wings like traditional airplanes. These wings provide lift as the UAV moves forward, typically powered by a propeller or jet engine. Fixed-wing drones are efficient for covering large distances and are often used in applications requiring long-endurance and high-altitude flights.

The applications of drones include long-range surveillance and reconnaissance, environmental monitoring, agricultural mapping, and military operations. These drones offer several advantages, such as longer flight times and greater range compared to rotorcraft, higher flight speeds, and being more energy-efficient for long-distance travel. However, there are some disadvantages, including the need for runways or catapults for takeoff and landing and limited hovering capability. An example of such drones is the Global Hawk, often used for military reconnaissance operations [17].

Rotary-wing UAVs Rotary-wing UAVs, also known as rotorcraft, include helicopters and multi-rotors (such as quadcopters, hexacopters, and octocopters). These UAVs use rotating blades to generate lift and thrust, allowing them to hover, take off, and land vertically.

Drones have a wide range of applications, including aerial photography and videography, infrastructure inspection, search and rescue operations, and delivery services. Their advantages include the ability to hover, perform vertical takeoff and landing (VTOL), high maneuverability, and stability, making them suitable for confined or urban environments. However, these drones also have disadvantages, such as shorter flight times and range compared to fixed-wing UAVs, as well as higher energy consumption. A well-known example of commercial drones used for these purposes is the DJI Phantom series [18].

Hybrid UAVs Hybrid UAVs combine the features of fixed-wing and rotary-wing designs. They typically have the vertical takeoff and landing capabilities of rotorcraft and the efficient forward flight characteristics of fixed-wing aircraft.

Hybrid drones are used in a variety of applications, including long-endurance missions with vertical takeoff and landing (VTOL) requirements, search and rescue operations in remote areas, and parcel delivery over varying terrains. These drones offer several advantages, such as versatility in takeoff and landing, and improved range and endurance compared to pure rotary-wing UAVs. However, they come with certain disadvantages, including a more complex design, higher maintenance requirements, and potentially higher costs. Examples of hybrid drones include the V-BAT and the X-Plane [19].

Flapping-wing UAVs Flapping-wing UAVs, which mimic the flight mechanics of birds or insects, generate lift through the flapping motion of their wings. These UAVs find applications in a variety of fields, including biological and environmental research, military reconnaissance in urban environments, and ornithological studies. Their primary advantages lie in their high maneuverability within complex environments and their low acoustic and visual signatures, making them less detectable. However, they also come with disadvantages, such as limited payload capacity and complex mechanical designs. An example of such UAVs can be found in biomimetic drones used for specific scientific studies, as noted by Lee et al. [20].

2.6 Understanding Throttle, Yaw, Pitch, and Roll in Drones

UAVs rely on four primary control inputs to maneuver: throttle, yaw, pitch, and roll. Each of these controls adjusts the drone's movement in specific ways, allowing for precise navigation and stability in flight.

Throttle Throttle controls the altitude of a drone by regulating the power sent to the motors, which directly influences its vertical movement. Increasing the throttle causes the drone to ascend, while decreasing the throttle leads to its descent. This control is used for key operations such as takeoff, landing, and maintaining a stable hover at a desired altitude.

Yaw Yaw controls the rotation of a drone around its vertical axis. It allows it to turn left or right while remaining at a fixed point. Moving the yaw control to the left rotates the drone counterclockwise, while moving it to the right rotates the drone clockwise. Yaw is used for changing the drone's direction while it is in flight. It is particularly useful for navigating around obstacles and for orientation adjustments during flight.

Pitch Pitch controls the tilt of the drone forward or backward. This input affects the drone's movement along the longitudinal axis (nose to tail). Pitch controls the forward and backward movement of a drone. Tilting the pitch control forward causes the drone to move forward, while tilting it backward causes the drone to move in reverse. Pitch is used to control the forward and backward movement of the drone. It is for navigating the drone in the desired direction, especially when covering ground or following a specific path.

Roll Roll controls the tilt of the drone to the left or right. This input affects the drone's movement along the lateral axis (wing to wing). Roll controls the lateral movement of a drone. Tilting the roll control to the left causes the drone to move left, while tilting it to the right makes the drone move right. Roll is for lateral movement and is particularly useful for side-stepping obstacles and making precise adjustments to the drone's position in the air.

Coordinated Control To effectively pilot a drone, a combination of throttle, yaw, pitch, and roll inputs is often required. Understanding how these controls interact is key to mastering drone flight. Here are a few scenarios where coordinated control is necessary.

Hovering involves maintaining a steady throttle to keep the drone at a constant altitude, while making small adjustments in yaw, pitch, and roll to ensure the drone remains stable and in one position. Flying forward requires a slight increase in throttle to gain altitude, followed by tilting the pitch control forward to move in the desired direction. Yaw and roll adjustments are necessary to maintain the drone's heading and avoid obstacles. Turning while moving involves using the yaw control to rotate the drone in the desired direction, while simultaneously adjusting pitch and roll to maintain speed and stability. We need the throttle balance to keep the drone at a consistent altitude. For landing, the throttle is gradually decreased to lower the drone toward the ground, with small adjustments in yaw, pitch, and roll guiding it to the landing spot for a smooth descent.

Each throttle, yaw, pitch, and roll has a specific role in maneuvering the drone, and understanding how to coordinate these inputs allows for precise and stable flight. Whether for recreational flying, aerial photography, or professional applications, proficiency in these controls improve the overall drone operation experience.

2.7 Drone Specification Fields

Drone specifications comprise a comprehensive range of features and parameters that define the performance, functionality, and usability of UAVs. These specifications cover various aspects, including the drone's general build, propulsion system, battery, flight controller, camera, communication, and remote controller. A pilot needs to understand these fields to evaluate a drone's suitability for specific applications, whether for professional use, recreational purposes, or research initiatives. In this section, we provide an in-depth overview of the components and features that make up drone specifications.

General Specifications The drone's specifications include several critical features. The Type specifies the drone's configuration, such as a "Quadcopter", which is equipped with four rotors and is renowned for its stability and ease of control. Dimensions indicate the size of the drone, measured diagonally from motor to motor. The Weight indicates the total weight of the drone, including the battery, which significantly influences its flight time and handling.

The Max Flight Time refers to the maximum duration the drone can remain airborne on a single battery charge, while the Max Speed represents the highest speed the drone can achieve, particularly beneficial for applications requiring rapid movement. The Max Service Ceiling is the highest altitude the drone can reach above sea level, and the Max Wind Resistance denotes the strongest wind speed the drone can endure while maintaining stable flight. Lastly, the Operating Temperature Range defines the range of temperatures within which the drone operates effectively.

Airframe The drone's structure and design incorporate several important elements. The Material used for its construction, such as carbon fiber composite, provides a balance of lightweight properties and strength to provide durability and performance. The Landing Gear supports safe landings and can be either fixed or retractable, often equipped with features like shock absorbers to protect the drone during touchdown. Additionally, the Color of the drone serves both practical and aesthetic purposes and assist in visibility and contributing to its overall appearance.

Propulsion System The performance of drones depends on the type and specifications of the motors utilized, with brushless motors being particularly efficient and durable. Similarly, the size and material of the propellers significantly affect the drone's lift and maneuverability. Another component is the Electronic Speed Controllers (ESCs), which regulate power to the motors and must be compatible with specific battery types to support optimal functionality.

Battery The battery type is typically Lithium-polymer (LiPo) due to its high energy density. The capacity, measured in milliampere-hours (mAh), determines the flight time of the drone. Charging time refers to the duration required to recharge the battery fully. Moreover, a battery life indicator, either visual or app-based, provides information about the remaining battery life.

Flight Controller The flight controller's performance is heavily influenced by the type of processor it uses. It relies on various sensors, including gyroscopes, accelerometers, magnetometers, barometers, and GPS, to maintain stability and accurate navigation. It supports multiple flight modes tailored to different conditions and purposes, such as GPS Hold, Altitude Hold, Manual control, Follow Me, Waypoint Navigation, Return to Home (RTH), and Headless Mode.

Camera The type of camera used in drones is often integrated and equipped with a gimbal for stabilization. Video resolution, such as 4K UHD, determines the quality of footage captured, while the field of view defines the width of the camera's viewing angle. Stabilization, typically achieved through a 3-axis gimbal, guarantees smooth video recording. The camera's photo resolution, measured in megapixels

(MP), impacts the quality of still images. Furthermore, the storage media, such as MicroSD cards, and their capacity determine how much footage and photos can be stored.

Communication The control range refers to the maximum distance from which the drone can be operated using the remote, while the video transmission range indicates the farthest distance for receiving a live video feed. Communication typically occurs over frequency bands such as 2.4 and 5.8 GHz, with a transmission system that is often digital and encrypted for secure and reliable control and video streaming. In addition, application compatibility allows users to control and configure the drone through mobile applications, which are usually supported on iOS and Android platforms.

Remote Controller The remote controller's form factor and design impact its usability and comfort. Several controllers feature a built-in display that provides live video feed and telemetry data to improve the user experience. The type and capacity of the controller's battery are important for ensuring long operational times. Moreover, many controllers offer extra features such as a telemetry display, customizable buttons for personalized control, and built-in GPS for additional functionality.

Safety Features Drones are equipped with obstacle avoidance sensors that detect and prevent collisions by identifying objects in their path. A fail-safe Return to Home feature automatically guides the drone back to its starting point when the battery is low or the signal is lost. Geofencing allows users to set configurable boundaries and makes sure the drone does not enter restricted areas. An emergency stop function enables an immediate shutdown of the motors in case of critical situations.

Additional Features The payload capacity of a drone determines the maximum weight it can carry in addition to its own weight for applications involving deliveries or specialized equipment. Several drones feature a foldable design, so their arms can fold for convenient transport and storage. Programmable LED lights improve orientation and visibility, particularly during night flights. Finally, the level of water resistance, often indicated by an IP rating, makes the drone can operate safely in various weather conditions.

Package Contents A standard drone package typically includes the main UAV unit, which serves as the core component of the system. It comes with a remote controller for operating the drone and one or more batteries to power it, along with a charger for recharging these batteries. Spare propellers are often provided for replacements in case of damage. Many packages also include a carrying case for safely transporting the drone and a user manual that provides detailed instructions and guidelines for operation. These fields cover the aspects of a drone's design, capabilities, and accessories and provide a comprehensive overview of its features and performance characteristics.

2.8 Drone Features Explained

Modern drones come equipped with a variety of advanced features designed to provide flight safety, ease of use, and overall functionality. These features include GPS technology for precise navigation and hovering capabilities, obstacle detection and collision avoidance systems to prevent accidents, and return-to-home functions that makes the drone can safely return to its starting point if it loses connection with the controller. They may also feature automatic flight modes such as follow-me or orbit mode, which allow the drone to autonomously track a subject or circle around a point of interest. In addition, many drones are equipped with high-resolution cameras and gimbal stabilization systems for capturing smooth and detailed aerial footage. These common drone features are explained to help readers understand the capabilities and advantages each provides.

Return to Home (RTH) Return to Home (RTH) is a safety feature that automatically brings the drone back to its takeoff point or a designated home location. The Return to Home (RTH) function can be activated in several ways. With Manual RTH, the pilot can initiate the return by pressing a button on the remote controller. Low Battery RTH triggers the drone to return automatically when the battery reaches a critically low level. Failsafe RTH guarantees the drone returns home automatically if it loses connection with the remote controller. The benefits of RTH include preventing the loss of the drone due to low battery or signal loss and securing a safe return in case of an emergency.

GPS Hold GPS hold allows the drone to maintain a stable position and altitude using GPS coordinates. The drone's GPS module works by locking onto satellites to maintain a fixed position, even in the presence of wind or minor disturbances. This feature provides stable flight, which is ideal for capturing smooth videos and photos, and also makes flying easier for beginners.

Obstacle Avoidance Obstacle avoidance feature uses sensors to detect and avoid obstacles in the drone's flight path. The drone's obstacle avoidance system works by using sensors such as ultrasonic, infrared, or visual sensors to scan the environment for obstacles. The flight controller processes the data from these sensors and adjusts the drone's path to prevent collisions. This system reduces the risk of crashes and enhances safety, particularly during autonomous flights.

Follow Me Mode Follow Me Mode allows the drone to automatically follow and film a subject. The drone follows a subject by using GPS or visual tracking to lock onto the target. It then maintains a set distance and angle while tracking the subject's movements. This feature is ideal for capturing action shots and filming moving subjects. It enables hands-free operation as the drone autonomously captures dynamic footage.

Waypoint Navigation Waypoint navigation lets users plan a flight path with specific points (waypoints) that the drone will follow autonomously. The process begins with the pilot setting waypoints on a map using a mobile application or a computer. The drone then flies from point to point and carries out pre-defined actions such as hovering or taking photos. This approach offers several benefits, including enabling precise and repeatable flight paths, which prove to be particularly useful for tasks like mapping, surveying, and inspections.

Altitude Hold Altitude hold maintains the drone at a constant altitude without the need for continuous throttle input from the pilot. The drone maintains its altitude by utilizing a barometer or GPS data. Therefore, the pilot can concentrate on directional control without needing to manage altitude adjustments. This simplifies control for beginners and improves stability and it is especially useful for aerial photography and videography.

Headless Mode Headless mode simplifies control by orienting the drone's movements relative to the pilot's position rather than the drone's front. In Headless mode, the drone's forward direction is always away from the pilot, and backward is towards the pilot, regardless of the drone's orientation. This feature makes control easier for beginners who may have difficulty with orientation and is particularly helpful when the drone is far away and hard to distinguish its front from its back.

First Person View (FPV) FPV allows the pilot to see from the drone's perspective in real-time, typically through a mobile device or FPV goggles. The drone's camera transmits live video to a screen or goggles, so the pilot can navigate as if they were onboard the drone. This setup provides the immersive flying experience and is particularly useful for precise flying and capturing specific shots.

Geofencing Geofencing uses GPS to create virtual boundaries that the drone cannot cross. The pilot establishes boundaries using a mobile app, and the drone's flight controller ensures that it does not fly beyond these set limits. This feature helps prevent the drone from entering restricted or hazardous areas. It guarantees safety and make sure the drones comply with regulations.

Follow Me Follow Me is a feature where the drone autonomously follows a moving object or person. The drone locks onto the GPS signal of the connected device or uses visual tracking to follow the target and maintain a set distance and angle. This functionality is well-suited for capturing footage of activities such as sports, hiking, or cycling, and offering a hands-free filming experience.

Gimbal Stabilization A gimbal stabilizes the camera to keep it level and reduce vibrations, resulting in smooth video footage. The gimbal uses motors and sensors to counteract the drone's movements. It provides stabilization across multiple axes, typically three. This ensures high-quality, stable video footage and is useful for professional aerial photography and videography.

2.9 Hands on Examples: DJI Mavic Air 2

As hands on examples, we discuss DJI Mavic Air 2 drone and its components.[2] A download centre page is also provided to check for DJI Fly application and the product manuals.[3] The DJI Mavic Air 2 drone comes with a set of cables designed to provide connectivity and power management as shown in Fig. 2.2. Included in the package is a USB-C to USB-C cable, it is for connecting devices with USB-C ports for data transfer or charging. For compatibility with older devices, a Micro USB to USB-C cable is provided, it supports connections between Micro USB and USB-C ports. Subsequently, a USB-A to USB-C cable facilitates connectivity between traditional USB-A ports and USB-C devices. It supports both charging and data transfer needs. To power the drone's batteries, an AC power cable and an AC power adapter are included. It will be safe conversion of AC power to the appropriate voltage for charging. These cables make sure that the Mavic Air 2 is fully equipped for both modern and legacy device compatibility, as well as reliable power supply.

Furthermore, Fig. 2.3 shows accessories included with the DJI Mavic Air 2 drone for operational convenience and maintenance. The remote controller is a key component for piloting the drone completed with ergonomic design and precise

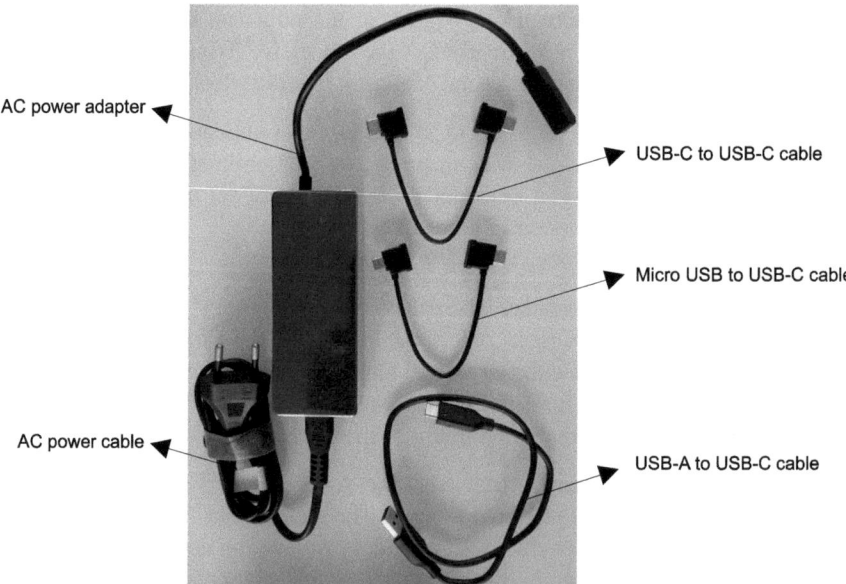

Fig. 2.2 Cables included in the DJI Mavic Air 2 drone

---

[2] https://www.dji.com/id/support/product/mavic-air-2.

[3] https://www.dji.com/id/downloads/products/mavic-air-2.

2.9 Hands on Examples: DJI Mavic Air 2

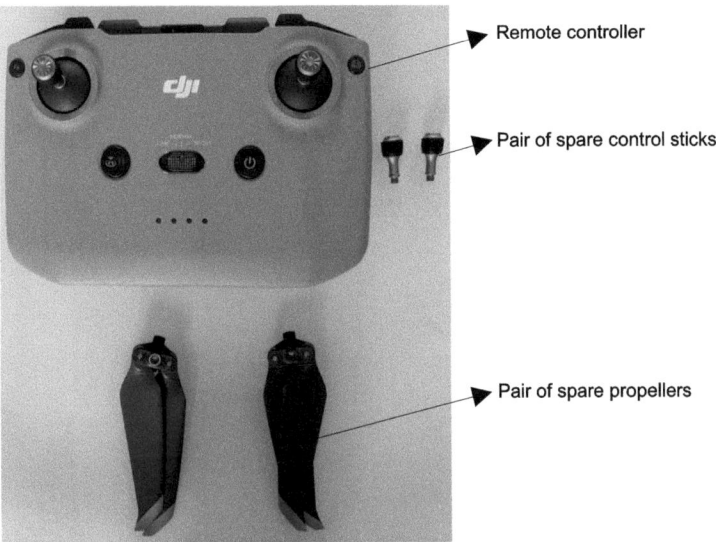

Fig. 2.3 Drone controller, spare control sticks, and spare propellers

control capabilities to enhance the flying experience. In addition, a pair of spare control sticks is provided to replace the controller's sticks if needed. The package also includes a pair of spare propellers. It can be used for quick replacement in the event of propeller damage. The remote controller, identified as Model RC231, operates on the same frequency bands and has a maximum transmission distance of 8 km (FCC), 5 km (CE), or 5 km (SRRC), depending on the region and conditions. It requires an operating current/voltage of 3.6 V at varying currents for Android and iOS devices. The controller also has an operating temperature range of 0° C–40° C (32° F–104° F).

Figure 2.4 shows other main components and accessories included with the DJI Mavic Air 2 drone to ensure functionality and protection. The aircraft represents the drone itself, equipped with advanced features for aerial operations. A battery is provided to power the drone for optimal flight time and efficient energy management. For data storage, a microSD card is included for the recording and saving high-resolution photos and videos captured by the drone. The gimbal cover serves as a protective shield for the drone's camera and gimbal system during transport and safeguards against damage. Finally, the propeller is an important component for flight to provide thrust and stability during operation.

The aircraft, identified as Model MA2UE3W, has a takeoff weight of 570 grams and offers a maximum flight time of 34 minutes under optimal conditions, flying at a consistent speed of 11.4 mph (18.4 kph) in zero wind. It operates efficiently within a temperature range of 0° C–40° C (32° F–104° F) and supports operating frequencies between 2.400–2.4835 and 5.725–5.850 GHz. The transmitter power varies based on frequency, adhering to regional standards.

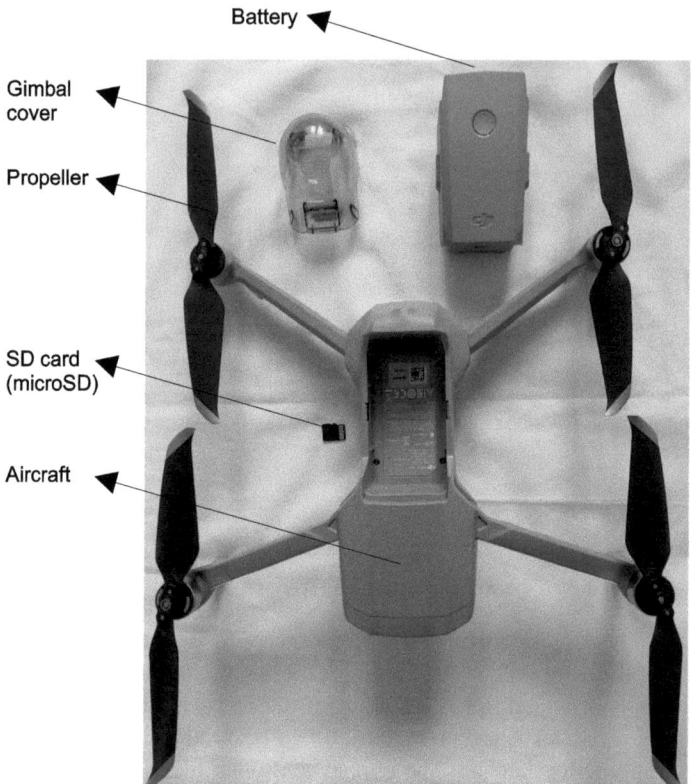

Fig. 2.4 Drone aircraft, battery, SD card, gimbal cover, and propeller

The DJI Mavic Air 2 drone is equipped with an Intelligent Flight Battery with a capacity of 3500 mAh, which determines the amount of charge it can hold and impacts the flight time. The battery operates at a nominal voltage of 11.55 V, which is typical for its 3S LiPo (Lithium Polymer) configuration. LiPo batteries are widely used in drones due to their high energy density and lightweight design. This particular battery has a total energy capacity of 40.42 Wh. It indicates the amount of energy it can deliver over an hour. For safe and efficient charging, the battery supports a maximum charging power of 38 W and should be charged within a temperature range of 5° C–40° C (41° F–104° F) to avoid potential damage or reduced performance. These specifications make the battery reliable and suitable for powering the Mavic Air 2. Proper handling and adherence to the recommended charging conditions can further extend the battery's lifespan.

Figure 2.5 depicts a drone camera equipped with advanced specifications for high-quality aerial photography and videography. The camera features a 1/2.0" CMOS sensor, which is capable of capturing detailed images and videos even in challenging lighting conditions. With effective pixels of 48 MP, the camera delivers

2.9 Hands on Examples: DJI Mavic Air 2

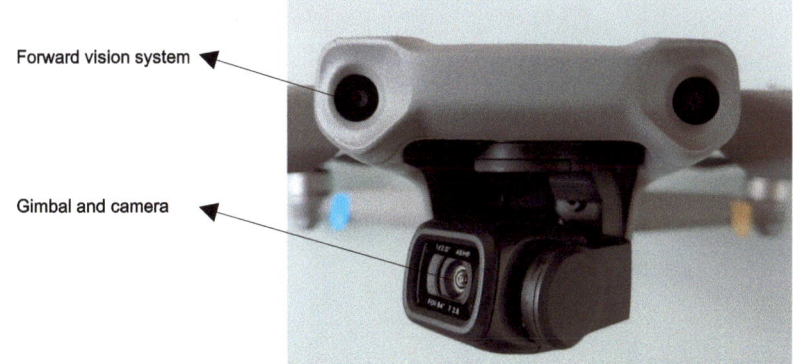

Fig. 2.5 The camera of DJI Mavic Air 2

ultra-high-resolution imagery, ideal for professional applications such as mapping, inspection, and cinematic recording. The ISO range varies between video and photo modes, with values between 100–6400 for video and an expanded range for photo modes. The electronic shutter speed is adjustable between 8 to 1/8000 seconds, and the maximum image size supported is 48 MP (8000×6000). The camera also supports 4K Ultra HD video recording and offers various resolutions and frame rates for flexibility.

The lens has a field of view (FOV) of 84° for wide-angle captures that includes more of the scene. It is suitable for expansive landscapes and aerial perspectives. The aperture of F/2.8 provides excellent light intake and contributes to brighter and clearer images, particularly in low-light environments. This combination of features makes the drone camera highly versatile and capable of delivering exceptional performance in various scenarios.

Figure 2.6 shows the bottom part of the DJI Mavic Air 2 drone and it displays several key components that contribute to its functionality and flight safety. The bottom section includes vision sensors and infrared sensors, which are integral for obstacle detection and precise hovering. These sensors supports stable flight, particularly during takeoff, landing, and indoor operations where GPS signals might be weak. The auxiliary LED light is another critical feature visible in this area. This light improves visibility in low-light conditions and aids in safe landing, especially at night or in poorly lit environments. The air vents on the bottom serves in the drone's cooling system to dissipate heat generated during operation and prevent overheating. In addition, the structural design and materials of the bottom part guarantee durability while maintaining the lightweight nature of the drone.

Figure 2.7 shows the SD card slot on the DJI Mavic Air 2 drone. It is designed to house a microSD card for storage purposes. The microSD card is an essential component for recording and saving high-resolution images and videos captured by the drone's camera. It supports large storage capacities and provides ample space for extended recording sessions and high-quality footage. The design of

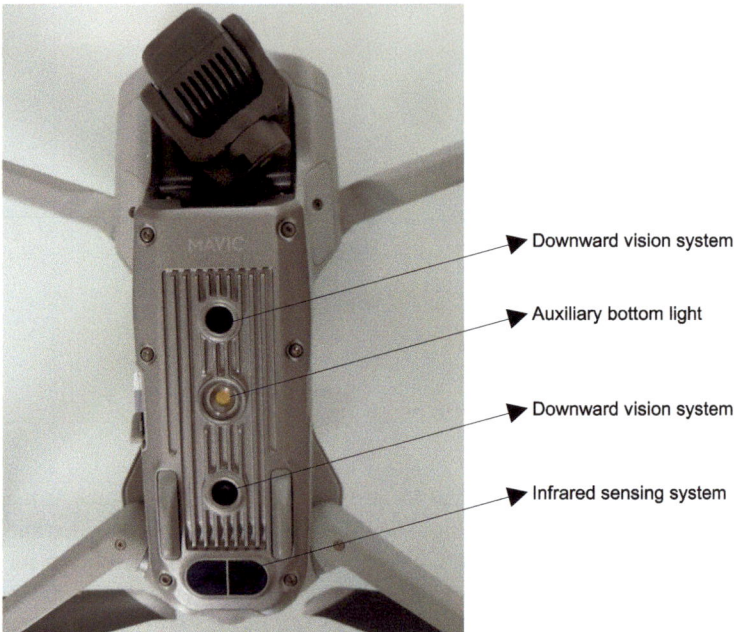

Fig. 2.6 Bottom part of the drone

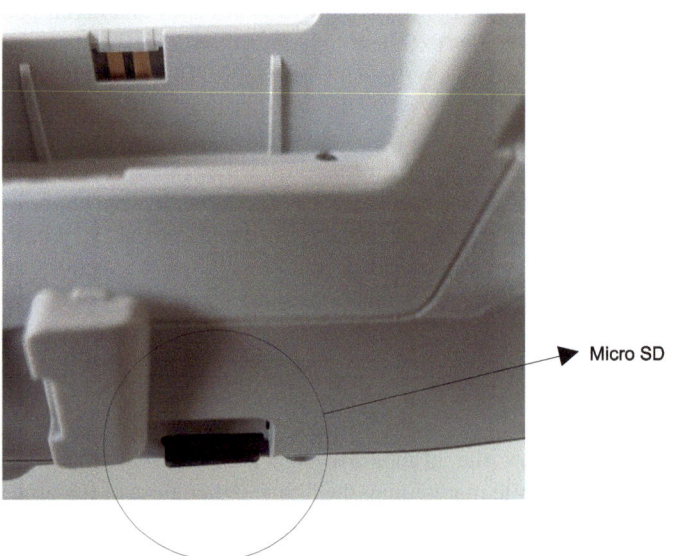

Fig. 2.7 SD card (microSD) as the drone external memory

2.9 Hands on Examples: DJI Mavic Air 2

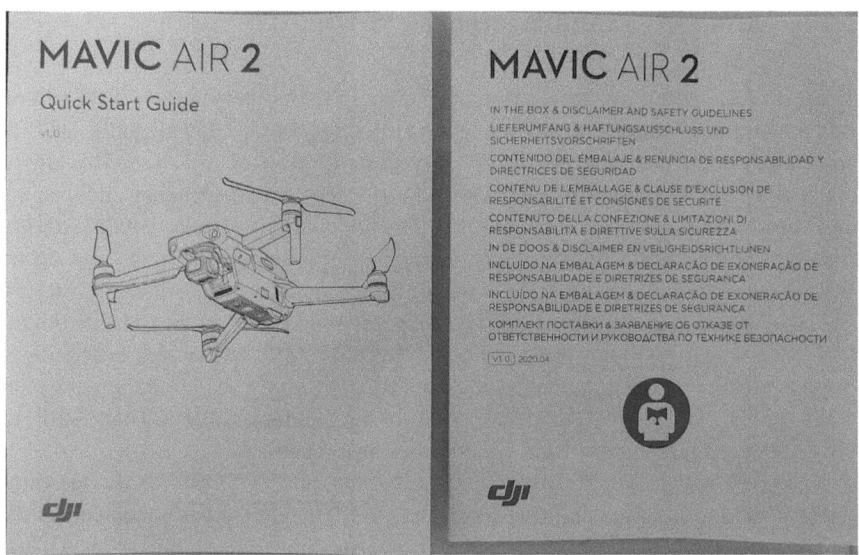

Fig. 2.8 DJI Mavic Air 2 manual book

the slot guarantees that the microSD card is held securely and minimizes the risk of unintended ejection while in use. The placement is easily accessible, so users can insert and remove the card conveniently. This feature enhances the drone's functionality by enabling users to expand storage capacity and easily transfer data to other devices for processing or sharing, including forensic analysis.

Furthermore, Fig. 2.8 shows the cover of the manual book for the DJI Mavic Air 2 drone. It is a key resource for users to understand the operation and safety guidelines of the drone. The Quick Start Guide (v1.0) on the left provides an overview of essential steps for setting up and operating the drone. It is ideal for first-time users to get started quickly. On the right, the section titled "In the Box & Disclaimer and Safety Guidelines" outlines the contents of the package and includes important information about responsibility disclaimers and safety instructions. The text is presented in multiple languages so it reflects DJI's effort to support a global audience. The version number v1.0 and the date 2020.04 indicate the guide's edition and publication date. This manual provides access to detailed instructions, safety tips, and operational recommendations to maximize the drone's performance while minimizing risks.

2.10 Hands on Examples: DJI FPV

The DJI FPV drone is a high-performance, immersive first-person view (FPV) drone designed for dynamic flying and cinematic video capture.[4] To download the DJI Fly application and manual book, users can access the official website.[5] This drone features a takeoff weight of approximately 795 grams, with compact dimensions that vary depending on whether the propellers are attached. Its aerodynamic design ensures stability and efficiency, with a diagonal distance of 245 mm. The drone supports multiple flight modes—Normal, Sport, and Manual—offering varying levels of speed and control. In Manual mode, the drone achieves a maximum horizontal speed of 39 m/s and accelerates from 0 to 100 kph in just 2 seconds. It can operate up to a maximum altitude of 6000 meters above sea level and withstand wind speeds of up to 13.8 m/s. Its operating temperature range is from $-10°$ to $40°$ C, so it is suitable for a wide variety of environments.

Equipped with a sophisticated gimbal system, the DJI FPV drone provides stable video footage even in fast-paced scenarios. The gimbal has a mechanical tilt range from $-65°$ to $+70°$ and provides precise single-axis stabilization. This is complemented by an electronic roll stabilization system, which strengthens footage quality and maintains smoothness. The gimbal can be controlled at speeds of up to $60°$ per second, with an angular vibration range of $\pm 0.01°$ in Normal mode.

The onboard camera is one of the standout features, boasting a 1/2.3" CMOS sensor with 12 MP resolution. Its wide $150°$ field of view and f/2.8 fixed aperture lens ensure detailed and vibrant imagery. The camera supports 4K video recording at 50/60 fps and Full HD (FHD) video recording at up to 120 fps, so it is ideal for creating high-quality video content. Features such as RockSteady electronic image stabilization and distortion correction further enhance the camera's capabilities. Media can be saved in JPEG (photos) and MP4/MOV (videos) formats, with a maximum video bitrate of 120 Mbps.

The drone is powered by an intelligent flight battery with a capacity of 2000 mAh. It provides approximately 20 minutes of flight time under optimal conditions. The battery charges fully in about 50 minutes. Alongside the drone, the FPV Goggles provide an immersive flying experience with dual 1440×810 resolution screens, a refresh rate of 144 Hz, and adjustable fields of view ranging from $30°$ to $54°$. They support live video transmission in Low-Latency and High-Quality modes, with a range of up to 10 km (FCC) and a maximum bitrate of 50 Mbps.

The remote controller, weighing 346 grams, offers ergonomic controls and a reliable transmission range matching the goggles. Both the controller and goggles support multiple operating frequencies and robust transmission power for seamless communication with the drone. Furthermore, the DJI FPV supports microSD cards

---

[4] https://www.dji.com/id/fpv.

[5] https://www.dji.com/id/support/product/dji-fpv.

2.10 Hands on Examples: DJI FPV

up to 256 GB for ample storage, and its obstacle sensing system ensures safety during operation by detecting forward and downward obstacles in Normal mode.

Figure 2.9 displays a DJI FPV drone AC cable adapter, which consists of two main components: the power adapter and the AC power cable. The power adapter, marked with the DJI logo, is designed to convert AC power from a standard wall outlet into the appropriate DC voltage needed to charge the DJI FPV drone battery efficiently and safely. The accompanying AC power cable features a European plug (Type C), suitable for regions with 220–240V AC outlets, and connects securely to the power adapter.

The front view of the DJI FPV drone remote controller shows controls designed for drone operation (Fig. 2.10). It includes the Pitch and Roll Stick to control the

Fig. 2.9 DJI FPV AC cable adapter

Fig. 2.10 DJI FPV remote controller (front)

Fig. 2.11 DJI FPV remote controller (top)

drone's forward/backward and side-to-side movements with precision. The Throttle and Yaw Stick is used for adjusting the drone's altitude and rotation to support aerial maneuvering. The controller is also equipped with an On/Off Button, which powers the device for a straightforward and quick setup process. In addition, the C1 Button provides customizable functionality, which allows the pilot to assign specific tasks or features for quick access. Note that this thoughtful layout makes the controller ergonomic and user-friendly.

The top view of the DJI FPV drone remote controller depicts its key controls and features designed for precise operation and convenience as shown in Fig. 2.11. It includes a Gimbal Dial, which allows the pilot to adjust the angle of the drone's camera for smooth and precise shots. The Shutter or Record Button enables the user to capture photos or start/stop video recording quickly. The Start/Stop Button is used to power the drone on or off. Furthermore, the controller features a Flight Pause or Return-to-Home (RTH) Button, which provides safety by allowing the drone to hover or return to its home location when needed. The Custom Mode (C2) Switch offers flexibility by enabling the pilot to assign specific functions based on personal preferences or operational requirements. This intuitive layout provides easy access to essential controls, making the DJI FPV remote controller highly functional and user-friendly for both beginners and experienced drone operators.

The DJI FPV drone battery system consists of two primary components: the intelligent flight battery and its protective casing. The intelligent flight battery, shown at the top in Fig. 2.12, is a high-capacity lithium-polymer (LiPo) battery that serves as the main power source for the DJI FPV drone. This battery is engineered to provide extended flight times and consistent energy output. It features built-in LED indicators that display the remaining charge level to monitor the battery's status at a glance. Furthermore, DJI incorporates advanced safety features into the battery, such as overcharge protection, over-discharge protection, and temperature regulation. The battery casing, displayed at the bottom of Fig. 2.12, offers robust protection for the intelligent flight battery. Its durable design safeguards the battery from

2.10 Hands on Examples: DJI FPV

Fig. 2.12 DJI FPV battery

Fig. 2.13 DJI FPV (front side)

potential physical damage during handling, storage, or minor impacts. The casing also includes a secure locking mechanism to ensure that the battery remains firmly in place when installed in the drone. It supposes to prevent accidental dislodgment during flight. Furthermore, the ergonomic design of the casing allows for easy installation and removal.

The front view of the DJI FPV drone (Fig. 2.13) reveals its advanced vision and sensing systems to support safe and stable flight. The forward vision system is

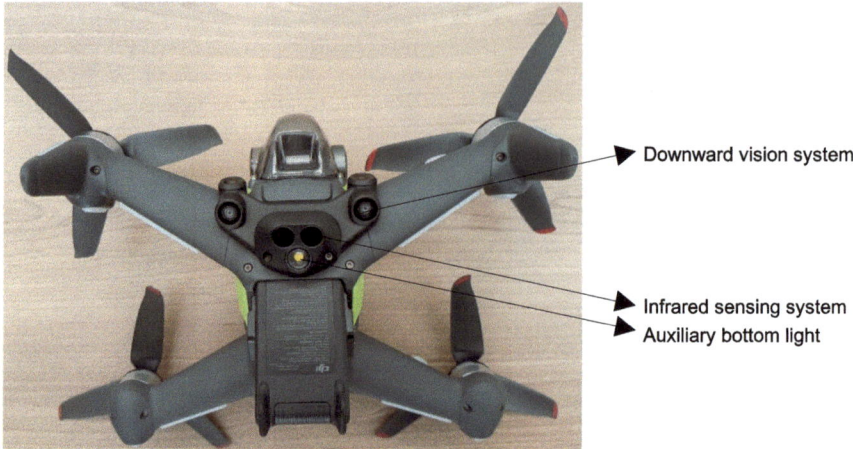

Fig. 2.14 DJI FPV (bottom side)

designed to detect obstacles in the drone's flight path. This functionality improves the drone's ability in navigating difficult environments and prevents collisions. It is ideal for both beginner pilots and advanced users operating in challenging scenarios. The bottom part of the drone has downward vision system, auxiliary bottom light, and infrared sensing system as shown in Fig. 2.14. The downward vision system, located on the bottom of the drone, assists in maintaining stable flight by detecting the ground or surfaces below. This system is particularly useful during takeoff, landing, and low-altitude operations.

The auxiliary bottom light provides additional illumination to improve visibility during low-light conditions. This light supports the downward vision system and maintains stability and positional accuracy, especially in dimly lit environments or at night. The infrared sensing system works in conjunction with the vision systems to detect obstacles and measure distances. By utilizing infrared sensors, the drone can perceive its surroundings with greater accuracy. Similar to other DJI drones, these features further improve its ability to avoid collisions and maintain safe flight.

The top view of the DJI FPV drone shows its sleek design and main components as depicted in Fig. 2.15. The most prominent feature is its quad-rotor configuration, consisting of four propellers positioned to provide stability, precise maneuverability, and efficient lift. Each propeller is marked with color-coded tips (red and black), so that users can easily identify and install them in the correct orientation.

At the center of the drone is the main body, which houses internal components such as the flight controller, battery compartment, and electronic speed controllers (ESCs). These elements are neatly integrated into the drone's compact frame. The top cover, designed in a vibrant green color, improves visibility and makes the drone easy to spot during operation. This feature is particularly important for maintaining a visual line-of-sight, which is a common requirement for safe drone operation in outdoor environments. The cooling vents, located near the central section, maintains

2.10 Hands on Examples: DJI FPV

Fig. 2.15 DJI FPV (top side)

the drone's performance by providing proper airflow. It prevents overheating of the internal components during extended flights or high-performance maneuvers.

The front view of the DJI FPV drone goggles demonstrates its robust design and advanced features tailored for an immersive flying experience as shown in Fig. 2.16. The most noticeable aspect is the sleek and ergonomic design to provide a comfortable fit for extended use. The goggles incorporate multiple antennas, as seen protruding from the sides and top to maintain a stable connection with the drone. These antennas support long-range video transmission to support a real-time and low-latency feed directly from the drone's camera. The goggles' front panel is equipped with large, shaded lenses that house high-resolution displays. It provides the pilot with a crystal-clear first-person view (FPV) of the drone's flight. This visual setup is useful for precise navigation, especially in high-speed or complex flying scenarios. The design also includes vents and subtle contours to ensure proper airflow, avoid fogging, and maintain user comfort during operation.

Figure 2.17 shows the back view of the DJI FPV drone goggles. It reveals its user-centric design and focuses on comfort and functionality for an immersive flying experience. The interior features two high-resolution eyepieces, which provide a clear and wide-angle view of the drone's live feed. These lenses are designed to deliver good image quality and allows pilots to navigate their drones with precision and confidence. The goggles are equipped with an adjustable head strap system to provide a secure and comfortable fit for users of all head sizes. The soft padding around the eyepieces enhances comfort, especially during prolonged usage, while also creating a snug fit to block out external light for an uninterrupted viewing experience. The presence of ventilation slots ensures proper airflow, prevents

Fig. 2.16 DJI FPV google (front)

Fig. 2.17 DJI FPV google (back)

2.10 Hands on Examples: DJI FPV 45

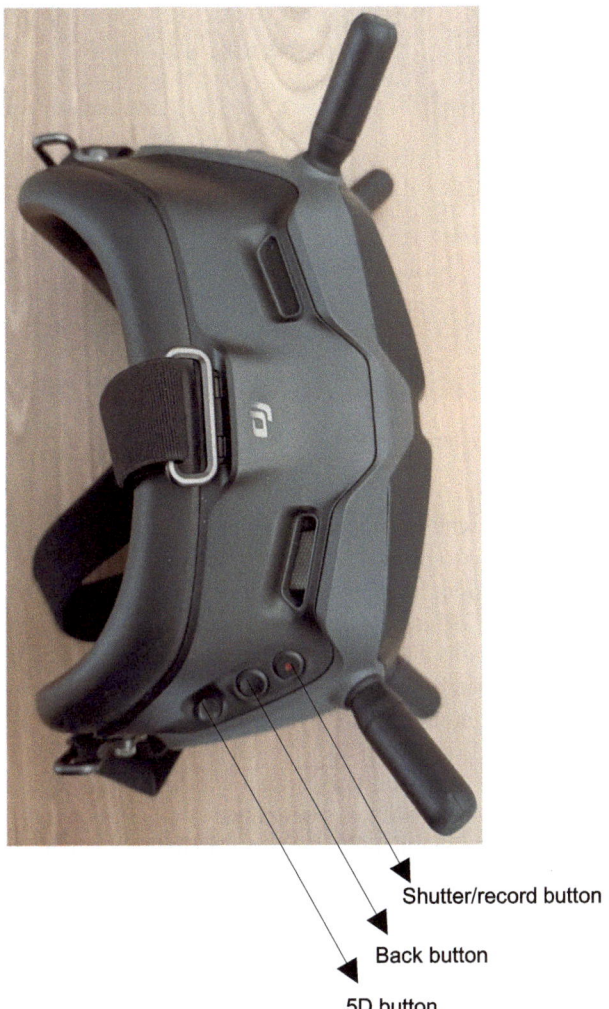

Shutter/record button
Back button
5D button

Fig. 2.18 DJI FPV google (top)

fogging of the lenses, and maintains clear visibility during operation. In addition, the goggles are designed to be lightweight and reducing fatigue for users.

The top view of the DJI FPV drone goggles shows its intuitive control interface (Fig. 2.18). Key controls include the 5D button as input mechanism allows users to navigate menus, make selections, and adjust settings with ease. This multi-directional button simplifies interaction with the goggles. In addition, the shutter/record Button is located for quick access to capture photos or start and stop video recordings easily. This feature is especially useful for capturing precise

Fig. 2.19 DJI Fly application for DJI FPV drone

moments during flight without interrupting the experience. The back button is another important control to navigate back to previous menu screens.

The DJI Fly application, as depicted in Fig. 2.19, is an platform designed to improve the experience of using DJI FPV drones. It offers a range of features to assist users in managing their drone operations effectively. At the forefront is the "Before You Fly" section, which provides information about nearby fly zones, including recommended areas for safe flying. This feature ensures users are aware of flying conditions and adhere to local regulations. The application includes a "QuickTransfer" feature that facilitates seamless transfer of media files, such as photos and videos, from the drone to a smartphone or other devices. This function is useful for users who want to quickly access and share their content. Moreover, the "Service" feature provides access to support resources, troubleshooting guides, and information about firmware updates.

For new users, the application offers a "DJI FPV Beginner Guide", which includes 11 lessons tailored to help beginners understand and operate their DJI FPV drone. This educational section is popular, with over 56.8K users already benefiting from it. Furthermore, the application displays notifications for firmware updates, such as the update for DJI FPV Goggles V2 to give information to users about the latest software. Another feature is the "Connection Guide", located at the bottom right of the screen, which provides step-by-step instructions for connecting the drone, controller, and goggles. This feature is especially useful for first-time users or those troubleshooting connectivity issues.

The application also includes additional sections like "Academy", which offers tutorials and educational resources to improve flying skills, and "SkyPixel", a social platform where users can share and explore creative content. We can see the app's "Album" and "Create" sections in the bottom part of the app. They allow users to manage their media files and create videos or images directly within the app. The "Profile" section enables users to manage account settings, view flight logs, and access personalized features.

Fig. 2.20 Flight Data Center feature in DJI Fly application

The Flight Data Center menu in the DJI Fly application provides a comprehensive overview of a user's drone flight activities as shown in Fig. 2.20. At the top, it confirms successful synchronization of flight records. It means that data is securely stored and accessible across devices or accounts. The menu displays key summary statistics, including the total distance flown (6.53 km), total flight time (48 minutes), and the number of flights (11). This menu gives users a quick snapshot of their overall flying activity. Below this, detailed flight logs are listed including information such as the date of each flight, the distance covered, the maximum altitude reached, and the duration of the flight. Using this information, users can review specific flights for performance analysis or troubleshooting. In addition, a filter for "All Aircraft" suggests the ability to sort data by different drones, which is useful for those managing multiple devices. The inclusion of map icons next to each flight log hints at geolocation data and can be used to visualize flight paths or locations. Finally, we hope that this hands-on section can provide an introduction to users before diving into UAV and drone forensics in following chapters in this book.

2.11 Summary

This chapter provides an overview of the basic components that make up UAVs and the broader ecosystem that supports their development and usage. It begins by exploring the fundamental elements of UAVs, including airframe structures, propulsion systems, sensors, cameras, navigation, control systems, communication systems, and power supplies. The chapter then extends the discussion to the classification of UAVs and detailed drone features, such as return to home, obstacle avoidance, and follow me mode. In addition, the chapter covers the integration of cloud services for UAV management, data processing, mission planning, and introduces the concept of the Internet of Drones (IoD). Finally, we provide a hands on tutorial for understanding drone components and use DJI Mavic Air 2 and DJI FPV as case studies.

2.12 Exercises

1. What are some of the fundamental components of UAVs mentioned in this chapter?
2. What systems are responsible for guiding UAVs during operation?
3. Name two key players in the UAV ecosystem.
4. What role do regulatory bodies play in the UAV ecosystem?
5. What is the "Internet of Drones" (IoD)?
6. Explain how cloud services are utilized in UAV management and mission planning.
7. Discuss the importance of the integration of sensors and cameras in UAV operations.
8. Analyze the relationship between UAV manufacturers and regulatory bodies in the development of drone technology.
9. Evaluate the potential benefits and challenges of implementing the Internet of Drones (IoD).
10. How do advancements in communication systems impact the functionality and applications of UAVs?

References

1. M.Y.-C. Jiang, et al., They believe students can fly: a scoping review on the utilization of drones in educational settings. Comput. Edu. 105113 (2024)
2. R. Díaz-Delgado, S. Mücher, Editorial of special issue 'Drones for biodiversity conservation and ecological monitoring. Drones **3**(2), 47 (2019)
3. D. Ventura, et al., Mapping and classification of ecologically sensitive marine habitats using unmanned aerial vehicle (UAV) imagery and object-based image analysis (OBIA). Remote Sensing **10**(9), 1331 (2018)
4. M. Moreira, et al., Precision landing for low-maintenance remote operations with UAVs. Drones **5**(4), 103 (2021)
5. A. Garcia-Rodriguez, et al., The essential guide to realizing 5G-connected UAVs with massive MIMO. IEEE Commun. Mag. **57**(12), 84–90 (2019)
6. S.A.H. Mohsan, et al., Towards the unmanned aerial vehicles (UAVs): A comprehensive review. Drones **6**(6), 147 (2022)
7. L. Kapustina, et al., The global drone market: Main development trends, in *SHS Web of Conferences*, vol. 129 (2021), p. 11004
8. R. Nouacer, et al., Framework of key enabling technologies for safe and autonomous drones' applications, in *2019 22nd Euromicro Conference on Digital System Design (DSD)* (2019), pp. 420–427
9. A. Otto, et al., Optimization approaches for civil applications of unmanned aerial vehicles (UAVs) or aerial drones: a survey. Networks **72**(4), 411–458 (2018)

10. G. Raja, et al., MLB-IoD: multi layered blockchain assisted 6G Internet of Drones ecosystem. IEEE Trans. Veh. Technol. **72**(2), 2511–2520 (2022)
11. W.-C. Chiang, et al., Impact of drone delivery on sustainability and cost: realizing the UAV potential through vehicle routing optimization. Appl. Energy **242**, 1164–1175 (2019)
12. M. Itkin, M. Kim, Y. Park, Development of cloud-based UAV monitoring and management system. Sensors **16**(11), 1913 (2016)
13. S. Sarkar, M.W. Totaro, K. Elgazzar, Leveraging the cloud to achieve near real-time processing for drone-generated data, in *2019 IEEE Women in Engineering (WIE) Forum USA East* (2019), pp. 1–6
14. G. Ermacora, S. Rosa, A. Toma, Fly4SmartCity: a cloud robotics service for smart city applications. J. Ambient Intell. Smart Environ. **8**(3), 347–358 (2016)
15. J. Moeyersons, et al., UAVs-as-a-Service: cloud-based remote application management for drones, in *2021 IFIP/IEEE International Symposium on Integrated Network Management (IM)* (2021), pp. 926–931
16. M. Rodrigues, D.F. Pigatto, K. RLJC Branco, Cloud-SPHERE: a security approach for connected unmanned aerial vehicles, in *2018 International Conference on Unmanned Aircraft Systems (ICUAS)* (2018), pp. 769–778
17. G. Cai, et al., A brief overview on miniature fixed-wing unmanned aerial vehicles, in *IEEE International Conference on Control and Automation (ICCA)* (2010), pp. 285–290
18. S. Rahman, D.A. Robertson, Multiple drone classification using millimeter-wave CW radar micro-Doppler data, in *Radar Sensor Technology XXIV*, vol. 11408 (2020), pp. 50–57
19. S. Krishnaraj, et al., Aerodynamic analysis of hybrid drone, in *IOP Conference Series: Materials Science and Engineering*, vol. 1012. 1 (2021), p. 012023
20. C. Lee, S. Kim, B. Chu, A survey: flight mechanism and mechanical structure of the UAV. Int. J. Precision Eng. Manuf. **22**(4), 719–743 (2021)

Chapter 3
Survey on Drone and UAV Forensics

Abstract Unmanned Aerial Vehicles (UAVs) are increasingly being utilized for various purposes, such as aerial photography, environmental monitoring, and military applications. As their usage grows, the likelihood of requiring a digital forensic investigation related to UAV activities—whether for accidents, incidents, or their presence in specific locations—also rises. In response, both industry and academia have developed guidelines and publications addressing the forensic investigation of UAV systems. This chapter focuses on exploring the academic contributions to UAV forensics. In this chapter, we conduct a systematic review of the existing UAV forensics literature to identify and document emerging research trends related to the digital forensic analysis of UAV incidents, accidents, and crimes. First, it provides a taxonomy categorizing previous work on UAV forensic artifacts, frameworks, models, forensic readiness, and tools employed in UAV investigations. Finally, this chapter outlines critical research themes that have yet to be fully explored. Furthermore, it offers a roadmap for advancing UAV forensic investigations in the future. This chapter is adopted from [1].

3.1 Introduction

Unmanned Aerial Vehicles (UAVs), also known as drones, are aircraft operated without an onboard pilot, ranging in size from tiny models as small as insects to larger ones comparable to commercial planes [2]. UAVs have seen a surge in popularity, reflected in increasing sales over the past year. For instance, the commercial UAV market is projected to reach $501.4 billion by 2028 [3]. As of May 2022, the U.S. Federal Aviation Administration reported 855,860 registered UAVs and issued 277,845 remote pilot certificates [4].

With technological progress, manufacturers have been able to integrate various accessories into UAVs, such as high-resolution cameras, thermal scanners, and even military-grade weaponry [2]. In military contexts, UAVs have become integral to the Internet of Battlefield Things (IoBT), which also includes networks of sensors, wearable devices, and other Internet of Things (IoT) technologies [5]. This interconnected system is designed to generate massive amounts of data to enhance battlefield decision-making and responsiveness [6].

Despite their benefits, UAVs raise significant cybersecurity concerns, as highlighted by both industry and academia [7, 8]. The U.S. Department of Homeland Security (DHS) has warned that certain UAVs may pose risks to organizational data security [9]. Various studies have demonstrated potential vulnerabilities in commercial UAV software, such as unauthorized takeovers and crashes [10]. One example includes GPS spoofing attacks that manipulate UAV positioning [11], while other research has shown how UAV sensors can be exploited in denial-of-service attacks [12].

These vulnerabilities could not only disrupt businesses and services but also be exploited by criminal organizations, terrorist groups, or adversarial governments. A compromised UAV could potentially be used to target critical infrastructure or public spaces, such as sports venues or government buildings. Furthermore, UAVs could be employed to disrupt communication systems during a coordinated attack, spreading fear and uncertainty. Consequently, multiple mitigation strategies have been developed to detect and address these vulnerabilities [13, 14].

In the event of UAV-related incidents or crimes, forensic investigations are often necessary to gather digital evidence from the UAV for criminal inquiries or civil lawsuits [15]. Organizations such as INTERPOL, EASA, and ISASI have provided guidelines on how to conduct UAV forensics in the context of incidents, accidents, or crimes, offering advice on identifying digital evidence sources and outlining investigative procedures [15–17]. Additionally, academic research has contributed to the development of best practices for UAV digital forensics, including evidence identification, preservation, examination, documentation, and reporting.

This chapter focuses on the academic perspective of UAV forensics by presenting a comprehensive review of the current literature on the subject. Its goal is to identify research trends and gaps related to UAV forensic investigations. Previous surveys on this topic have either overlooked key aspects, such as event reconstruction, or are outdated [18, 19]. After analyzing the literature, we developed as taxonomy to assist with the discussion of our findings and to highlight potential gaps in the literature. This taxonomy is presented in Fig. 3.1.

The taxonomy of UAV forensics serves as a structured framework to address various aspects of investigating incidents involving unmanned aerial vehicles (UAVs). This taxonomy divides UAV forensics into distinct categories. By organizing these categories, investigators and researchers can systematically approach the complexities of UAV-related incidents. One key component of the taxonomy is UAV forensics artifacts, which represents the tangible evidence collected during investigations. These artifacts are further classified into subcategories. UAV device artifacts refer to physical or digital evidence retrieved directly from the UAV, such as hardware components or onboard storage. Flight logs form another subcategory, as they contain detailed records of flight paths, timestamps, and coordinates to assist investigators to reconstruct the UAV's movements. In addition, UAV identification focuses on identifying the specific UAV involved in the incident, including its make, model, serial number, and potentially its operator.

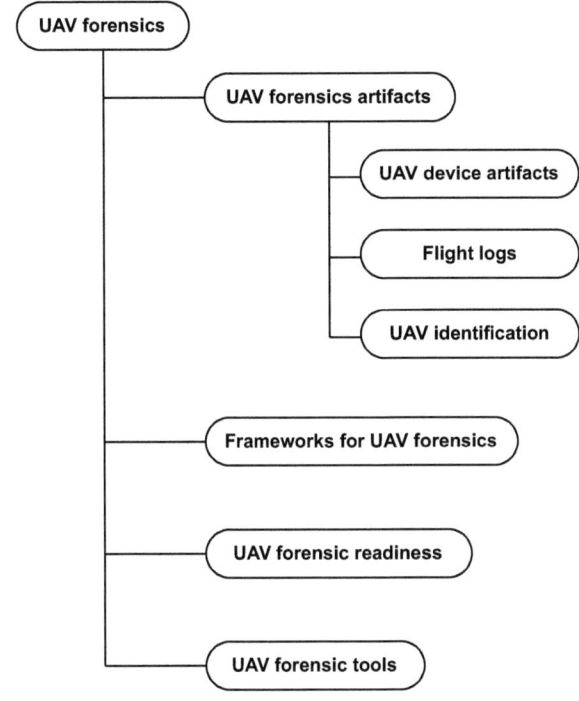

Fig. 3.1 Taxonomy of UAV forensics [1]

Another category is frameworks for UAV Forensics, which provides standardized methodologies for conducting investigations. These frameworks motivates forensic processes to be systematic, reliable, and legally defensible. They guide investigators through evidence collection, preservation, analysis, and reporting to provide consistent practices across different cases and jurisdictions. Complementing these frameworks is the concept of UAV forensic readiness, which focuses on the importance of preparedness in handling UAV-related incidents. Readiness involves having the necessary tools, protocols, and trained personnel in place to respond swiftly to forensic challenges. Finally, the taxonomy highlights the role of UAV forensic tools, which are specialized software and hardware used in analyzing UAV-related data. These tools assist in extracting information from UAV devices, interpreting flight data, and recovering deleted or hidden information. Examples include GPS analysis software, data recovery tools, and specialized hardware for accessing UAV components.

3.2 UAV Forensic Artifacts

A review of the literature indicates that there are four primary sources of potential digital evidence that investigators should consider during a UAV forensic examination: the flight controller, ground control station, the smartphones used to control

Table 3.1 Artifact path and type of UAV

Reference	UAV	Location	Type
[23]	A.R Drone	/data/syslog.bin	System logs
		/data/config.ini	Configuration file
		/data/emergency.bin	n/a
		/data/custom.configs/sessions/	GPS data
		/data/custom.configs/profiles/	Controller footprint
[27]	DJI Phantom 3 Advanced	/root/FLY*.DAT	GPS data
[28]	Parrot AR	/data/video/boxes/flight_*/	Flight data
		/data/video/boxes/tmp_flight_*/	Media (pictures)
		/data/video/usb/media_*/	Media (videos)
		/data/custom.config/	Configuration file
[29]	Cheerson CX-20	/logs/QUADROTOR/1	Flight data
[30]	DJI Mavic Air	/DCIM/100MEDIA/	Media
[31]	Parrot Bebop 2	internal_000/Bebop_2/media/	Media
		/internal_*/log/	System logs
		/internal_*/Debug/crash_reports/	Crash logs
[32]	Yuneec Typhoon H	/1/vol_*/DCIM/100MEDIA/	Media
[33]	DJI Mini 2	/DCIM/100MEDIA	Media
[34]	DJI Phantom 4	/DCIM/100MEDIA/	Media
	DJI Mavic Pro	/flyctrl/FLY*.DAT	Flight data
		/NO NAME/DCIM/100MEDIA/	Media
	Yuneec Mantis Q	/NO NAME/DCIM/100MEDIA/	Media

UAVs, and the UAV itself [20–22]. These artifacts are discussed in Chap. 6 in details. Table 3.1 summarize the artifacts and their respective locations that can be recovered from UAVs and their ground controllers (Table 3.2). Several key insights can be drawn from Table 3.1. For example, in their forensic investigation of a DJI Phantom 3 Professional UAV, Barton and Azhar [23] found that Android smartphones controlling this UAV could be used to recover flight data and media files. Later, [24] identified several open-source tools—such as dd, mount, dmesg, and file—that can be used for forensic analysis of UAVs, tools generally available on Linux systems. Their analysis recovered session activities, GPS data, email messages, media, and flight information [24]. In addition, Hamdi et al. [25] discussed flight data from the DJI Phantom 4 UAV, reporting that credentials required to control the UAV can be retrieved from smartphone ground stations, specifically from the following file paths: Library/Preferences/com.dji.go.plist and Documents/.Dji.configs.

In their investigation of the DJI Spark, Kao et al. [26] detailed the recovery of flight data in .DAT format from an internal SD card, along with pictures and videos from a memory card. They also noted that additional pictures and videos could be recovered from an external memory card, though it was unclear if these files

3.2 UAV Forensic Artifacts   55

Table 3.2 Artifact path and type of a UAV ground controller [1]

UAV	Location of artifact(s)	Type
DJI Phantom 3 Professional [23, 38]	/media/0/DJI/dji.pilot/LOG/CACHE	Flight data
	/media/0/DJI/dji.pilot/LOG/CACHE/NFZ	Flight data
	/media/0/DJI/dji.pilot/LOG/ERROR_POP_LOG	Flight data
	/media/0/DJI/dji.pilot/DJI_RECORD	Media
	/media/0/DJI/dji.pilot/FlightRecord	Flight data
	/media/0/DJI/dji.pilot/CACHE_IMAGE	Media
DJI Phantom 3 Advanced [27]	/root/DJI/dji.pilot/FlightRecord/DJIFlightRecord_*.txt	Flight data
Parrot A.R Drone 2.0 [24]	userdata/data/com.parrot.freeflight	Sessions activity
	userdata/com.parrot.freeflight/shared_prefs/Preferences.xml	GPS data, email
	userdata/media/0/DCIM	Media
DJI Phantom III Professional [39]	data/dji.pilot/DJI/FlightRecords	Flight data
DJI Phantom 4 [25]	Apps/DJIGo/FlightRecord (iOS)	Flight data
	/DJI/dji.pilot/FlightRecord (Android)	Flight data
DJI Spark [26]	DJI/dji.go.v4/CACHE_IMAGE	Media
	DJI/dji.go.v4/DJI_RECORD	Media
	DJI/dji.go.v4/FlightRecord	Flight data
DJI Phantom [21]	/Media01/apps/dji.go.v4/	Controller artifacts
DJI Phantom 4 Pro [40]	/DJI/dji.pilot/LOG/CACHE/NFZ	NFZ data
DJI Phantom 3 [41]	Apps/DJIGo/FlightRecord (iOS)	Flight data
	DeviceStorage/DJI/dji.pilot/-FlightRecord (Android)	Flight data
DJI Phantom 4	Apps/DJIGo/FlightRecord (iOS)	Flight data
	DeviceStorage/DJI/dji.pilot/FlightRecord (Android)	Flight data
Bebop Parrot Drone 1	/Applications/com.parrot.freeflight3/documents/academy (iOS)	Flight data
	/data/data/com.parrot.freeflight3/files/academy (Android)	Flight data
DJI Mini 2 [33]	/private/var/mobile/Containers/Data/Application*/Documents/FlightRecords/	Flight data
	/Application/*/Library/.space_db/flysafe_dji_flight_dynamic_areas.db	Flight data
	Application/*/Library/Preferences/com.dji.golite.plist	Controller artifacts
	Application/*/Documents/Tmp/DJISyncLog*.txt	Sync data
	private/var/mobile/Library/Caches/com.apple.routined/Cache.sqlite	Cache data
	data/media/0/DJI/dji.go.v5/FlightRecord	Flight data
	/data/data/dji.go.v5/databases/	Controller artifacts

(continued)

Table 3.2 (continued)

UAV	Location of artifact(s)	Type
	/data/media/*/DJI/dji.go.v5/CACHE_IMAGE/ImageCaches/	Media
DJI Phantom 4 [34]	/mobile/Containers/Data/Application/com.dji.go/Library/Preferences/ (iOS)	Controller Artifacts
	/mobile/Containers/Data/Application/com.dji.go/Library/ApplicationSupport/ApplicationSupport/	Application artifacts
	/mobile/Containers/Data/Application/com.dji.assistant/Library/Preferences/	User credentials
DJI Mavic Pro	/mobile/Containers/Data/Application/com.dji.go/Documents/FlightRecords/	Flight data
	/mobile/Containers/Data/Application/com.dji.go/Documents/.mediaLibrary.Cache/	Media
	data/Root/media/0/DJI/dji.go.v5/DJI FLY/Video/	Media
	data/Root/media/0/DJI/dji.go.v5/DJI FLY/Photo/	Media
Yuneec Mantis Q	/data/data/com.yuneec.android.z/shared_prefs/	Controller artifacts
	data/com.yuneec.android.z/databases/yuneec.db	Flight data

duplicated those found on the internal card. The research [26] also analyzed the temporal relationship between artifacts to explore interactions between the UAV and its Android smartphone controller.

Another work identified encrypted flight logs from the DJI Phantom 4 Pro, DJI Mavic Pro, and Yuneec Typhoon H UAVs [32]. While decrypting these logs improves their integrity as digital evidence, the encryption used on the Yuneec Typhoon H was found to be weak, potentially allowing an attacker to alter the logs before analysis. Furthermore, another study conducted an in-depth analysis of the DJI Mavic 2 Pro and DJI Phantom 4 UAVs, discovering media files such as images and videos, along with EXIF metadata embedded in the images [35]. When examining the backup generated from an iPhone used to control the UAVs, they found further details, including the serial number and platform of the specific UAV. However, flight data could not be recovered in either case due to encryption.

Salamh et al. [36] evaluated various digital forensic tools—such as Autopsy, Cellebrite Universal Forensic Extraction Device (UFED), and Magnet AXIOM—for examining UAVs. One notable finding was that timestamps retrieved using Autopsy differed from those recovered using Cellebrite UFED due to differences in the algorithms used to decrypt flight logs. Separately, [37] demonstrated a brute force attack on a UAV's camera, enabling an intruder to access the file system and create or delete files or directories. This led to the proposal of a UAV Kill Chain as a method for advanced intrusion detection. Focusing on cyberattacks, Iqbal et al. [31] investigated forensic methods for detecting de-authentication and integrity attacks on UAVs. They identified plist files and flight logs as critical sources of evidence, recovering de-authentication attack evidence from log files and integrity attack evidence from files uploaded to the UAV. To provide more insight into the

3.2 UAV Forensic Artifacts

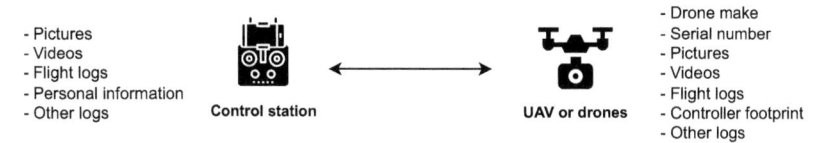

Fig. 3.2 Common UAV or drone forensic artifacts [1]

types of data that can be recovered from UAVs, the paper proposed a data context model.

Another research [42] explored different sources of information for investigating cybercrimes involving UAVs and recommended using traceback solutions to guide UAV investigations. They also proposed a neural network-based model for identifying real-time cyberattacks on UAVs. Forensic reconstruction involves examining the sequence of events leading to a security incident or accident [43]. One can reconstruct events forensically using a directed graph-based search, extracting events from UAV log messages, and employing depth-first search to trace the sequence from one event to another [44]. In addition to flight logs, UAV forensic investigations also focus on video files captured by the UAV's camera, which may provide visual evidence of incidents such as collisions. Editya et al. [45] aimed to automatically analyze UAV videos to determine object movement across frames. They find that the Lucas-Kanade method offered the most accurate estimates of object direction by calculating specific motion vectors.

Figure 3.2 summarizes the most common artifacts recoverable during UAV forensic investigations. Drone or UAV artifacts from the controller often contain vital information linking the drone's activities to its operator. The controller, acting as the intermediary between the operator and the drone, stores a variety of data types that can provide insights into the drone's usage, operator behavior, and potential intent. One of the key artifacts found on the controller includes pictures and videos. These media files, which may have been captured during flights, can provide evidence of the drone's purpose, whether for recreational, commercial, or malicious use. These files might include aerial images, videos of target locations, or other recordings that could aid investigators in reconstructing the context of the flight.

Another artifact is the flight logs stored in the controller. These logs record critical data about the drone's operation, including timestamps, GPS coordinates, altitude, and speed. By analyzing flight logs, investigators can reconstruct the drone's flight path and determine its activities during specific periods. Additionally, these logs might reveal any unauthorized or suspicious activities carried out by the drone. The controller can also contain personal information about the operator. This could include user profiles, account details, or even linked mobile devices that provide clues about the operator's identity. Such information is valuable in identifying the individual responsible for the drone's operation.

Other artifacts include the controller footprint and other logs. The controller footprint refers to data remnants left behind during usage, such as settings configurations, network connections, and communication protocols with the drone. These remnants can reveal how the controller was set up and connected to the drone, offering further insights into the operation's technical details. In addition, other logs might include metadata, error reports, or operational records, all of which can provide supplementary evidence during forensic analysis.

Although UAVs and their controllers are interconnected, the literature reveals a lack of research correlating events between these devices. This presents several challenges and opportunities for advancing event correlation and reconstruction between UAVs and their controllers.

3.3 UAV Flight Logs Analysis

Flight logs in UAV forensic investigations provide valuable insights into flight history, including specific latitude, longitude, and altitude coordinates. This information assists investigators in determining where an accident, incident, or cybercrime involving a UAV occurred. In a comprehensive study, Renduchintala et al. [22] analyzed and visualized flight log data using open-source tools like DroneForensicsSoftware.[1] This software allows visualization of UAV flight routes via Google Maps and facilitates flight control analysis through line graphs. Clark et al. [46] developed another open-source tool, the Drone Open Source Parser (DROP), which is used to extract flight logs from proprietary DAT file formats recovered from UAVs. DROP supports file correlation to establish relationships between DAT files containing flight logs and artifacts from Android controller devices.

Moreover, another research focused on the GPS data found in flight logs, identifying three flight scenarios related to GPS signals: P-GPS (full GPS signal), P-OPTI (weak but available GPS signal), and P-ATTI (no GPS signal) [27]. Their findings indicated that available GPS signals are logged in flight logs and can be effectively used to trace the UAV's flight history. Further studies by Barton and Azhar [38] demonstrated that flight logs could be retrieved from both UAV storage devices and their ground stations or controllers. The study employed various tools such as dd, adb, exiftool, fsstat, and mount to conduct forensic investigations of flight logs, which included GPS coordinate data.

In another study, Mantas et al. [47] explored the analysis of dataflash and telemetry logs from UAVs using a tool called GRYPHON.[2] It is developed to analyze logs from the Ardupilot platform. GRYPHON helps extract flight data, identify anomalies, map GPS coordinates, detect altitude variations, analyze commands issued to the UAV, and perform timeline analysis. When a UAV crash or accident occurs, determining the cause and preventing future incidents becomes a priority [21]. However, as noted by researchers [48–50], it can be challenging to ascertain whether a crash was caused by a crime, malfunction, external factors,

---

[1] https://github.com/ankitrlps/DroneForensicsSoftware.
[2] https://github.com/emantas/GRYPHON_dft.

or human error. Consequently, several research efforts have developed tools and methodologies to assist with UAV crash investigations. For example, a sequence mining technique is applied to detect critical flight log patterns, which can support postmortem analysis or accident investigation [49]. Separately, the application of natural language processing on flight logs to enhance forensic investigations of UAV accidents is introduced [50]. Note that a more detailed discussion of flight logs forensics is provided in Chap. 7.

3.4 UAV Identification

Several researchers have concentrated on the challenge of UAV identification, aiming to determine either the UAV pilot or the UAV itself. For instance, a study developed a pilot identification method using machine learning techniques [51]. They collected flight behavior data from twenty pilots, including signals related to pitch, roll, yaw, thrust, and their derivatives. Machine learning algorithms such as random forests, support vector machines, and linear discriminant analysis were then applied to identify pilots based on their flying styles. However, a limitation of this approach is that it requires prior knowledge of a pilot's flying behavior, which may not be accessible in real-world forensic investigations. Therefore, this method less practical for pilot identification in such scenarios.

Moreover, concerns have been raised about the possibility of malicious pilots spoofing traditional UAV identification methods, such as RF communication and ID markers. To address this issue, Li et al. [52] proposed a UAV identification technique based on the unique hardware characteristics of the UAV itself, such as parasitic electronic components detected through RF interrogation, which are specific to each UAV. Building on this concept, they designed and implemented DroneTrace, a comprehensive and passive identification system capable of identifying UAVs with an error rate of less than 5%.

3.5 UAV Forensic Tools and Datasets

UAV forensics is a field that requires specialized tools and datasets to address the unique challenges associated with investigating UAV-related incidents. These tools and datasets are required to extract, analyze, and interpret forensic artifacts from UAVs and their controllers. By using such resources, investigators can identify evidences, reconstruct UAV activities, and evaluate anomalies or malicious behavior. This section provides an overview of forensic tools and publicly available datasets that support UAV forensic investigations.

Table 3.3 List of UAV forensic tools

Tool	Description	Reference
DatCon	DatCon parses .DAT file containing flight logs data and output a CSV file	[53]
CsvView	CsvView provides a visualization from a CSV file generated from DatCon	[54]
Gryphon	A tool to perform an investigation on telemetry and flight logs	[47]
DROP	An open source parser for UAV data	[46]
Magnet AXIOM	A generic tool which supports UAV forensics	[55]
Cellebrite UFED and physical analyzer	A tool for collecting evidence and performing generic investigations	[56]

3.5.1 Tools

To conduct a digital forensic examination of a UAV, investigators have access to various tools, both proprietary and open-source, developed by industry and academic sources, as shown in Table 3.3. Several insights can be gathered from this table. One effective method for reading, parsing, and visualizing UAV flight logs stored in .DAT format is the combination of DatCon [53] and CsvView [54]. DatCon is used to read and parse flight log files. It produces a CSV output, which can then be input into CsvView to generate visualizations of the flight data.

Other noteworthy tools include Magnet Axiom and Cellebrite's Universal Forensic Extraction Device (UFED), both of which are designed to process UAV data and extract relevant artifacts for analysis. Similarly, DROP [46] is an open-source tool which can parse encrypted DJI flight logs data. Another open-source tool, GRYPHON, offers multiple features for UAV investigators, such as integrity checking, trajectory analysis, error detection, command verification, and timeline analysis [47].

Additional tools mentioned in the literature include DroneForensicsSoftware [22], which is capable of extracting, parsing, and visualizing UAV data. It is also an open-source tool that can parse and extract information from various UAV artifacts, including flight data, GPS coordinates, control device information, battery levels, altitudes, and flight statuses. In addition, PhantomHelp LogViewer[3] can be used for parsing and viewing flight logs from DJI UAVs. Subsequently, GPSVisualizer[4] can generate visualizations of UAV flight paths; and AirData,[5] an online tool which provides insights such as flight data, battery health, and control responses. While there are many tools available for UAV forensic investigations, Al-Dhaqm et al. [19] stress the importance of using "proper and trusted" tools. However, they provide

---

[3] https://www.phantomhelp.com/LogViewer/Upload/.
[4] https://www.gpsvisualizer.com/.
[5] https://airdata.com/.

3.5 UAV Forensic Tools and Datasets

Table 3.4 List of UAV public datasets

Dataset	Description	Reference
VTO Drone data set	Data sets from VTO's drone forensics program	[62]
UAV Attack data set	Contains data related to attacks on UAVs including GPS spoofing and jamming	[58]
ALFA	AirLab failure and anomaly detection data set for UAVs	[59]
VisDrone data set	A data set for object detection and object tracking from UAV images and videos	[60]
Drone detection data set	A data set to detect flying objects such as UAVs	[61]
VisDrone-data set	Object detection and tracking data set captured from UAVs	[60]

limited guidance on how to evaluate whether a specific UAV forensic tool meets this standard. Several of the tools, such as DROP and GRYPHON discussed in this section will be put into practice in subsequent chapters in this book.

3.5.2 Datasets

As the field of UAV forensics advances, it becomes important to guarantee that the tools and methodologies designed for this domain are dependable and yield consistent, reproducible outcomes. [57]. To evaluate these metrics, researchers need access to publicly available data sets containing UAV data. Such data sets allow for objective comparisons of proposed tools and techniques against other methods. In addition, they are valuable resources for training and educational purposes in digital forensics. Table 3.4 presents a summary of the public UAV forensic data sets identified in the literature.

The VTO Drone data set,[6] compiled by VTO, Inc., includes data from 79 UAVs across 30 different models, such as the Phantom 3, Phantom 4, Phantom 4 Pro, Inspire 1, Inspire 2, Mavic Pro, and Mavic Air. This data set was created by flying the UAVs and collecting data from both the UAVs and the mobile devices used to control them. Furthermore, for UAVs with memory card capabilities, forensic copies of both internal and external SD cards are included. Whelan et al. [58] introduced the UAV attack data set, which contains flight logs from both a benign UAV flight and a UAV subjected to GPS spoofing and jamming attacks. The experiment was conducted using a PX4 Autopilot v1.11.3 UAV, equipped with a Pixhawk 4 flight controller and Pixhawk GPS receiver.

The ALFA data set [59] provides flight information on fixed-wing UAVs, including anomalous events and records. It is suitable for UAV fault detection

---
[6] https://cfreds.nist.gov/all/SteveWatson%2FVTOInc./Dronedataset.

and anomaly isolation research. This data set consists of information from 47 autonomous flights, including 23 scenarios of full engine failure and 24 other types of faults. It also includes hours of raw data from fully autonomous, autopilot-assisted, and manual flights. Moreover, Zhu et al. [60] released the VisDrone data set, containing UAV flight data collected over various urban and suburban areas across 14 cities in China. This data set supports research on visual analysis algorithms for UAV platforms. From a forensic perspective, it could be used to develop machine learning models for detecting and tracking objects captured by UAV cameras. Lastly, Ozel [61] published a data set containing approximately 1400 UAV images and labeled files, intended for research on UAV collisions, including both images and videos related to such events.

3.6 UAV Forensic Readiness

Forensic readiness is generally understood as optimizing the use of digital evidence while minimizing the costs associated with digital forensic investigations [63]. Researchers have explored different strategies to improve UAV forensic readiness. For example, [64] suggest using electromagnetic watermarking as a way to enhance forensic readiness in UAVs. This technique involves storing encrypted information, such as a random number and a timestamp, with a secret key. When decrypted during an investigation, this information helps identify the UAV involved.

In a study by Yu et al. [65], they introduced a blockchain-based solution to ensure tamper-proof storage of UAV flight logs. Instead of storing logs locally, they are uploaded to a cloud-based blockchain system. This approach allows investigators to access flight logs from the cloud, even if the UAV is unrecoverable following an incident or accident, enabling the continuation of the investigation. To accommodate readiness-based framework, Alotaibi et al. [66] proposed the Drone Forensics Readiness Framework (DRFRF), which is divided into two stages: a proactive forensic stage and a reactive forensic stage. The DRFRF emphasizes centralized logging of all UAV-related events in anticipation of potential future investigations. By doing so, the time required to gather evidence is reduced, allowing investigators to proceed more quickly to the examination and analysis phase in any subsequent forensic investigation.

3.7 Conceptual Drone Forensics Framework (CDFF)

Several researchers have proposed frameworks aimed at guiding UAV forensic investigations. However, upon reviewing these frameworks, it is evident that many either omit critical phases or standard practices in traditional digital forensics. To address these shortcomings, we introduce the Conceptual Drone Forensics Framework (CDFF) as a more comprehensive solution designed to assist in UAV

3.7 Conceptual Drone Forensics Framework (CDFF)

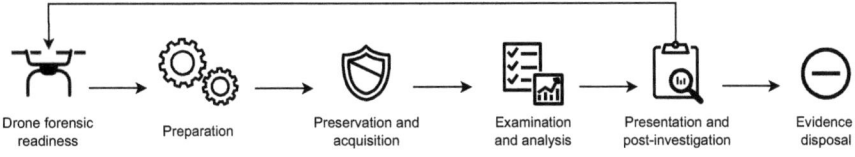

Fig. 3.3 Conceptual Drone Forensics Framework (CDFF) [1]

forensic investigations. The CDFF is structured into six distinct phases: (1) UAV forensic readiness, (2) Preparation, (3) Preservation and acquisition, (4) Examination and analysis, (5) Presentation and post-investigation, and (6) Evidence disposal. Figure 3.3 provides a visual overview of the framework, while the subsequent sections explain the activities associated with each phase in detail.

UAV Forensic Readiness Phase This phase focuses on ensuring that investigators can efficiently collect potential evidential data while minimizing the cost of the UAV forensics process [67]. The framework incorporates and builds upon the forensic readiness activities introduced by [63]:

- Defining scenarios that require digital evidence. The first step is to identify scenarios where digital evidence from a UAV might be necessary. These scenarios could involve traditional crimes where a UAV is used as a tool (e.g., transporting illegal substances), digital crimes (e.g., video surveillance), crashes or accidents, or use in a military context as a weapon. Investigators must identify what information is needed in each case to determine the incident and responsible parties.
- Identifying sources and types of evidence. According to the literature [20–22], key sources of potential evidence in the UAV ecosystem include the flight controller, ground control station, controlling smartphones, and the UAV itself. Cloud storage could also be relevant [65]. Investigators should be aware of the data generated at each source, its format, duration of storage, who holds responsibility for it, and any legal or technical support needed to obtain it.
- Determining evidence collection requirements. UAVs are not typically designed for digital forensics, so investigators must consider how to retrieve data from the identified sources. If data is stored in the cloud, for instance, maintaining the chain of custody and ensuring the evidence is preserved is crucial.
- Establishing secure storage policies for evidence. Once evidence is collected, it must be securely stored following digital forensics standards [68]. A centralized, secure database may be used to store UAV data, with regular audits for accountability.
- Training staff in incident awareness. All individuals involved in a UAV forensics investigation should be well-trained in handling digital evidence and understanding its legal implications. Proper training ensures evidence is preserved and handled correctly, minimizing the risk of contamination or accidental deletion.

Preparation Phase In this phase, forensic investigators must ensure they are equipped to handle a UAV forensics investigation. This step involves having the necessary hardware and software for acquiring and analyzing UAV data. It should establish policies and standards for UAV forensics, and creating a protocol to maintain the chain of custody throughout the investigation. Investigators should also have mechanisms in place to be alerted to potential or ongoing investigations, either through manual reporting or automated detection.

Preservation and Acquisition Phase Once an investigation is identified, the investigator proceeds with the preservation and acquisition phase, which involves preserving and collecting both volatile and non-volatile data from relevant evidence sources. The process starts with assessing the condition of the evidence sources (e.g., powered on/off, damaged/undamaged), followed by the appropriate acquisition method. For instance, data retrieval from a damaged UAV might require advanced techniques like "chip-off" acquisition, while an undamaged UAV may allow for conventional approaches such as accessing an onboard port or memory card. Throughout this process, it is important to use forensically sound tools that do not alter or destroy the evidence before it is analyzed.

Examination and Analysis Phase During this phase, the collected artifacts are examined to determine their relevance to the investigation. Investigators may form hypotheses, gather evidence to support or refute these hypotheses, and establish a timeline of events. Any discovered evidence should undergo validation to ensure it was not altered during the examination process. A key activity in this phase is reconstructing the UAV's flight path, which can provide further insight into the incident under investigation.

Presentation and Post-incident Phase At the end of the investigation, the findings are summarized in reports or briefings that can be presented to courts, regulatory bodies, or even UAV manufacturers. The investigation may also serve as a learning opportunity, leading to improvements in UAV operations or the UAV forensics process. Any lessons learned should be integrated into the UAV forensic readiness phase to enhance future investigations.

Evidence Disposal Phase Once the investigation concludes, evidence may need to be properly disposed of according to legal, regulatory, and privacy guidelines. Evidence disposal involves securely removing or destroying data that is no longer needed. The following steps are proposed for evidence disposal:

1. Documentation and cataloging of the evidence, including its source, chain of custody, and any analysis techniques used.
2. Sanitization of sensitive or confidential information using approved methods.
3. Verification and certification that the disposal process was successful.

Table 3.5 summarizes the phases of the CDFF, comparing it with other frameworks found in the literature. The CDFF stands out for incorporating UAV forensic readiness and evidence disposal. Future work will focus on validating the

3.8 Challenges and Future Research Directions

Table 3.5 Comparison of existing frameworks with CDFF framework

Framework	Readiness	Preparation	Acquisition and preservation	Examination and analysis	Presentation and post-investigation	Evidence disposal
Jain et al. [69]	×	✓	✓	✓	✓	×
Peruzzi et al. [70]	✓	✓	✓	✓	✓	×
Bouafif et al. [28]	×	×	✓	✓	✓	×
McAteer et al. [29]	×	✓	✓	✓	✓	×
Gulatas et al. [39]	×	✓	✓	✓	✓	×
Roder et al. [71]	✓	✓	✓	✓	✓	×
Renduchintala et al. [22]	✓	✓	✓	✓	✓	×
Salamh et al. [72]	×	✓	✓	✓	✓	×
Thornton et al. [34]	×	✓	✓	✓	✓	×
Alhussan et al. [73]	×	✓	✓	✓	✓	×
CDFF [1]	✓	✓	✓	✓	✓	✓

CDFF through empirical studies involving UAVs from different manufacturers and scenarios.

3.8 Challenges and Future Research Directions

Our review of UAV forensics literature has identified several areas requiring further investigation, which supports the need for a comprehensive research agenda in UAV forensic investigations. This section outlines key areas for future research, including: forensic investigations of UAV ecosystems, the role of artificial intelligence and machine learning in UAV forensics, and the quality of data used in UAV forensic investigations.

3.8.1 Forensic Investigation of UAV Ecosystems

Current literature shows that UAV pilots use controllers or smartphones to operate UAVs. Much of the existing research focuses on recovering and analyzing artifacts from UAVs or their controllers [26, 41]. However, with the increasing use of cloud

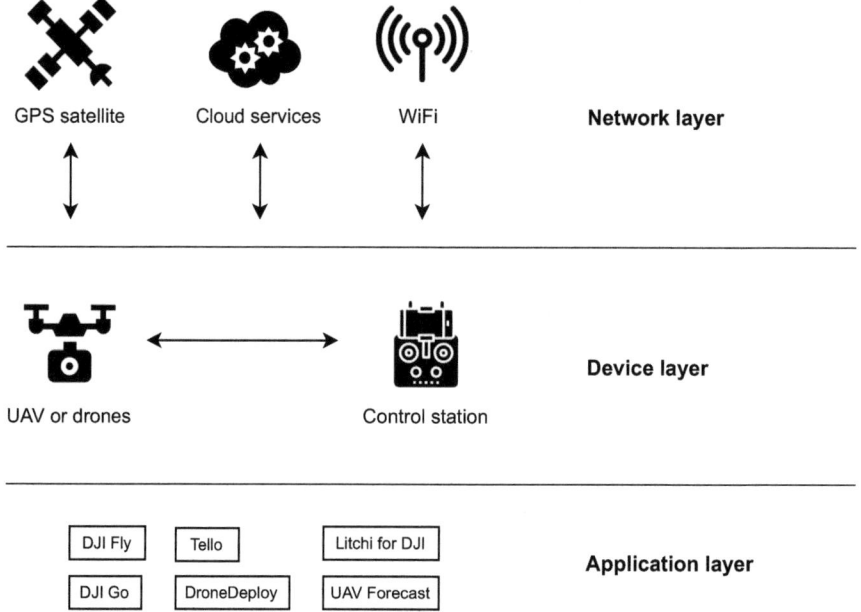

Fig. 3.4 The UAV ecosystem [1]

storage for UAV data [65], forensic investigators must consider the cloud as a potential source of evidence. Moreover, some UAVs interact with GPS satellites, leaving residual data that could be valuable for investigations [23, 27]. Thus, the inclusion of cloud data, GPS interactions, smartphone apps, controllers, and UAVs has led to the creation of a broader UAV ecosystem as depicted in Fig. 3.4.

Within this ecosystem, potential evidence may reside in the application, device, or network layers [74, 75]. For example, the application layer may contain data from applications such as DJI Fly[7] or DroneDeploy,[8] the device layer might hold firmware and UAV operating system data, and the network layer could reveal GPS, Wi-Fi information, and cloud-stored data. Future research should focus on identifying which parts of the ecosystem provide the most comprehensive evidence and explore methods for retrieving it, especially from cloud and GPS sources. Legal, technical, and organizational challenges in accessing and analyzing this evidence also need to be addressed [76, 77].

In addition, cloud service providers such as Flytbase[9] and Airdata[10] are gaining popularity for UAV data management. Flytbase allows for uploading flight

---

[7] https://www.dji.com/id/dji-fly.
[8] https://www.dronedeploy.com/.
[9] https://www.flytbase.com/.
[10] https://www.airdata.com/.

3.8 Challenges and Future Research Directions

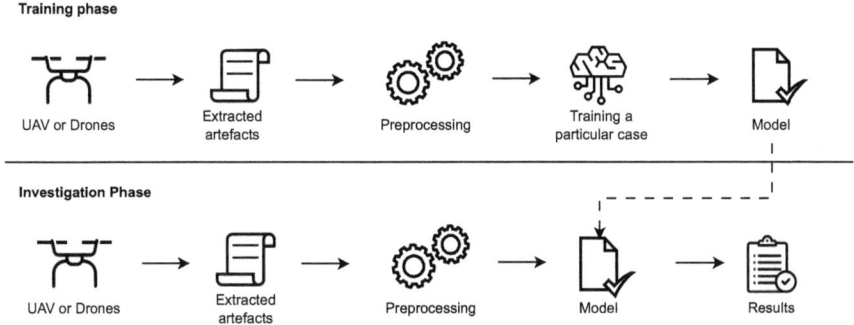

Fig. 3.5 Application of AI and ML in UAV forensics [1]

telemetry, while Airdata collects flight data, engine information, and environmental conditions. Future research should investigate how data from these third-party providers can be accessed to provide deeper insights into UAV incidents or accidents.

3.8.2 UAV Forensics Support by Artificial Intelligence and Machine Learning

Another promising area for future research is integrating machine learning into UAV forensic investigations as illustrated in Fig. 3.5. Previous studies have shown that artificial intelligence, including machine learning and deep learning techniques, can assist in digital forensic investigations [78–81]. For example, deep learning has been used to detect anomalies in forensic timelines [40, 82], but there has been limited exploration of its application in UAV forensics.

For machine learning to be effectively applied, there is a need for publicly available, comprehensive data sets of UAV activities, which can be used for training models. This reinforces the call for the UAV community to publish data sets to aid in model development. With such data sets, machine learning models could be developed to automate the detection of UAV failures or suspicious flight paths. These models could analyze data from the UAV ecosystem to determine the root cause of an incident.

3.8.3 Quality of Data for UAV Forensics Investigations

A third potential topic for future research involves examining the quality of data used in UAV forensic investigations. Data quality can be defined as "fitness for purpose" [83], and in UAV forensics, it can be evaluated based on accuracy, timeliness, completeness, and consistency [84]. Future studies should investigate whether the data and artifacts recovered from UAV ecosystems meet these standards and are suitable for enhancing forensic investigations.

From an accuracy standpoint, researchers need to determine whether the recovered artifacts truly reflect what occurred during the UAV flight. In terms of timeliness, the timestamps associated with forensic artifacts should match the actual dates and times of events. If there are errors, inaccurate timestamps could lead to incorrect investigative conclusions. Lastly, in terms of completeness and consistency, investigators need sufficient information, presented in a standardized format, to conduct thorough and reliable UAV forensic investigations.

3.9 Summary

The chapter provides a comprehensive review of the field of Unmanned Aerial Vehicle (UAV) forensics. It highlights the increasing use of UAVs in areas such as aerial photography, environmental monitoring, and military applications, which in turn heightens the need for digital forensic investigations related to UAV activities. These investigations may focus on accidents, incidents, or unauthorized UAV presence.

The chapter offers a systematic review of the current literature, focusing on key academic contributions to UAV forensics. It categorizes previous work into various areas, including UAV forensic artifacts, frameworks, models, forensic readiness, and tools used in UAV forensic investigations. Moreover, it identifies and outlines three critical research themes that remain underexplored, suggesting potential directions for future innovations in UAV forensics.

3.10 Exercises

1. What are some of the common applications of UAVs mentioned in this chapter?
2. Why is there an increasing need for forensic investigations related to UAV activities?
3. What does this chapter aim to explore within the field of UAV forensics?
4. What types of events often require UAV forensic investigations?
5. What does the chapter's taxonomy categorize in the field of UAV forensics?
6. Discuss the significance of academic contributions to the field of UAV forensics in addressing incidents and crimes involving UAVs.
7. How can the taxonomy provided in this chapter aid in organizing research on UAV forensic artifacts and frameworks?
8. Analyze the challenges faced in conducting forensic investigations for UAV incidents and accidents.

9. Evaluate how existing frameworks and models contribute to improving the forensic analysis of UAV systems.
10. Propose additional categories or considerations that could enhance the taxonomy of UAV forensics introduced in this chapter.

References

1. H. Studiawan, G. Grispos, K.-K.R. Choo, Unmanned Aerial Vehicle (UAV) forensics: the good, the bad, and the unaddressed. Comput. Secur. 103340 (2023)
2. T. Matiteyahu, Drone regulations and fourth amendment rights: the interaction of state drone statutes and the reasonable expectation of privacy. Colum. JL Soc. Probs. **48**, 265 (2014)
3. G. Newswire, *Commercial Drone Market Size, Share & Trends Analysis Report By Product, By Application, By End-use, By Region And Segment Forecasts, 2021–2028* (2021). Available Online: https://www.globenewswire.com/news-release/2021/05/05/2223128/0/en/Commercial-Drone-Market-Size-Share-Trends-Analysis-Report-By-Product-By-Application-By-End-use-By-Region-And-Segment-Forecasts-2021-2028.html
4. Philly By Air, *16 Eye-Opening Drone Stats for 2022* (2022). Available Online: https://www.phillybyair.com/blog/drone-stats/
5. J. Park, et al., Cyber deception in the internet of battlefield things: techniques, instances, and assessments, in *International Workshop on Information Security Applications* (Springer, 2019), pp. 299–312
6. L. Cameron, Internet of things meets the military and battlefield: connecting gear and biometric wearables for an IoMT and IoBT (2018). Retrieved from URL https://publications.computer.org/cloud-computing/2018/03/22/internet-ofmilitary-battlefield-things-iomt-iobt
7. A. Fotouhi, et al., Survey on UAV cellular communications: practical aspects, standardization advancements, regulation, and security challenges. IEEE Commun. Surv. Tutor. **21**(4), 3417–3442 (2019)
8. Y. Zhi, et al., Security and privacy issues of UAV: A survey. Mobile Netw. Appl. **25**(1), 95–101 (2020)
9. D. Shortell, *DHS warns of 'strong concerns' that Chinese-made drones are stealing data* (2019). Available Online: https://edition.cnn.com/2019/05/20/politics/dhs-chinese-drone-warning/index.html.2019.
10. J. Burns, *Johns Hopkins Team Hacks, Crashes Hobby Drones To Expose Security Flaws* (2016). Available Online: https://www.forbes.com/sites/janetwburns/2016/06/13/johns-hopkins-team-hacks-crashes-hobby-drones-to-expose-security-flaws/
11. S.P. Arteaga, et al., Analysis of the GPS spoofing vulnerability in the drone 3DR solo. IEEE Access **7**, 51782–51789 (2019)
12. L. Watkins, et al., Defending against consumer drone privacy attacks: a blueprint for a counter autonomous drone tool, in *Workshop on Decentralized IoT Systems and Security (DISS)* (2020)
13. M.S. bin Mohammad Fadilah, et al., DRAT: a drone attack tool for vulnerability assessment, in *Proceedings of the Tenth ACM Conference on Data and Application Security and Privacy* (2020), pp. 153–155
14. H. Choi, et al., Cyber-physical inconsistency vulnerability identification for safety checks in robotic vehicles, in *Proceedings of the 2020 ACM SIGSAC Conference on Computer and Communications Security* (2020), pp. 263–278
15. INTERPOL, *Framework For Responding to a Drone Incident: For First Responders and Digital Forensics Practitioners* (2019)
16. The European Union Aviation Safety Agency, *Drone Incident Management at Aerodromes, Part 1: The Challenge of Unauthorised Drones in the Surroundings of Aerodromes* (2021)

17. International Society of Air Safety Investigators, *Unmanned Aircraft System Handbook and Accident/Incident Investigation Guidelines* (2015)
18. H. Al Hosani, et al., State of the art in digital forensics for small scale digital devices, in *2020 11th International Conference on Information and Communication Systems (ICICS)* (2020), pp. 72–78
19. A. Al-Dhaqm, et al., Research challenges and opportunities in drone forensics models. Electronics **10**(13), 1519 (2021)
20. C.-C. Yang, H. Chuang, D.-Y. Kao, Drone forensic analysis using relational flight data: a case study of DJI spark and mavic air. Proc. Comput. Sci. **192**, 1359–1368 (2021)
21. R. Kumar, A.K. Agrawal, Drone GPS data analysis for flight path reconstruction: a study on DJI Parrot & Yuneec make drones. Forensic Sci. Int. Digital Investigation **38**, 301182 (2021)
22. A. Renduchintala, et al., A comprehensive micro unmanned aerial vehicle (UAV/Drone) forensic framework, in Digital Investigation **30**, 52–72 (2019)
23. T.E.A. Barton, M.A.H. Azhar, Forensic analysis of popular UAV systems, in *2017 Seventh International Conference on Emerging Security Technologies (EST)* (2017), pp. 91–96
24. M.A.H. Azhar, T. Barton, T. Islam, Drone forensic analysis using open source tools. J. Digital Forensics, Secur. Law **13**, 7–30 (2018)
25. D.A. Hamdi, et al., Drone forensics: a case study on DJI Phantom 4, in *2019 IEEE/ACS 16th International Conference on Computer Systems and Applications (AICCSA)* (2019), pp. 1–6
26. D.-Y. Kao, et al., Drone forensic investigation: DJI spark drone as a case study. Proc. Comput. Sci. **159**, 1890–1899 (2019)
27. S.E. Prastya, I. Riadi, A. Luthfi, Forensic analysis of unmanned aerial vehicle to obtain GPS log data as digital evidence. Int. J. Comput. Sci. Inf. Secur **15**(3) (2017)
28. H. Bouafif, et al., Drone forensics: challenges and new insights, in *2018 9th IFIP International Conference on New Technologies, Mobility and Security (NTMS)* (2018), pp. 1–6
29. I. McAteer, et al., Forensic analysis of a crash-damaged cheerson CX-20 auto pathfinder drone. J. Digital Forensics Secur. Law **13**(4) (2018)
30. M. Yousef, F. Iqbal, Drone forensics: a case study on a DJI Mavic Air, in *2019 IEEE/ACS 16th International Conference on Computer Systems and Applications (AICCSA)* (2019), pp. 1–3
31. F. Iqbal, et al., Drone forensics: examination and analysis. Int. J. Electron. Secur. Digital Forensics **11**(3), 245 v
32. F.E. Salamh, U. Karabiyik, M.K. Rogers, RPAS forensic validation analysis towards a technical investigation process: a case study of yuneec typhoon H. Sensors **19**(15), 3246 (2019)
33. M. Stanković, M.M. Mirza, U. Karabiyik, UAV forensics: DJI mini 2 case study. Drones **5**(2), 49 (2021)
34. G. Thornton, P.B. Zadeh, An investigation into Unmanned Aerial System (UAS) forensics: data extraction & analysis. Forensic Sci. Int. Digital Investigation **41**, 301379 (2022)
35. M. Yousef, F. Iqbal, M. Hussain, Drone forensics: a detailed analysis of emerging DJI models, in *2020 11th International Conference on Information and Communication Systems (ICICS)* (2020), pp. 66–71
36. F.E. Salamh, et al., A comparative UAV forensic analysis: static and live digital evidence traceability challenges. Drones **5**(2), 42 (2021)
37. F.E. Salamh, M.M. Mirza, U. Karabiyik, UAV forensic analysis and software tools assessment: DJI phantom 4 and matrice 210 as case studies. Electronics **10**(6), 733 (2021)
38. T.E.A. Barton, M.A.H. Azhar, Open source forensics for a multi-platform drone system, in *Lecture Notes of the Institute for Computer Sciences, Social Informatics and Telecommunications Engineering* (2018), pp. 83–96
39. İ. Gülataş, S. Baktir, Unmanned aerial vehicle digital forensic investigation framework. J. Naval Sci. Eng. **14**, 32–53 (2018)
40. H. Studiawan, F. Sohel, Anomaly detection in a forensic timeline with deep autoencoders. J. Inf. Secur. Appl. **63**, 103002 (2021). ISSN: 2214–2126
41. K. Al-Room, et al., Drone forensics. Int. J. Digital Crime Forensics **13**(1), 1–25 (2021)

References

42. H. Alsulami, Implementation analysis of reliable unmanned aerial vehicles models for security against cyber-crimes: attacks, tracebacks, forensics and solutions. Comput. Electr. Eng. **100**, 107870 (2022)
43. G. Grispos, K. Bastola, Cyber autopsies: the integration of digital forensics into medical contexts, in *2020 IEEE 33rd International Symposium on Computer-Based Medical Systems (CBMS)* (IEEE, 2020), pp. 510–513
44. H. Studiawan, et al., Forensic event reconstruction for drones, in *2021 4th International Seminar on Research of Information Technology and Intelligent Systems (ISRITI)* (2021), pp. 41–45
45. A.S. Editya, T. Ahmad, H. Studiawan, Direction estimation of drone collision using optical flow for forensic investigation, in *2022 10th International Symposium on Digital Forensics and Security (ISDFS)* (IEEE, 2022), pp. 1–6
46. D.R. Clark, et al., DROP (DRone Open source Parser) your drone: forensic analysis of the DJI phantom III. Digital Investigation **22**, S3–S14 (2017)
47. E. Mantas, C. Patsakis, GRYPHON: drone forensics in dataflash and telemetry logs. Adv. Inf. Comput. Secur. 377–390 (2019)
48. H. Moon, et al., Digital forensic methodology for detection of abnormal flight of drones. J. Inf. Secur. Cyber. Res. **4**(1), 27–35 (2021)
49. S. Silalahi, T. Ahmad, H. Studiawan, Drone flight logs sequence mining, in *2022 IEEE International Conference on Cybernetics and Computational Intelligence (CyberneticsCom)* (2022), pp. 107–111
50. S. Silalahi, T. Ahmad, H. Studiawan, Named entity recognition for drone forensic using BERT and DistilBERT, in *2022 International Conference on Data Science and Its Applications (ICoDSA)* (2022), pp. 53–58
51. A. Shoufan, et al., Drone pilot identification by classifying radio-control signals. IEEE Trans. Inf. Forensics Secur **13**(10), 2439–2447 (2018)
52. Z. Li, et al., Reliable digital forensics in the air: exploring an RF-based drone identification system. Proc. ACM Interact. Mob. Wearable Ubiquitous Technol. **6**(2), 1–25 (2022)
53. DatFile, *DatCon* (2021). https://datfile.net/DatCon/intro.html
54. DatFile, *CsvView* (2021). https://datfile.net/CsvView/intro.html
55. M. Forensics, *Magnet Axiom: Recover and Analyze Your Evidence in One Case* (2021). https://www.magnetforensics.com/products/magnet-axiom/
56. Cellebrite, *Cellebrite: Digital Intelligence Solution Suite* (2021). https://www.cellebrite.com/en/product/
57. G. Horsman, J.R. Lyle, Dataset construction challenges for digital forensics. Forensic Sci. Int. Digital Investigation **38**, 301264 (2021)
58. J. Whelan, et al., *UAV Attack Dataset* (2020). https://dx.doi.org/10.21227/00dg-0d12
59. A. Keipour, M. Mousaei, S. Scherer, ALFA: a dataset for UAV fault and anomaly detection. Int. J. Robot. Res. **40**(2–3), 515–520 (2021)
60. P. Zhu, et al., Detection and tracking meet drones challenge. IEEE Trans. Pattern Anal. Mach. Intell. **44**(11), 7380–7399 (2022)
61. M. Ozel, *Drone Detection Dataset* (2020). https://github.com/dasmehdix/drone-dataset
62. VTO Inc., *Drone Forensics Datasets* (2018). https://cfreds-archive.nist.gov/drone-images.html
63. R. Rowlingson, et al., A ten step process for forensic readiness. Int. J. Digital Evidence **2**(3), 1–28 (2004)
64. J.L. Esteves, Electromagnetic watermarking: exploiting IEMI effects for forensic tracking of UAVs, in *2019 International Symposium on Electromagnetic Compatibility - EMC EUROPE* (2019), pp. 1144–1149
65. Y. Yu, et al., LiveBox: a self-adaptive forensic-ready service for drones. IEEE Access **7**, 148401–148412 (2019)
66. F.M. Alotaibi, A. Al-Dhaqm, Y.D. Al-Otaibi, A novel forensic readiness framework applicable to the drone forensics field, in Computational Intelligence and Neuroscience, ed. by V. Kumar (2022), pp. 1–13

67. G. Grispos, W.B. Glisson, K.-K. Raymond Choo, Medical cyber-physical systems development: a forensics-driven approach, in *2017 IEEE/ACM International Conference on Connected Health: Applications, Systems and Engineering Technologies (CHASE)* (IEEE, 2017), pp. 108–113
68. E. Casey, *Digital Evidence and Computer Crime: Forensic Science, Computers, and the Internet* (Academic Press, New York, 2011)
69. U. Jain, M. Rogers, E.T. Matson, Drone forensic framework: sensor and data identification and verification, in *2017 IEEE Sensors Applications Symposium (SAS)* (2017), pp. 1–6
70. R.O. Peruzzi, Forensic engineering analysis of quadcopter drone personal injury. J. Natl. Acad. Forensic Eng. **34**(2) (2017)
71. A. Roder, K.-K. Raymond Choo, N.-A. Le-Khac, Unmanned aerial vehicle forensic investigation process: DJI Phantom 3 drone as a case study, in *Proceedings of the Annual ADFSL Conference on Digital Forensics, Security and Law* (2018), pp. 55–70
72. F.E. Salamh, et al., Drone disrupted denial of service attack (3DOS): towards an incident response and forensic analysis of remotely piloted aerial systems (RPASs), in *2019 15th International Wireless Communications & Mobile Computing Conference (IWCMC)* (2019), pp. 704–710
73. A.A. Alhussan, et al., Towards development of a high abstract model for drone forensic domain. Electronics **11**(8), p. 1168 (2022)
74. M. Chernyshev, et al., Internet of Things (IoT): research, simulators, and testbeds. IEEE Int. Things J. **5**(3), 1637–1647 (2017)
75. L. Chettri, R. Bera, A comprehensive survey on Internet of Things (IoT) toward 5G wireless systems. IEEE Int. Things J. **7**(1), 16–32 (2019)
76. K. Ruan, et al., Cloud forensics, in *Advances in Digital Forensics VII: 7th IFIP WG 11.9 International Conference on Digital Forensics* (2011), pp. 35–46
77. B. Cusack, M. Simms, Evidential recovery from GPS devices. J. Appl. Comput. Inf. Technol. **15**(1) (2011)
78. L.F. Sikos, AI in digital forensics: ontology engineering for cybercrime investigations. Wiley Interdiscip. Rev. Forensic Sci. **3**(3), e1394 (2021)
79. S.W. Hall, A. Sakzad, K.-K.R. Choo, Explainable artificial intelligence for digital forensics. Wiley Interdiscip. Rev. Forensic Sci. **4**(2), e1434 (2022)
80. P. Sharma, U. Siddanagaiah, G. Kul, Towards an AI-based after-collision forensic analysis protocol for autonomous vehicles, in *2020 IEEE Security and Privacy Workshops (SPW)* (2020), pp. 240–243
81. M. Negrão, P. Domingues, SpeechToText: an open-source software for automatic detection and transcription of voice recordings in digital forensics. Forensic Sci. Int. Digital Investigation **38**, 301223 (2021)
82. H. Studiawan, F. Sohel, C. Payne, Sentiment analysis in a forensic timeline with deep learning. IEEE Access **8**, 60664–60675 (2020)
83. R.Y. Wang, D.M. Strong, Beyond accuracy: what data quality means to data consumers. J. Manag. Inf. Syst. **12**(4), 5–33 (1996)
84. G. Grispos, W. Glisson, T. Storer, How Good is Your Data? Investigating the quality of data generated during security incident response investigations, in *Proceedings of the 52nd Hawaii International Conference on System Sciences* (2019), pp. 7156–7165

Chapter 4
Setting Up a Drone Forensics Laboratory

Abstract This chapter serves as a guide for establishing a forensic laboratory specialized in drone investigations. It explains in detail the requirements for physical infrastructure, equipment, and software for data examination. This chapter describes the workflow of a forensic investigation from installing softwares to analysis. It also highlights the necessity of maintaining evidence integrity. This chapter equips readers with the knowledge needed to create a drone forensics laboratory. It also addresses both the technical aspects for the successful analysis of drones. By the end of the chapter, readers are equipped with a foundation for setting up a drone forensics laboratory. The setup is designed to meet the demands of modern forensic science and contribute to the resolution of cases involving drones and UAVs.

4.1 Introduction

This chapter is designed as a guide for forensic professionals, students, and cybersecurity experts interested in establishing a laboratory equipped to handle a drone forensic case. We discuss the components of a drone forensics laboratory, including the physical infrastructure, necessary equipment, and software tools for effective investigation. From the initial steps of installing specialized forensic software to the comprehensive processes involved in data examination and analysis, this chapter outlines a systematic approach to setting up the laboratory.

The integrity of evidence is important of any forensic investigation, and the procedures for handling, analyzing, and storing data from drones are no exception. This chapter also provides insights into best practices for maintaining evidence integrity. We need to make sure that the findings of an investigation can withstand the scrutiny of legal proceedings. By equipping readers with a thorough understanding of both the technical and procedural aspects necessary for a drone forensics laboratory, this chapter lays the groundwork for the following chapters and building a facility capable of contributing to the resolution of drone-related cases. As the drone use and misuse continues to expand, the insights offered in this chapter will prepare students and forensic professionals to meet the demands of modern forensic science and play an important role in upholding security and accountability in the age of drones.

In this section, we explain the hardware-related preparation hardware in Sects. 4.2 and 4.3. Subsequently, we describe the software needed to perform UAV and drone forensics in Sects. 4.4–4.6. Finally, we discuss the public forensic dataset for hands-on experiments in this book.

4.2 Preparing the Forensic Workstation

Setting up a digital forensics computer or workstation involves several considerations to ensure it is capable of handling the demands of forensic analysis. Common hardware requirements are as follows:

1. Processor (CPU): A powerful CPU is crucial for processing large amounts of data. Consider a multi-core processor (i.e., Intel i7, i9, or AMD Ryzen 7, 9) for multitasking and handling compute-intensive tasks.
2. Memory (RAM): A minimum of 8 GB of RAM is recommended, although 16 GB or more is preferable for handling large datasets and running multiple virtual machines.
3. Storage: SSDs (Solid State Drives) are recommended for the operating system and applications due to their speed. In addition, have multiple high-capacity HDDs (Hard Disk Drives) for data storage and evidence preservation.
4. Network interface: Multiple network interfaces may be required, including both wired and wireless, to facilitate network forensic activities and isolated investigations.
5. Graphics card (GPU): A dedicated GPU can be beneficial for tasks that require graphic-intensive processing and password cracking.
6. Write blocker: A write blocker is needed to prevent unintentional write operation to the evidence such as a drone SD card. In this book, we employ Tableau T35U Forensic Bridge as the write blocker [1]. The details of this hardware can be accessed on its official website: https://www.opentext.com/products/tableau-forensic.
7. SD card reader: An SD card reader is an external device that allows computers to access data from SD cards when built-in slots are unavailable, incompatible, or lack advanced features. In addition, external readers support a wide range of SD card formats and provide greater portability and compatibility. This device is needed as many drones are equipped with an external memory to store the flight data.

The Tableau Forensic T35u/T35u-RW SATA/IDE Bridge is a purpose-built hardware write blocker designed to assist in the secure acquisition of digital evidence from SATA and IDE drives. Its primary function is to ensure that the integrity of the source data is preserved during the forensic process by preventing any modifications to the original storage medium. This is achieved through its read-only mode, which is controlled by a DIP switch. When the read-only mode is

4.2 Preparing the Forensic Workstation

enabled, all write operations to the source drive are blocked. Thus, it is a critical tool for maintaining evidentiary value in digital forensics.

The device supports both SATA and IDE drives and provides flexibility for handling various types of storage media commonly encountered during forensic investigations. Connection to the source drive is accomplished using dedicated Tableau cables for power and data transfer. The T35u/T35u-RW is then connected to a forensic workstation via a USB 3.0 interface to provide fast and reliable data transfer while preventing any unintentional or intentional alterations to the source drive. The device also includes clear LED indicators and DIP switch settings that allow forensic investigators to verify the write protection status and configure other operational preferences, such as error reporting.

Another feature of the Tableau Forensic Bridge is its ability to maintain data integrity throughout the acquisition process. By allowing only read operations, the device ensures that the original data on the source drive remains untouched to preserve it for further analysis or presentation in court. The device also includes a "safe removal" procedure by unmounting the source drive before disconnection, further safeguarding against data corruption. Moreover, DIP switch settings offer flexibility to hide or report write errors as needed. For scenarios that require write capabilities, the T35u-RW variant is available. This version, distinguished by its yellow casing, is pre-configured for read/write operations and is useful in cases where data needs to be modified during the investigation. The visual differentiation between the read-only T35u and the read/write T35u-RW reduces the risk of accidental misuse.

In drone forensics, the Tableau Forensic Bridge's role is to acquire data from storage devices, such as SD cards used in drone systems. By connecting the storage media through the device to a forensic workstation, investigators can create a forensic image, which is an exact bit-by-bit copy of the original data. This approach allows for in-depth analysis of flight logs, images, videos, and other operational data without compromising the integrity of the original evidence.

Furthermore, software requirements are as follows:

1. Operating system: A stable and secure operating system such as Windows for commercial forensic tools, Linux (e.g., Kali Linux, which includes numerous forensic and hacking tools), macOS, depending on our preference and tool compatibility.
2. Forensic tools: Install forensic software tools based on our needs, including disk and data capture tools (e.g., FTK Imager [2]), analysis tools (e.g., Autopsy [3]), and specialized tools for mobile forensics, such as Android Debug Bridge (adb).
3. Virtualization software (optional): Tools such as VMware Fusion or VMware Workstation [4] for creating and managing virtual machines. This is useful for safely examining malware or running different operating systems. However, virtualization software is rarely necessary for drone forensics unless testing involves malware that specifically targets drone communication protocols or software.

Note that in this book we only use free and/or open-source software tools. The reason is that these tools can be downloaded at no cost and it is affordable for all groups, especially students or academics. Moreover, the open-source nature also allows us to learn the internals of the tools.

For the network configuration, we recommend to make sure that the forensic workstation is isolated from the internet and the organization's main network to prevent contamination or data leakage. It is advisable to use a dedicated network for forensic analysis or a virtual private network (VPN) for secure connections when necessary. Furthermore, we need to implement a robust firewall and security protocols to protect the workstation from unauthorized access and malware.

4.3 Forensic Workstation Used in this Book

In this book, we will utilize a range of forensic workstations to demonstrate different tools and techniques. Each type brings unique specifications that enable effective handling of various forensic tasks. The first workstation is a Windows-based laptop running Microsoft Windows 11 Pro. It features an 11th Generation Intel Core i7-1165G7 processor, 16 GB of RAM, and a 1 TB solid-state drive (SSD). This configuration provides ample processing power and memory. It assists investigators to handle demanding forensic software and large data files typically encountered in drone analysis. Windows is highly suitable for free and open-source tools such as Autopsy and FTK Imager used in this book, as these tools are designed to run smoothly on this platform. In contrast, based on our experiments, they may not perform as seamlessly as on Linux or macOS.

The second workstation we use is a MacBook Pro, which operates on macOS Sonoma 14.5 and is powered by Apple's M1 chip. With 8 GB of RAM and a 256 GB SSD, the MacBook Pro offers efficient performance for lighter forensic tasks and compatibility with Mac-specific forensic tools. Although it has a smaller storage capacity, it is well-suited for scenarios where flexibility and compatibility within the Mac ecosystem are required. For example, we use the MacBook to create a backup of an iOS drone controller. This backup can be further analyzed using the provided forensic tools. Finally, we include a Mac Mini M2 as another forensic workstation. Running on macOS Sequoia 15.1, this device is equipped with Apple's M2 processor, 8 GB of RAM, and a 256 GB SSD. Compact yet powerful, the Mac Mini M2 is ideal for Mac-based forensic analyses and tasks that benefit from Apple's processing capabilities.

Each of these workstations has a different purpose in our configuration of forensic analysis. They range from so many scenarios to exercise different tools on varied operating systems and even hardware configurations. Such variation allows us also to show how forensic techniques can be applied on several platforms to give an overview of workflows in drone forensics. The reader can use any device to run hands-on practice in this book based on the recommendation in Sect. 4.2.

4.4 Installation of Anaconda

Anaconda simplifies Python programming, especially for data science, machine learning, and scientific computing, by bundling libraries, tools, and an environment manager in one package [5]. However, it can also be used for forensic analysis as there are many Python-based forensic tools. It includes the conda package manager, which helps users quickly install and manage packages without dependency issues and supports creating isolated environments for different projects. This isolation prevents conflicts between package versions. Therefore, investigators can seamlessly switch between projects with different dependencies or Python versions.

Moreover, Anaconda comes with tools such as Jupyter Notebook and Spyder. These tools support easier experiment, visualize data, and document work in an interactive environment. For beginners and professionals alike, Anaconda provides a complete, consistent setup across Windows, macOS, and Linux. Therefore, it will reduce the time and complexity involved in configuring Python environments from scratch. To install Anaconda, we can follow these steps.

1. Download the Anaconda installer
   Visit the Anaconda Download website.[1] Click on the "Download" button or a link "64-Bit Graphical Installer (912.3M)" in the Windows section.
2. Launch the installer
   Locate the downloaded installer file. In Windows, we need to double-click the .exe file. Follow the on-screen instructions to proceed. It will first show an Anaconda installer welcome window as shown in Fig. 4.1.
3. Accept the license agreement
   During the installation, we will be prompted to review the license agreement. We need to read through it, then accept to continue (Fig. 4.2).
4. Choose installation location
   Select the installation directory where Anaconda will be installed. In Windows By default, it installs in (Fig. 4.3). C:\Users\YourUsername\Anaconda3. Choose whether to make Anaconda available to all users or just our own user account (Fig. 4.4).
5. Configure advanced options
   Add Anaconda to PATH (Windows): It is recommended to leave this option unchecked to avoid conflicts with other software. Instead, use the Anaconda Prompt to manage Anaconda environments. Register Anaconda as the default Python: This option is checked by default. It makes Anaconda's Python interpreter the default on the system. This window is shown in Fig. 4.5.
6. Complete the installation
   Click the "Install" button to start the installation. Wait for the installation to complete. This may take several minutes as shown in Fig. 4.6.

---

[1] https://www.anaconda.com/download/.

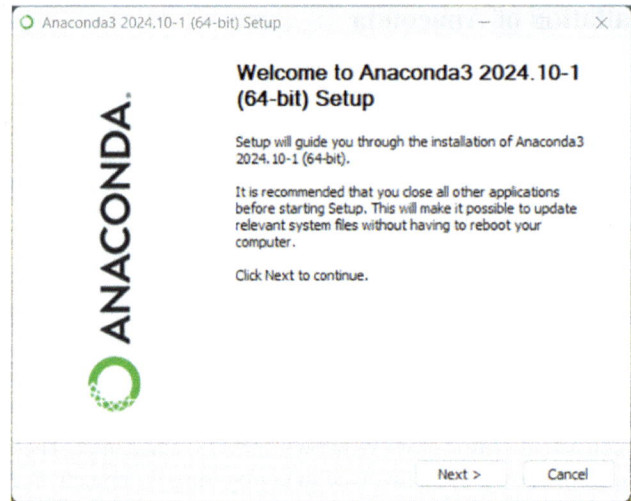

Fig. 4.1 Anaconda installer welcome window

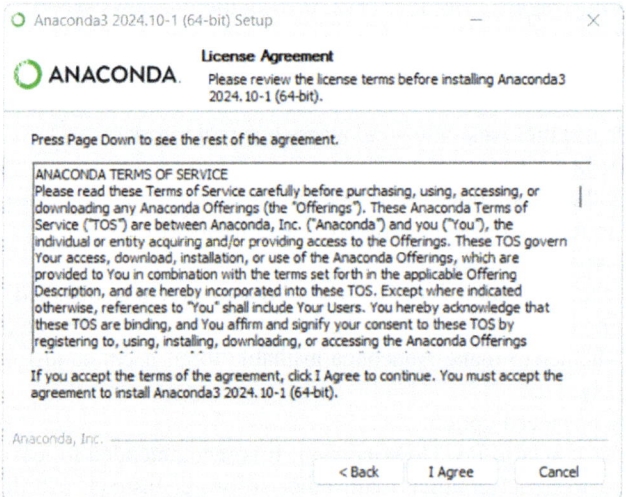

Fig. 4.2 Anaconda installer license agreement

4.4 Installation of Anaconda 79

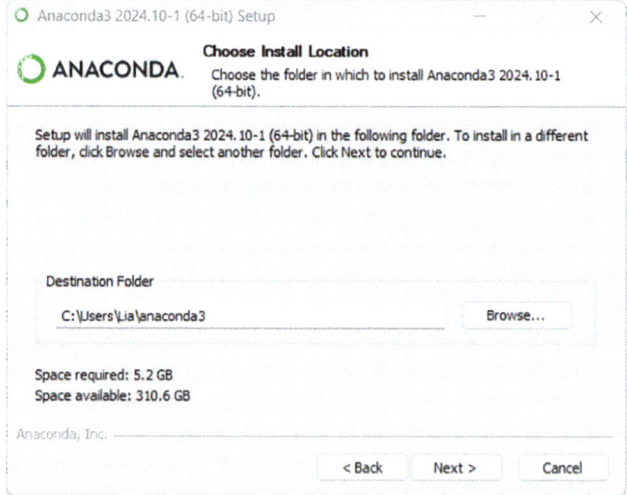

Fig. 4.3 Anaconda install location

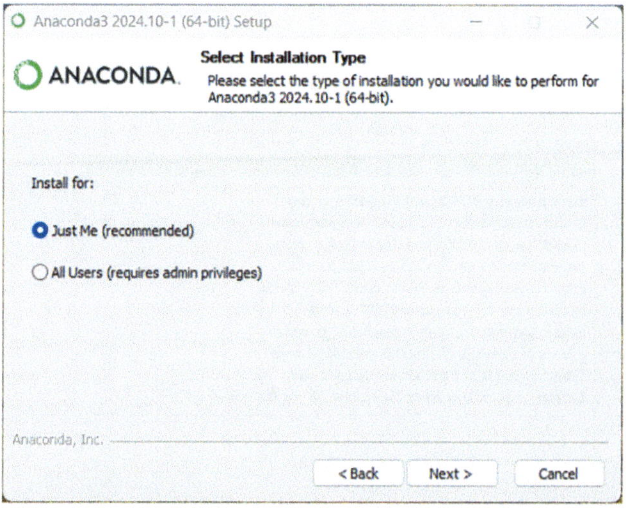

Fig. 4.4 Anaconda installation type

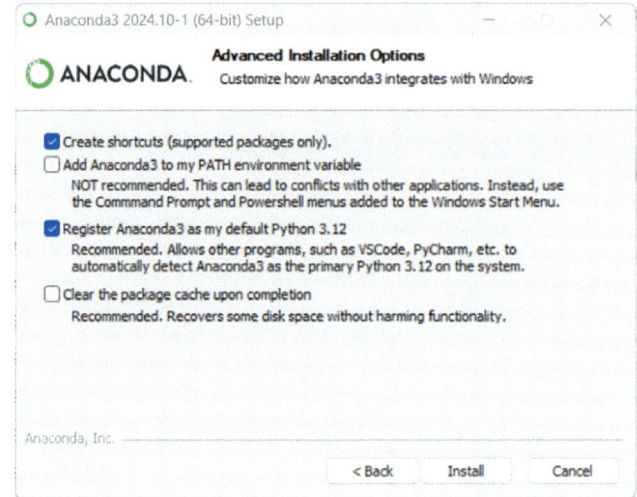

Fig. 4.5 Anaconda advance installation options

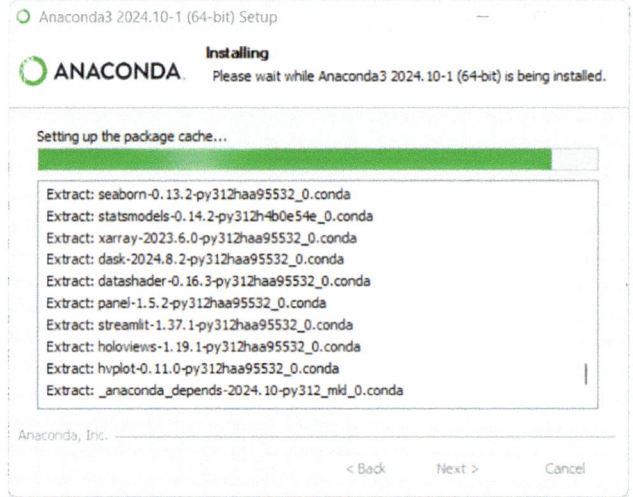

Fig. 4.6 Anaconda installation in progress

4.4 Installation of Anaconda

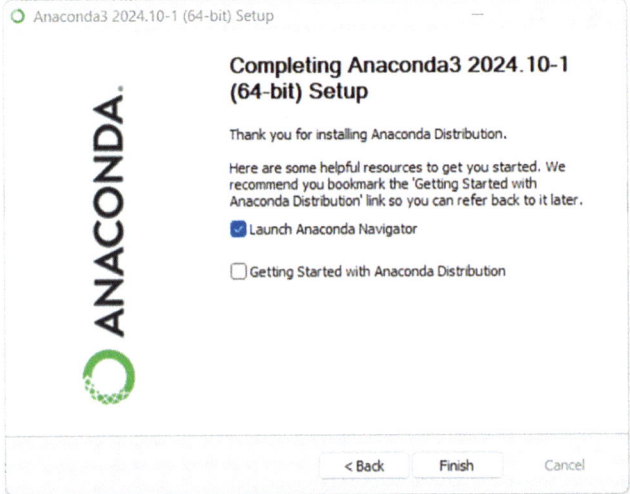

Fig. 4.7 Anaconda installation is complete

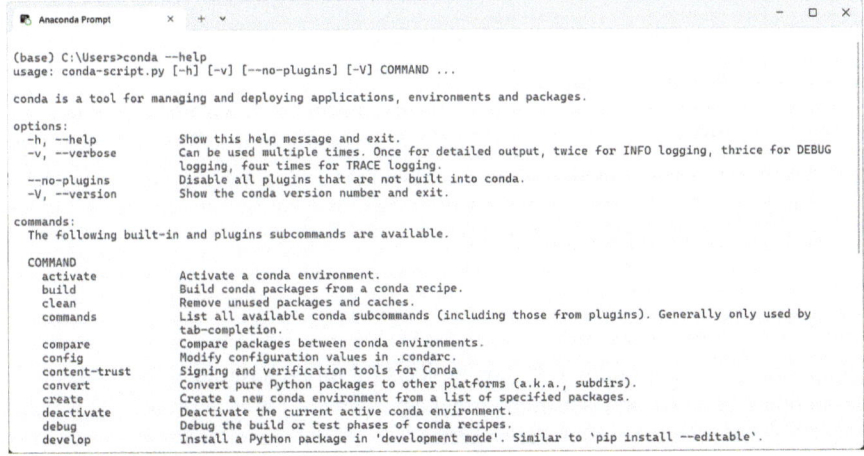

Fig. 4.8 Anaconda help prompt

7. Verify the installation

After the installation completes (Fig. 4.7), close the installer. Open a terminal (macOS/Linux) or the Anaconda Prompt (Windows). Type the following command to verify the installation (Fig. 4.8):

```
conda --version
conda --help
```

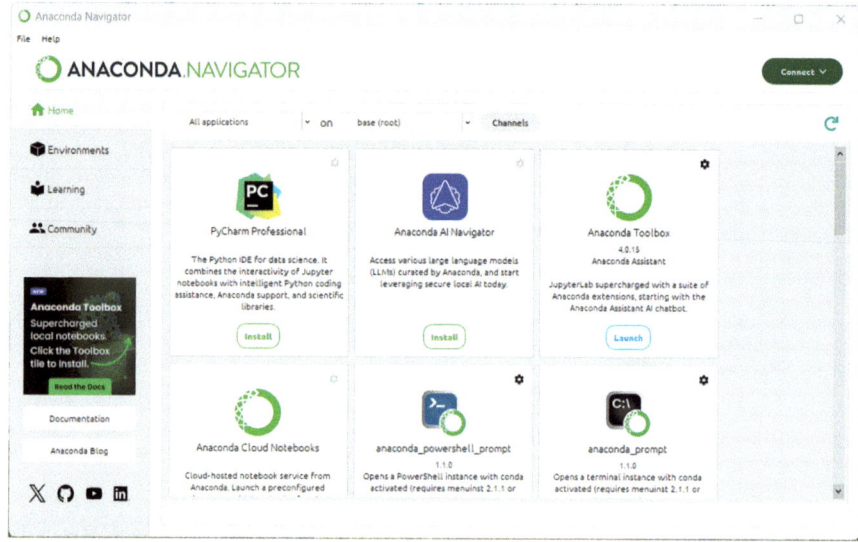

Fig. 4.9 Anaconda Navigator

8. Launch Anaconda Navigator (Optional)

To use the graphical interface, open Anaconda Navigator. In Windows, Search for "Anaconda Navigator" in the Start menu. In macOS and Linux, type anaconda-navigator in the terminal. Anaconda Navigator provides a GUI for managing environments, installing packages, and launching Jupyter Notebook or other tools as depicted in Fig. 4.9.

> **! Attention**
>
> Since we will install multiple Python projects for drone forensic tools in this book, it is important to set up a separate virtual environment for each project. This helps prevent any conflicts between the projects.

4.5 Installation of FTK Imager

FTK Imager (Forensic Toolkit Imager) is a free and widely used forensic imaging tool developed by Exterro [2]. It is primarily used for creating forensic images of storage devices, files, or other digital evidence while preserving the integrity of the evidence. FTK Imager is one of basic tools in the digital forensic investigator's toolkit.

4.5 Installation of FTK Imager

FTK Imager offers several key features that make it a valuable tool in digital forensics. It excels in data acquisition by creating exact, bit-for-bit forensic images of various storage devices, including hard drives, USB drives, SD cards, and optical media. The tool supports multiple imaging formats such as E01 (EnCase), a common forensic format with metadata and compression; DD (Raw), a raw format for forensic preservation; and AFF (Advanced Forensic Format), known for its flexibility and extensibility. FTK Imager maintains integrity verification by automatically calculating and verifying cryptographic hashes such as MD5 and SHA-1. It provides hash reports to confirm that the original evidence remains unaltered.

The software also facilitates live evidence collection by capturing of volatile data, such as memory (RAM), from a live system. It is useful in scenarios where powering down a system might lead to evidence loss. Investigators can preview files and directories before imaging, and the tool supports the recovery and previewing of deleted files unless they have been overwritten. FTK Imager is compatible with various file systems, including NTFS, FAT, exFAT, EXT, and HFS+. We can use it to perform the acquisition and analysis of both logical and physical drives. In addition, it supports logical data collection and supports the acquisition of specific files or folders without requiring the creation of a full disk image. It will save time when only a subset of data is necessary.

FTK Imager provides several advantages that make it a popular choice among forensic professionals. First, it is free to use, it provides robust forensic imaging, and preview capabilities without any cost. Its intuitive interface makes it easy to navigate, even for users with limited experience in digital forensics. Most importantly, FTK Imager is forensically sound, it means that data integrity is preserved throughout the acquisition process. To install FTK Imager, we can follow the steps below:

1. Open a web browser and go to the official AccessData website to download FTK Imager.[2] Look for the latest version of FTK Imager.
2. Some AccessData pages require users to register before downloading their tools. Fill in basic details (name, email, organization, etc.). Submit the form to receive access to the download link.
3. Depending on our system, we can choose the appropriate file, specifically Windows 64-bit: For most modern systems and Windows 32-bit: For older or specific systems.
4. Save the installer file (usually named FTKImager.exe) to an easily accessible location on the computer, such as the Desktop or Downloads folder.
5. Locate the installer

    Navigate to the location where we saved the file (FTKImager.exe). Double-click the file to begin the installation process. We may see a prompt asking for administrative permissions—click Yes to proceed.

---

[2] https://www.exterro.com/ftk-product-downloads.

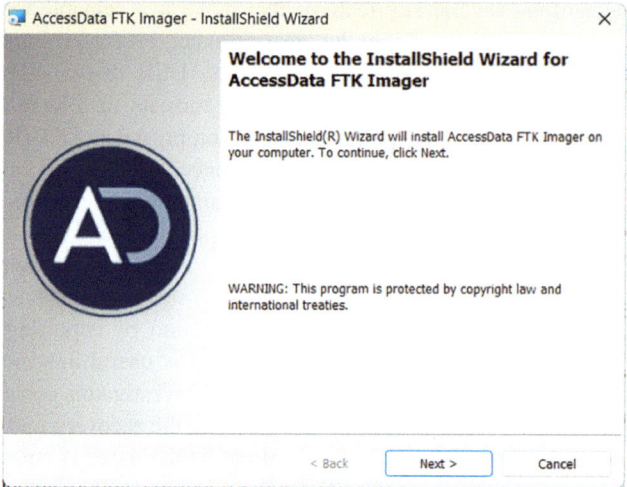

Fig. 4.10 Welcome window of FTK Imager installer

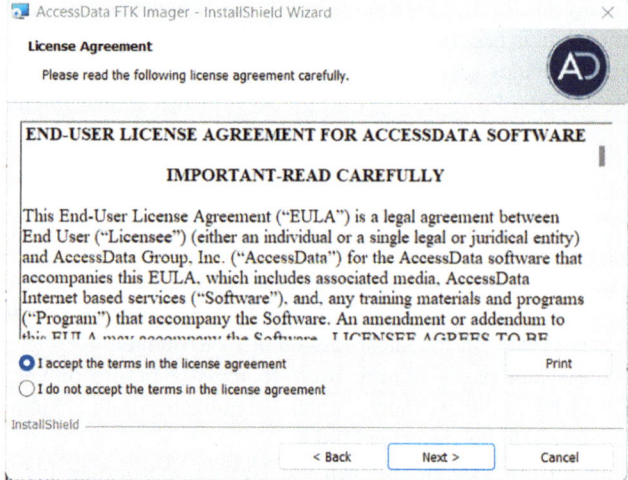

Fig. 4.11 License agreement of FTK Imager installer

6. Read the welcome screen

    The installer will display a welcome screen or a brief description of the software as shown in Fig. 4.10. Click Next to continue.

7. Accept the license agreement

    Carefully read the license agreement (if shown) and select I Accept the Terms of the License Agreement to proceed (Fig. 4.11). Click Next to move to the next step.

4.5 Installation of FTK Imager

Fig. 4.12 Select destination folder for FTK Imager installation

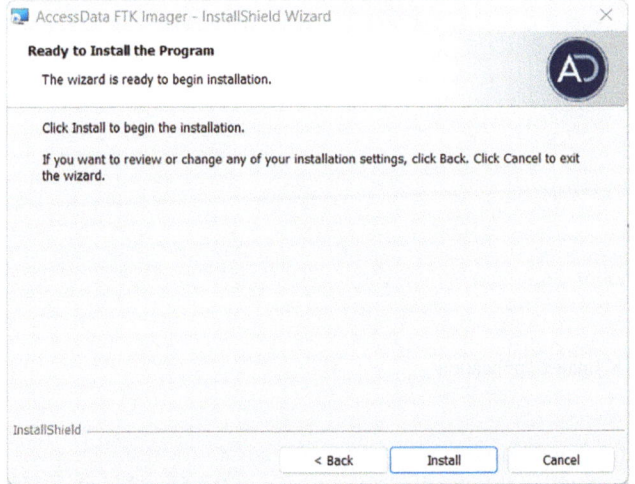

Fig. 4.13 Install the FTK Imager

8. Select the installation directory

   The installer will ask where to install the software as shown in Fig. 4.12. By default, it will install in C:\Program Files\AccessData\FTK Imager. We can keep this or choose a different location by clicking Browse.

9. Start the installation

   Click Install or Next to begin the installation (Fig. 4.13). The progress bar will show the installation process. Once the installation is complete, we will see a confirmation message. Click Finish to exit the installer as shown in Fig. 4.14.

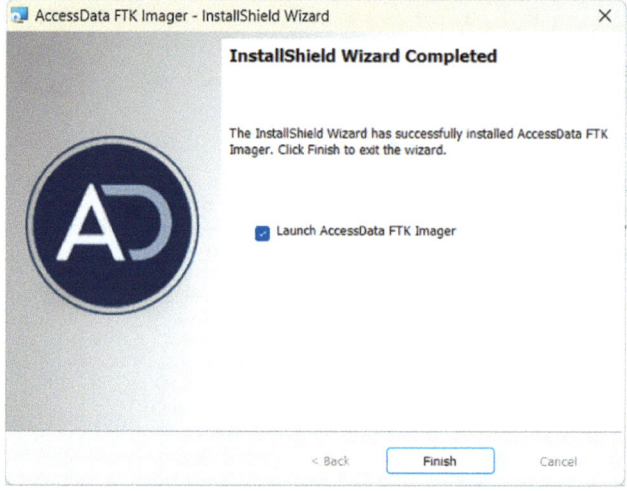

Fig. 4.14 FTK Imager has been successfully installed

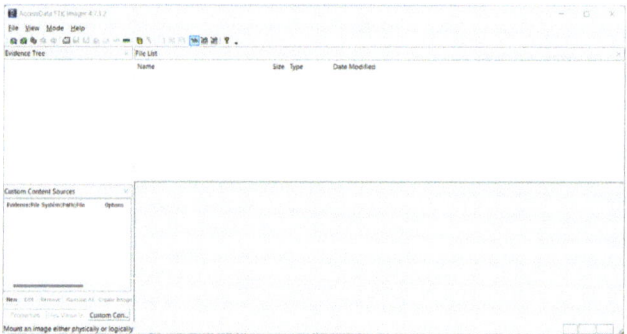

Fig. 4.15 FTK Imager application

10. Verify the installation

    If we installed the standard version, Go to the Start Menu or search for FTK Imager in the Windows search bar. Click the application to open it and we can see the installed application as shown in Fig. 4.15.

4.6 Installation of Autopsy

Autopsy is an open-source digital forensics tool used to investigate and analyze digital devices, especially storage media such as hard drives, USB drives, and memory cards [3]. It is built on top of The Sleuth Kit (TSK), a library of forensic tools, and provides an easy-to-use graphical interface for forensic investigators.

Autopsy offers a wide range of features that make it a powerful tool for digital forensic investigations. One key feature is File System Analysis to recover deleted files and browse through various file systems, including NTFS, FAT, EXT, and HFS+. In addition, the tool provides Keyword Search functionality to search for specific terms, phrases, or patterns within files and metadata. This feature makes it easier to pinpoint critical evidence. The Timeline Analysis feature is used to assist event reconstruction by visualizing file activity over time. For web-related investigations, Web Artifacts Analysis allows users to extract and analyze browser history, cookies, cached files, and bookmarks to provide insights into online activity. Similarly, Email Analysis supports the investigation of email data in formats such as PST and MBOX to review of communication records.

Autopsy also includes Registry Analysis, which focuses on extracting and analyzing Windows registry information to uncover key system details and user activities. For investigations involving live memory, the Memory Analysis feature enables forensic examination of RAM dumps to identify active processes, malware, or user activity at the time of the dump. With Hash Set Matching, investigators can quickly identify known good or bad files by comparing them against hash sets like MD5 and SHA256. The tool's Media Analysis capabilities extend to analyzing pictures and videos, including extracting metadata for further insights. Finally, Case Management allows users to handle multiple forensic cases. It also provides logging and reporting to ensure thorough documentation of the investigation process. To install Autopsy follow these steps:

1. Download the installer

   Visit the official Autopsy website[3] and download the latest Windows installer. Make sure to save the file to an easily accessible location on the computer.

2. Run the installer

   Locate the downloaded installer file (e.g., autopsy-x.y.z.exe) and double-click it to begin the installation process. The Autopsy installer will launch and display a Welcome Message (Fig. 4.16). This message introduces the software and highlights key features and benefits of using Autopsy. Click Next to proceed.

3. Select destination folder

   The installer will prompt users to choose a destination folder for the installation. By default, the software will be installed in the Program Files directory. If we wish to install it elsewhere, click Browse and select a different location as depicted in Fig. 4.17. Once satisfied with the choice, click Next.

4. Ready to install

   Before proceeding with the installation, the installer confirms whether or not we want to change the installation settings. Click the Install button to begin the installation process as shown in Fig. 4.18.

---

[3] https://www.autopsy.com/download/.

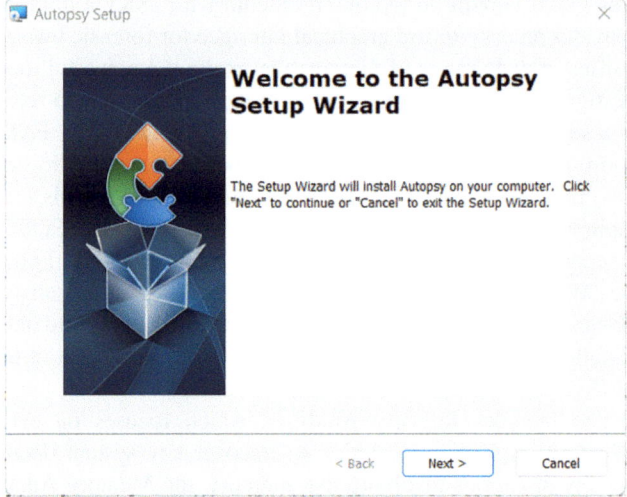

Fig. 4.16 Welcome window of Autopsy wizard installation

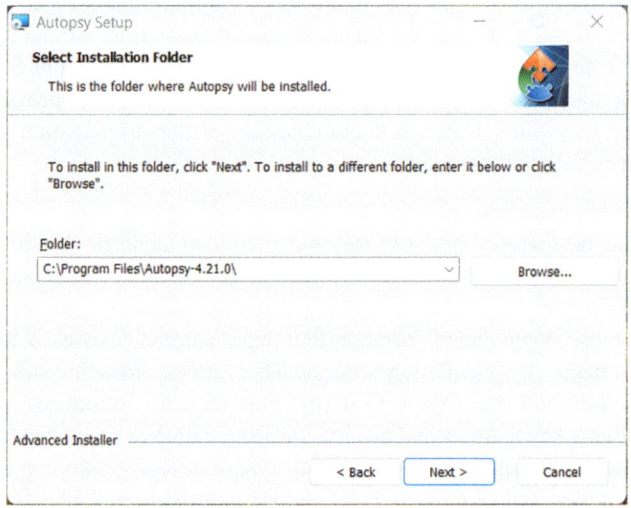

Fig. 4.17 Select installation folder for Autopsy

5. Install and wait for progress

   The installer will begin copying files and configuring the application. A progress bar will appear and show the installation's status (Fig. 4.19). This step may take a few minutes depending on the system's performance. Once completed, we will see a confirmation screen indicating the installation was successful.

4.6 Installation of Autopsy

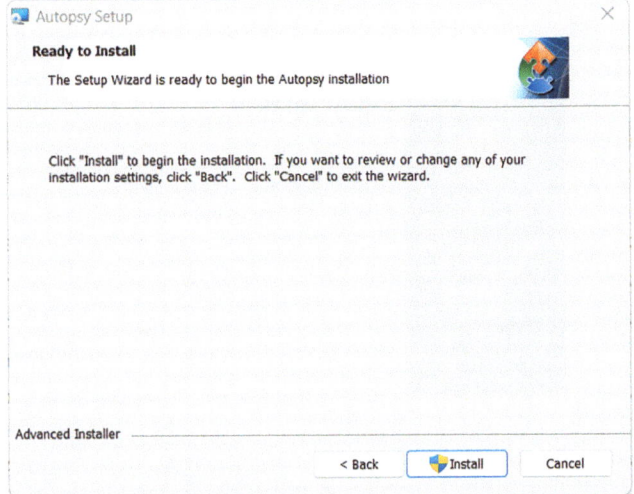

Fig. 4.18 Autopsy is ready to install

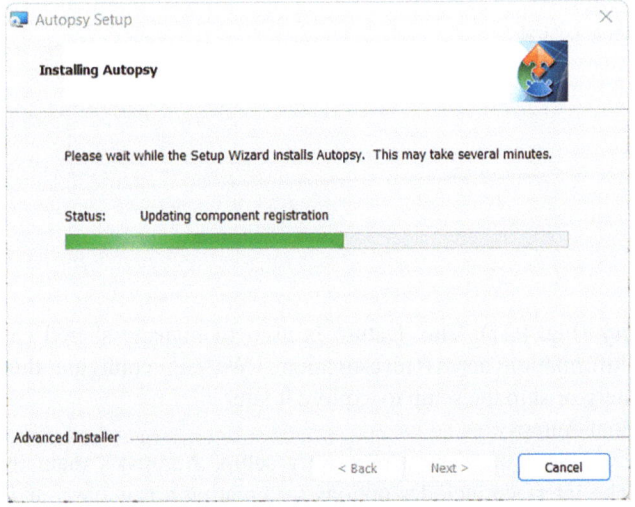

Fig. 4.19 Installing Autopsy and wait for a few minutes

6. Finish installation

   Click the Finish button to close the installer (Fig. 4.20). We can choose to launch Autopsy immediately by selecting the corresponding option before finishing.
7. Open the Autopsy program

   After installation, open Autopsy from the Start Menu or the desktop shortcut. Upon launching the program, a window will appear about the Autopsy Central

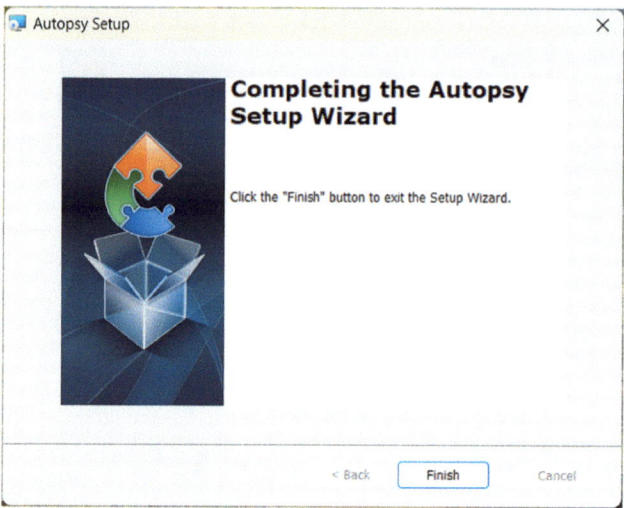

Fig. 4.20 Autopsy installation is complete

Fig. 4.21 Autopsy information about central repository

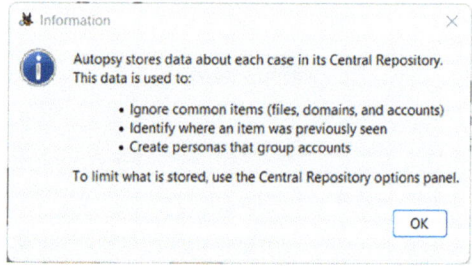

Repository (Fig. 4.21). This feature is used to manage shared case data and artifact information across investigations. We can configure this repository immediately or skip the setup to explore it later.

8. First look at Autopsy

Once we close the central repository setup, Autopsy's main interface will appear. The interface includes options for creating a new case, and opening an existing case as shown in Fig. 4.22. At this point, we are ready to start a forensic investigation.

4.7 Downloading Datasets for Experiments

Datasets is used in both advancing research and practical applications in drone and UAV forensics. They provide a foundation for testing, validating, and benchmarking forensic methodologies and tools. By utilizing real-world and simulated data,

4.7 Downloading Datasets for Experiments

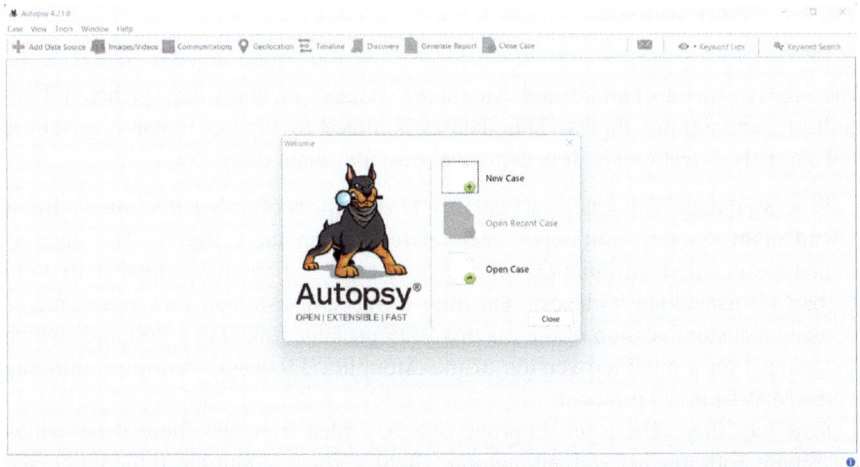

Fig. 4.22 First look at Autopsy

researchers can explore various scenarios, investigate anomalies, and develop techniques to extract, analyze, and interpret digital evidence from drones and their associated controller devices. This section presents two datasets—VTO Labs and the ALFA dataset—that have contributed to the field by offering diverse data for forensic investigations and anomaly detection in drone systems.

4.7.1 VTO Labs dataset

VTO Labs. is recognized for its pioneering research in digital forensics relating to both consumer and professional drones. They aim to uncover and analyze digital forensic evidence from these drones and provide supporting datasets to law enforcement and governmental agencies during investigations. The datasets mentioned are available through the U.S. National Institutes of Standards and Technology (NIST) on the Computer Forensic Reference Datasets (CFReDS) website [6]. For access to these datasets, we can visit the following website: https://www.cfreds.nist.gov/.

In the framework of their Drone Forensics Program, VTO Labs. acquired a total of 79 drones including 30 different models with approximately three units per model. These drones were set up and flown within a restricted and geofenced area. The next step involved efforts to capture and create images of the data storage sections found in each drone, as well as the drones' controllers, any connected mobile devices, and computers. To pinpoint the locations of data storage, each drone was dismantled and disassembled. We will use these datasets throughout many chapters in this book.

4.7.2 ALFA dataset

The ALFA (AirLab Failure and Anomaly)[4] dataset includes data gathered from multiple autonomous flights. This dataset is aimed to support research in failure and anomaly detection [7]. It is organized into four main categories:

1. Processed data: This category contains 47 sequences of fully autonomous flights and eight distinct fault types encountered during these flights. The datasets include detailed information on the fault type and timing, available in ROS .bag format, along with .csv and .mat formats. These .bag files were created using a custom version of the mavros ROS package linked to a Pixhawk, which operated on a modified version of the Ardupilot 3.9.0beta1 firmware, utilizing the MAVLink 2.0 protocol.
2. Raw bag files: These are unprocessed .bag files from the flight data, which capture both manual and autonomous flight segments. Similar to the processed data, these files are generated through a custom mavros ROS package that interfaces with the Pixhawk on the modified Ardupilot 3.9.0beta1 firmware and using the MAVLink protocol.
3. Telemetry logs: These logs are from the NVidia TX2 onboard computer, which is connected to the Pixhawk autopilot system.
4. Dataflash logs: These are the records maintained by the Pixhawk throughout the flights. Alongside the logs, the dataset includes supplementary code for dataset manipulation in C++, Python, and MATLAB.[5] This code is designed to be independent of ROS or any other external libraries and is compatible across Linux, macOS, and Windows platforms.

The dataset was compiled in the course of developing a new real-time method for detecting anomalies in autonomous aerial vehicles. Further details about this research can be found on the project's webpage. We will use ALFA dataset in Sect. 8.

4.8 Summary

This chapter serves as a guide for setting up a forensic laboratory tailored to drone investigations. It addresses the requirements for infrastructure, equipment, and software needed for thorough data examination. It begins with an introduction to the components of a drone forensics laboratory. The chapter highlights the nature of maintaining evidence integrity throughout the forensic investigation process, from software installation to analysis.

---

[4] https://theairlab.org/alfa-dataset/.

[5] https://github.com/castacks/alfa-dataset-tools.

In detailing the setup of a forensic workstation, the chapter outlines specific hardware requirements, including a powerful processor, RAM, fast and high-capacity storage options, multiple network interfaces, and a dedicated graphics card for intensive processing tasks. Software requirements include a stable operating system, a suite of forensic analysis tools, and virtualization software for managing virtual machines and safely examining potential threats.

The chapter also advises on network configuration and recommend the isolation of the forensic workstation from external networks to prevent contamination and advocating for robust firewall and security measures to protect against unauthorized access. By providing a understanding of the technical aspects of establishing a drone forensics laboratory, this chapter prepares readers to perform hands-on approach in UAV and drone forensics. This preparation will make sure users are equipped to meet the demands of forensic science in the context of drone and UAV investigations.

4.9 Exercises

1. What is the primary purpose of a drone forensics laboratory?
2. What are the three main requirements for establishing a drone forensics laboratory?
3. Why is maintaining evidence integrity important in forensic investigations?
4. What are the key steps in the workflow of a forensic investigation described in this chapter?
5. What tools do investigators need for setting up a drone forensics laboratory?
6. Explain the importance of having a specialized physical infrastructure for a drone forensics laboratory.
7. Discuss the role of open-source software tools in the forensic analysis workflow.
8. Analyze the challenges involved in the context of integrity of evidence during the forensic process.
9. Evaluate the procedural aspects critical to the successful setup and operation of a drone forensics laboratory.
10. Why do we need a separate virtual environment for each Python-based drone forensic tool?

References

1. Open Text, *Quick Reference Quide: Tableau Forensic T35u/T35u-RW SATA/IDE Bridge* (2021). https://www.opentext.com/file_source/OpenText/en_US/PDF/opentext-t35u-t35u-rw-quick-referenceguide-en.pdf

2. Exterro, *Create forensic images with Exterro FTK Imager* (2024). https://www.exterro.com/digital-forensics-software/ftk-imager.
3. Autopsy, *Autopsy: Digital Forensics* (2024). https://www.autopsy.com/
4. VMware by Broadcom, *Desktop Hypervisor* (2024). https://www.vmware.com/products/desktop-hypervisor/workstation-andfusion
5. Anaconda Inc., *Anaconda: The Operating System for AI* (2024). https://www.anaconda.com/
6. VTO Inc., *Drone Forensics Datasets* (2018). https://cfreds-archive.nist.gov/drone-images.html
7. A. Keipour, M. Mousaei, S. Scherer, ALFA: a dataset for UAV fault and anomaly detection. Int. J. Robot. Res. **40**(2–3), 515–520 (2021)

Chapter 5
Data Acquisition from Drone and UAV

Abstract This chapter will discuss technical procedures involved in collecting and securing data on UAVs and drones for forensic analysis. A review of methods tailored to work with iOS, Android, macOS, and Windows operating systems will follow. The chapter steps through the forensic images acquisition from iPhones, Android devices, and drone SD cards in detailed steps using open-source and native tools. This chapter concerns itself with the logical and physical methods of imaging and equips the users with practical skills to execute data extraction and analysis on drones while keeping digital evidence intact.

5.1 Introduction

Collecting data from UAVs and drones is one of important parts of digital forensics. It is all about preserving and analyzing the data these devices capture during their flights. The goal is to extract this data from the drone's storage while keeping its integrity for forensic investigations. If we are using macOS, the Finder [1] feature makes the initial steps of data acquisition straightforward for acquiring iOS-based drone controller. By connecting our UAV or drone controller to our Mac with the right cables, we can easily browse through the device's storage using Finder application. Users can explore and extract files directly from the device.

For those dealing with Android-based UAVs or drone controllers, the Android File Transfer tool is a utility for accessing and transferring files between the Android device and a macOS [2]. By connecting the Android device to a computer via USB, the Android File Transfer tool allows for the extraction of user-generated content and data stored on the device. While it may not capture all system-level files or create complete disk images like other specialized tools, it is a straightforward solution for retrieving accessible files. On the other hand, it is important to maintain the integrity of the data throughout this process. This is where a write blocker, such as the Tableau TK8U USB 3.0 Forensic Bridge Kit [3], comes into play. This device guarantees that no changes are made to the original data on the UAV or drone SD card and it keeps the evidence intact and admissible in judicial processes.

5.2 Acquisition of iPhone Drone Controller

Creating a forensic image from an iPhone using open-source tools or native macOS features presents unique challenges due to Apple's robust security architecture. It includes features such as data encryption, secure boot, and hardware-based security measures. These security implementations are designed to protect user data and makes forensic acquisition more complex. Despite these challenges, there are methods and tools available for extracting data from an iPhone for forensic purposes, particularly when dealing with iOS-based drone controllers.

This section provides a guide to acquire data from iOS-based drone controllers using macOS native features and free and/or open-source tools. The focus is on utilizing macOS's built-in tools such as Finder, which can detect and access connected iOS devices, and Terminal, which can be used to interact with the device via command-line utilities. Additionally, open-source tools such as libimobiledevice–a cross-platform software library that allows communication with iOS devices–can be used to extract logical data without jailbreaking the device.

5.2.1 Using macOS Features

To guarantee all relevant data is securely backed up and ready for forensic analysis, we can create a full backup of the iPhone used as a drone controller. Follow these steps:

1. Prepare the iPhone for backup

    Begin by charging the iPhone so it has sufficient power for the entire backup process, as interruptions can corrupt the backup file. In addition, close any active applications on the iPhone that might interfere with the data transfer.
2. Connect the iPhone to the MacBook

    Use a certified lightning data cable to connect the iPhone to the MacBook. If any prompts appear on the iPhone, such as Trust This Computer, tap Trust and enter the device passcode if required.
3. Access the backup interface

    For macOS Catalina and later, open Finder and look for the iPhone under Locations in the sidebar. We need to click "Trust" to connect and synchronize with the iPhone (Fig. 5.1).
4. Configure backup settings

    In the iPhone's Summary or General tab, confirm that "Encrypt local backup" is unchecked as shown in Fig. 5.2. This setting enables access to non-encrypted data that may be necessary for evidence acquisition. If encrypted data is specifically required, enable "Encrypt local backup", choose a secure password, and document it carefully for future access.
5. Initiate the backup process

5.2 Acquisition of iPhone Drone Controller

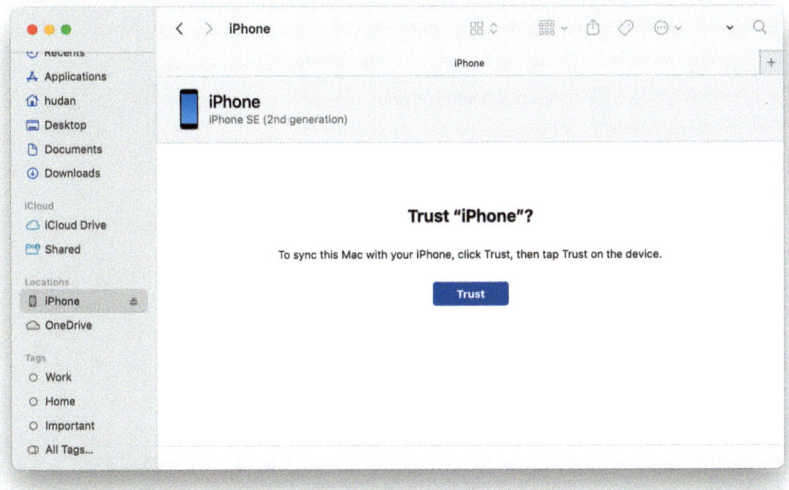

Fig. 5.1 Trust iPhone to connect and synchronize MacBook with the iPhone

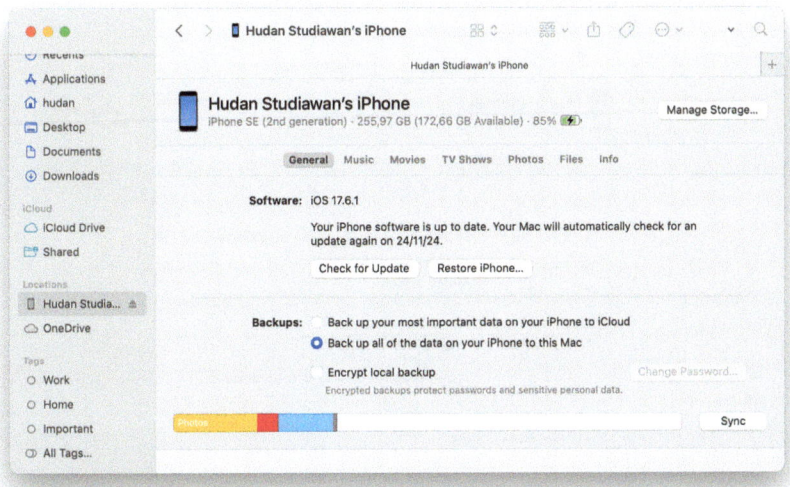

Fig. 5.2 iPhone summary

Click "Back Up Now" button to start creating a local backup of the iPhone (Fig. 5.3), which may take several minutes to an hour depending on the device's data volume. Check the progress in Finder to monitor the backup completes without interruptions as shown in Fig. 5.4.

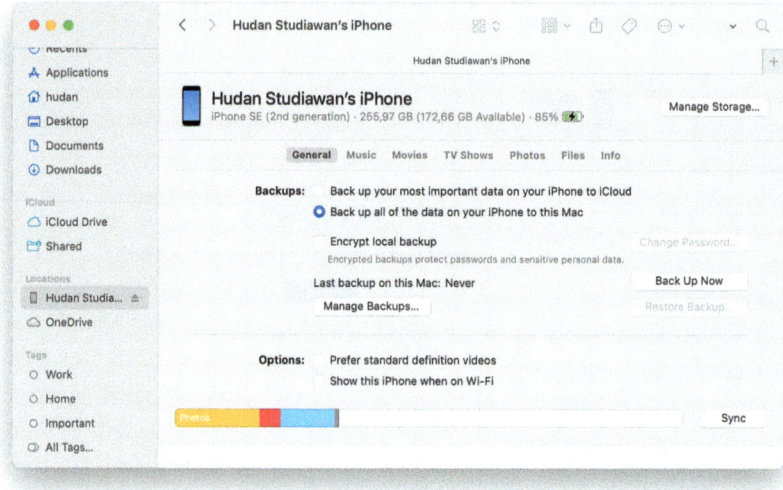

Fig. 5.3 Back Up Now button to start backup

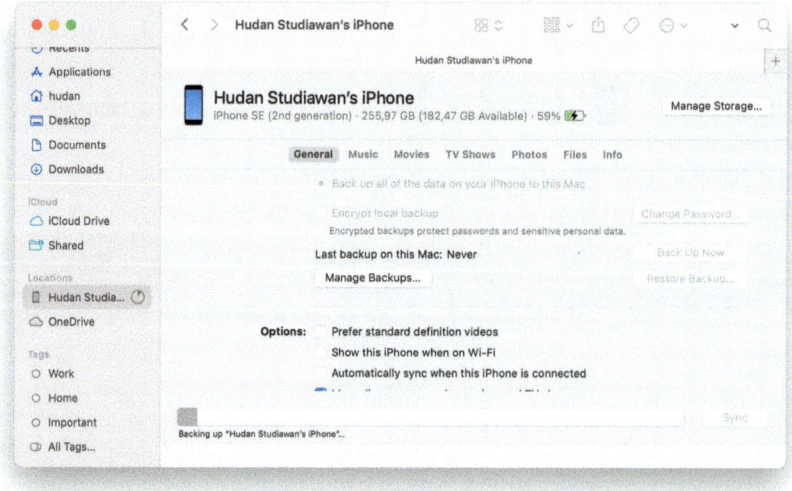

Fig. 5.4 iPhone Back up is running

6. Verify backup completion

After completion, we can locate the backup file in macOS by navigating to the backup directory. It is typically found at ~/Library/Application Support/MobileSync/Backup/.

7. Store the backup securely

    Once the backup process is complete, locate the stored backup file on our Mac. By default, iPhone backups are saved in the directory ~/Library/Application Support/MobileSync/Backup/. Navigate to this folder to access the newly created backup.

    To preserve the data for forensic analysis, copy the entire backup folder to a secure and isolated storage location. This could be an external hard drive or a dedicated forensic storage device. This step is performed to assure that no changes or corruption occur to the original data during transfer. It is recommended to use a write blocker or to set the destination drive to read-only mode if feasible, as this further maintains the integrity of the backup for later investigation. Consider creating additional copies of the backup as part of a backup management plan. We need to keep an archival copy while working on the primary data in a controlled forensic environment.

5.2.2 Using Open-source Tools

To interact with iOS devices and manage backups with open-source tools, we need to install the libimobiledevice suite of tools [4]. This installation can be done quickly using apt, a popular package manager for Debian-based Linux that simplifies the installation of software and libraries. Note that we do not use libimobiledevice on macOS as we encounter a protocol error which we are unable to resolve. Therefore, we Windows Subsystem for Linux (WSL) on Windows operating system. We can run the following steps.

1. Begin by opening the Terminal or WSL on Windows.
2. Use apt to install libimobiledevice[1] by entering the command (Fig. 5.5):

    ```
    sudo apt-get install usbmuxd libimobiledevice6
        ↪libimobiledevice-utils
    ```

3. After the installation is complete, verify that libimobiledevice is installed by running:

    ```
    ideviceinfo --help
    ```

    This command should display a help message with options for ideviceinfo. It confirms that the tools and device are ready for use as depicted in Fig. 5.6.

    With libimobiledevice successfully installed, we will have access to a suite of tools, such as ideviceinfo, idevicebackup2, and others. These tools can be used to retrieve device information, create and manage backups, and perform various interactions with iOS devices. These tools are especially useful for extracting

---

[1] https://github.com/libimobiledevice/libimobiledevice.

```
hudan@DESKTOP-NAPT0TL:~$ sudo apt-get install usbmuxd libimobiledevice6 libimobiledevice-utils
Reading package lists... Done
Building dependency tree... Done
Reading state information... Done
The following additional packages will be installed:
  libplist3 libusbmuxd6
Suggested packages:
  libusbmuxd-tools
The following NEW packages will be installed:
  libimobiledevice-utils libimobiledevice6 libplist3 libusbmuxd6 usbmuxd
0 upgraded, 5 newly installed, 0 to remove and 2 not upgraded.
Need to get 260 kB of archives.
After this operation, 998 kB of additional disk space will be used.
Do you want to continue? [Y/n] Y
Ign:1 http://archive.ubuntu.com/ubuntu jammy/main amd64 libplist3 amd64 2.2.0-6build2
Get:2 http://archive.ubuntu.com/ubuntu jammy/main amd64 libusbmuxd6 amd64 2.0.2-3build2 [20.4 kB]
Get:3 http://archive.ubuntu.com/ubuntu jammy/main amd64 libimobiledevice6 amd64 1.3.0-6build3 [71.1 kB]
Get:4 http://archive.ubuntu.com/ubuntu jammy/universe amd64 libimobiledevice-utils amd64 1.3.0-6build3 [93.7 kB]
Get:5 http://archive.ubuntu.com/ubuntu jammy/main amd64 usbmuxd amd64 1.1.1-2build2 [42.8 kB]
Get:1 http://archive.ubuntu.com/ubuntu jammy/main amd64 libplist3 amd64 2.2.0-6build2 [32.1 kB]
Fetched 260 kB in 40s (6546 B/s)
Selecting previously unselected package libplist3:amd64.
(Reading database ... 42572 files and directories currently installed.)
Preparing to unpack .../libplist3_2.2.0-6build2_amd64.deb ...
Unpacking libplist3:amd64 (2.2.0-6build2) ...
Selecting previously unselected package libusbmuxd6:amd64.
Preparing to unpack .../libusbmuxd6_2.0.2-3build2_amd64.deb ...
Unpacking libusbmuxd6:amd64 (2.0.2-3build2) ...
Selecting previously unselected package libimobiledevice6:amd64.
Preparing to unpack .../libimobiledevice6_1.3.0-6build3_amd64.deb ...
```

Fig. 5.5 Installation of `ilibmobiledevice` package

```
hudan@DESKTOP-NAPT0TL:~$ ideviceinfo
ActivationState: Activated
ActivationStateAcknowledged: true
BasebandActivationTicketVersion: V2
BasebandCertId: 524245983
BasebandChipID: 104
BasebandKeyHashInformation:
  AKeyStatus: 64
  SKeyStatus: 2
BasebandMasterKeyHash: 1B41607650EBF11C6B39F41CB267DC64C121A9BCF44DBA5D28F55ACC86361BBA366554CD57B4C466055803E1EF81C870
BasebandRegionSKU: AAAAAAAAAAAAAAAAAAAAAAAAAAAAAAAAAAAAAAAAAAAAAAAAAAAAAAAAAAAAAAAAAAAAAAAAAAAAAAAAAAAAAA==
BasebandSerialNumber: ieXv0tAAdLQAAAAA
BasebandStatus: BBInfoAvailable
BasebandVersion: 6.00.00
BluetoothAddress: 44:35:83:2c:9b:17
BoardId: 16
BootSessionID: ED7F0D02-4BC3-4496-8961-4F7632BA959B
BrickState: false
BuildVersion: 22B91
CPUArchitecture: arm64e
CarrierBundleInfoArray[0]:
CertID: 524245983
ChipID: 32816
ChipSerialNo: ieXv0tAAdLQAAAAA
DeviceClass: iPhone
DeviceColor: 1
DeviceName: Hudan Studiawan's iPhone
DieID: 6315805482565678
EthernetAddress: 44:35:83:41:41:4f
FirmwareVersion: iBoot-11881.40.163
FusingStatus: 3
HardwareModel: D79AP
HardwarePlatform: t8030
HasSiDP: true
HostAttached: true
HumanReadableProductVersionString: 18.1.1
InternationalMobileEquipmentIdentity: 356795118301030
InternationalMobileEquipmentIdentity2: 356795118297667
MLBSerialNumber: FG3042701DCMWQRAK
MobileEquipmentIdentifier: 35679511830103
MobileSubscriberCountryCode:
```

Fig. 5.6 Checking device information with `ideviceinfo` command

forensic data from iPhone backups and other iOS artifacts. To perform a full forensic backup of an iOS device, `idevicebackup2` from the `libimobiledevice` suite offers options for both creating and analyzing backups. This command-line tool allows investigators to safely back up and interact with iPhone data for forensic acquisition.

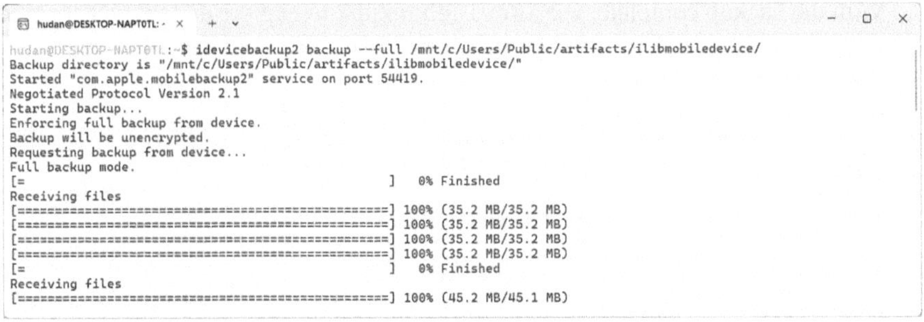

5.2 Acquisition of iPhone Drone Controller

```
hudan@DESKTOP-NAPT0TL:~$ idevicebackup2 backup --full /mnt/c/Users/Public/artifacts/ilibmobiledevice/
Backup directory is "/mnt/c/Users/Public/artifacts/ilibmobiledevice/"
Started "com.apple.mobilebackup2" service on port 54419.
Negotiated Protocol Version 2.1
Starting backup...
Enforcing full backup from device.
Backup will be unencrypted.
Requesting backup from device...
Full backup mode.
[=                                                 ]    0% Finished
Receiving files
[==================================================] 100% (35.2 MB/35.2 MB)
[==================================================] 100% (35.2 MB/35.2 MB)
[==================================================] 100% (35.2 MB/35.2 MB)
[==================================================] 100% (35.2 MB/35.2 MB)
[=                                                 ]    0% Finished
Receiving files
[==================================================] 100% (45.2 MB/45.1 MB)
```

Fig. 5.7 Backup iOS device using `idevicebackup2` command

To create a secure backup of the iOS device, use the `idevicebackup2 backup` command. This command saves a full backup of the device to a specified directory on our computer.

1. Launch Terminal on WSL
2. Enter the following command to create a backup as shown in Fig. 5.7. Note that we need to replace `/path/to/backup/directory` with the desired destination for storing the backup. For example in Fig. 5.7, the path for the iOS drone controller backup is `/mnt/c/Users/Public/artifacts/ilibmobiledevice`.

   `idevicebackup2 backup --full /path/to/backup/directory`

   This command will copy all accessible data from the iOS device to the specified location. The backup may take some time, depending on the amount of data on the iPhone.
3. Once the backup is complete, check the backup directory to verify that all data files have been successfully saved as shown in Fig. 5.8.

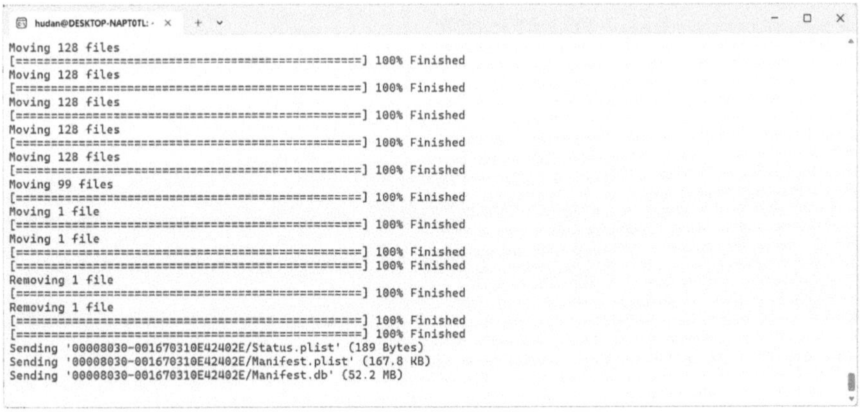

```
Moving 128 files
[==================================================] 100% Finished
Moving 128 files
[==================================================] 100% Finished
Moving 128 files
[==================================================] 100% Finished
Moving 128 files
[==================================================] 100% Finished
Moving 128 files
[==================================================] 100% Finished
Moving 99 files
[==================================================] 100% Finished
Moving 1 file
[==================================================] 100% Finished
Moving 1 file
[==================================================] 100% Finished
[==================================================] 100% Finished
Removing 1 file
[==================================================] 100% Finished
Removing 1 file
[==================================================] 100% Finished
[==================================================] 100% Finished
Sending '00008030-001670310E42402E/Status.plist' (189 Bytes)
Sending '00008030-001670310E42402E/Manifest.plist' (167.8 KB)
Sending '00008030-001670310E42402E/Manifest.db' (52.2 MB)
```

Fig. 5.8 Backup is complete

With the backup complete, use `idevicebackup2` to explore and extract specific data types or gain detailed information about the backup contents. Here are some useful commands for analysis:

1. To view metadata about the backup, such as backup date, device name, and software version, run the `info` command:

   `idevicebackup2 info /path/to/backup/directory`

   This command is for verifying the backup details and understanding the context of the data stored within it.

2. To view a directory-style listing of files within the backup, use the ls command:

   `idevicebackup2 ls /path/to/backup/directory`

   This command outputs a list of all files and directories in the backup, allowing we to identify specific files of interest. Alternatively, we can use File Explorer to view the results as shown in Fig. 5.9.

3. If we need to extract certain files for detailed analysis, use the `idevicebackup2` restore command with the specific file path. For example:

   `idevicebackup2 restore -d /path/to/backup/directory`
   `↪path/to/file`

Forensic investigators can analyze and retrieve data from an iOS backup by integrating these commands and facilitate a comprehensive examination of the drone controller contents.

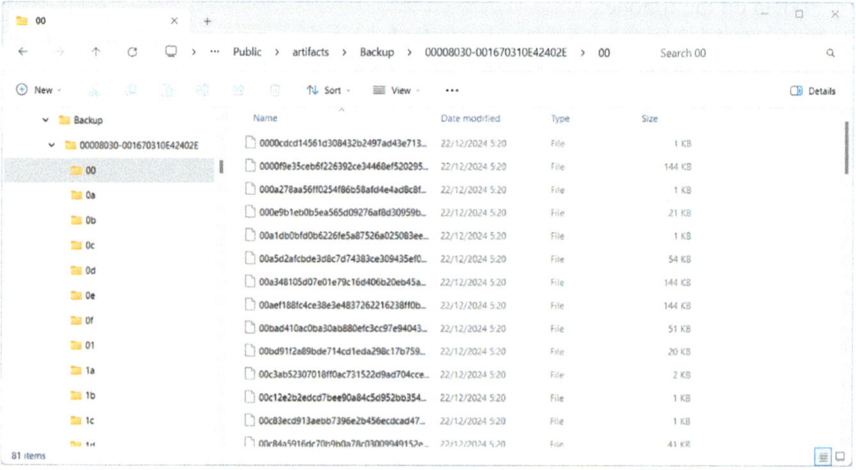

Fig. 5.9 Results of iOS backup in Windows File Explorer

5.2 Acquisition of iPhone Drone Controller

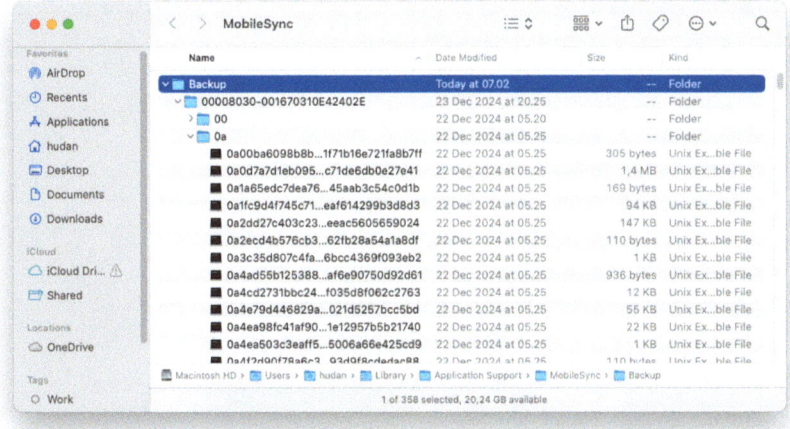

Fig. 5.10 Results of iOS backup in ~/Library/Application Support/MobileSync/Backup/ directory

5.2.3 Verifying iOS Controller Acquisition

Parsing and analyzing an iPhone backup involves systematically examining the files within the backup directory to identify relevant data, such as messages, call logs, app data, and other artifacts that may be of interest in forensic investigations. This process can be conducted through both manual browsing and automated tools, depending on the depth and scope of the analysis.

The backup files can be manually explored using Finder (Fig. 5.10) or Autopsy (Fig. 5.11). Figure 5.10 displays the contents of an iOS backup directory located at /Users/hudan/Library/Application Support/MobileSync/Backup. This directory is where macOS stores iPhone backups created using Finder. Each folder within the Backup directory corresponds to a unique backup of a device. It is identified by an alphanumeric string (00008030-001670310E42402E in this case). The internal structure of these backups includes subdirectories like 00 and 0a, which contain hashed files representing data from the iPhone. The hashed filenames are generated from the SHA-1 hashes of the original file paths. Within the backup directory, the stored iPhone backup contains various folders and files that can be browsed to locate specific data files. Each file is labeled with a unique alphanumeric identifier. It requires cross-referencing with iOS file structure documentation to identify common data types. Examples include sms.db for SMS and iMessages or Manifest.db, which serves as a complete index of the backup contents.

Furthermore, files can be examined in detail using text or hex editors. Text editors such as VS Code or Sublime Text allow for the inspection of raw data and metadata within structured files, while hex editors like Hex Fiend are suited for analyzing

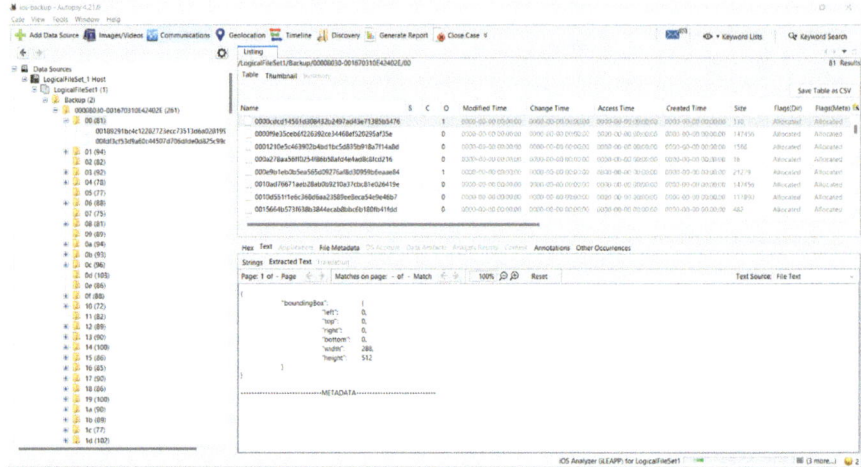

Fig. 5.11 File listing from iOS backup of a drone controller

binary files [5]. For example, the `sms.db` file can be explored in a SQLite DB Browser [6] to access SMS and iMessage content, whereas Hex Fiend is used for checking additional metadata or embedded information in binary files that may be relevant to an investigation.

Parsing and analysis of iOS backups can be automated using scripts and open-source forensic tools, which are designed to interpret backup structures and present the data in an organized and readable format. This approach reduces the time and effort required for manual analysis. Tools such as Autopsy provide automated data extraction from iPhone backups. They will categorize files into groups such as messages, contacts, call history, and media files.

The Manifest.db file, as seen in the Autopsy tool interface (Fig. 5.12), is a SQLite database to decode iOS backups. It maps hashed filenames, which iOS uses to store backup files, to their original file paths and metadata. We need this mapping because hashed filenames do not directly convey any meaningful information about the files they represent. Manifest.db bridges this gap, so investigators is able to reconstruct the original filenames, file paths, and their associated contexts within the iOS filesystem.

This database also includes valuable metadata about the files, such as the application or system domain (e.g., which application created the file), the file's relative path within the iOS file system, and specific flags that describe file properties or states. Forensic analysts rely on this information to understand the purpose of each file, its location, and its relation to the user or the system. The inclusion of metadata makes it possible to identify not only where a file was located but also how it was used within the context of the iOS environment.

The file also provides the organization of iOS backup data during forensic analysis. In tools such as Autopsy, Manifest.db allows investigators to browse

5.2 Acquisition of iPhone Drone Controller

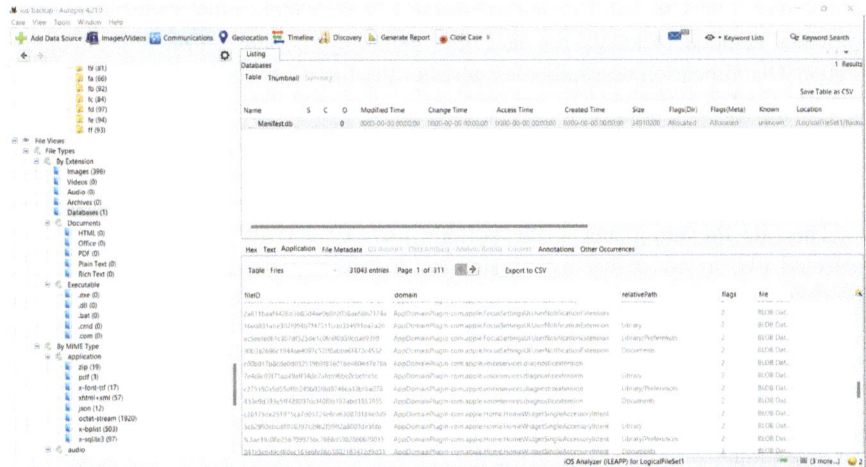

Fig. 5.12 Manifest.db contains mapping the real file names to hashed ones

and export a structured view of the backup's contents, with columns for hashed filenames (fileID), domains, relative paths, and file types. This functionality helps investigators analyze large amounts of data, such as in the example shown in the interface (Fig. 5.12), where over 31,000 entries are displayed.

Another case for Autopsy is shown in Fig. 5.13. In this case, the investigator is focusing on files categorized by their MIME type, specifically image/png, which represents PNG image files. On the left side, the File Views panel organizes the data into categories based on file extensions or MIME types. The selected MIME type,

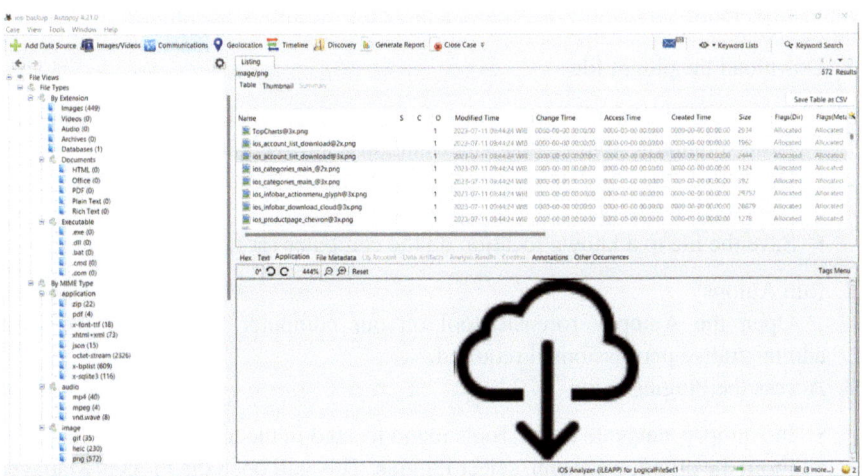

Fig. 5.13 View a file based on MIME type

image/png, filters the files displayed in the Listing panel on the right. This panel provides detailed metadata for each file, including its name, size, and allocation status. The allocation flag indicates whether the file is still active in the backup or has been deleted. In addition, timestamps such as Modified Time, Change Time, and Access Time are shown, although in this instance, these fields are populated with default values of 00:00:00, which could suggest missing metadata or issues during the extraction of the backup.

The File Preview panel at the bottom provides a visual representation of the selected file, to assist the investigator in identifying its content. Tabs below the preview allow access to more technical details, such as the file's hexadecimal structure, plain text content, or detailed metadata. In the selected example, the preview displays an image of a cloud download icon, indicating that this file could be part of an app's interface or cached data. Additional metadata and contextual information about the file can also be explored through this panel to enhance the investigation.

From a forensic perspective, this analysis provides valuable insights. By focusing on specific MIME types, the investigator can quickly filter relevant data, such as images, videos, or documents, that may be important to the case. The ability to examine metadata allows for tracing user activities and understanding device behavior. For instance, the filenames such as ios_account_list_download@2x.png or TopCharts@3x.png suggest these images might be associated with an iOS application. They are potentially containing evidence of user interactions or app-generated content. Furthermore, the allocation status of the files indicates whether they were actively stored in the backup at the time of its creation.

To provide a more meaningful iOS backup viewer, we use the iOS Device Data Extractor plugin [7]. It is an Autopsy module designed to create encrypted or unencrypted iOS backups for iPhones or iPads. This tool also extracts specific files from these backups and provides a view of iOS backup conveniently without hashed file names. Outlined below are the procedures for installing this plugin.

1. Download the plugin file

    - Visit the official GitHub repository for the iOS Device Data Extractor: https://github.com/ernestbies/iOSDeviceDataExtractor/tree/master/build.
    - Download the file named `org-gbies-iosdevicedataextractor.nbm` from the repository.
    - Save the file to a known location on the computer for later use.

2. Run Autopsy

    Open the Autopsy forensic tool on our computer. Make sure we have administrative permissions if required.

3. Access the Plugins menu

    - In Autopsy, navigate to the Tools menu located in the top navigation bar.
    - From the dropdown menu, select Plugins. This will open the Plugins Manager as shown in Fig. 5.14.

5.2 Acquisition of iPhone Drone Controller

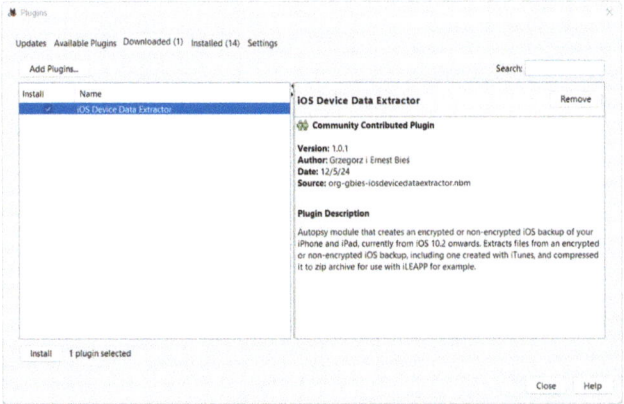

Fig. 5.14 iOS Device Data Extractor: Add plugin

4. Add the plugin
   - In the Plugins Manager window, switch to the Downloaded tab.
   - Click on the Add Plugins button.
5. Select the plugin file
   - In the file selection dialog that appears, browse to the location where we saved the `org-gbies-iosdevicedataextractor.nbm` file.
   - Select the file and click Open.
6. Install the plugin
   - Once the plugin file is added, click on the Next button as depicted in Fig. 5.15.
   - Follow the on-screen installation instructions. This process might include accepting terms and conditions in License Agreement window (Fig. 5.16) or confirming a validation warning (Fig. 5.17).
7. Restart Autopsy
   - After the installation is complete (Fig. 5.18), we will be prompted to restart Autopsy.
   - Ensure that we save any ongoing work, then close and reopen Autopsy to finalize the installation process.
8. Verify installation
   To confirm that the iOS Device Data Extractor has been successfully installed, go to the Tools menu and check for the module in the available list of plugins or modules.

To analyze an iOS backup file using the iOS Device Data Extractor plugin in Autopsy, follow these steps:

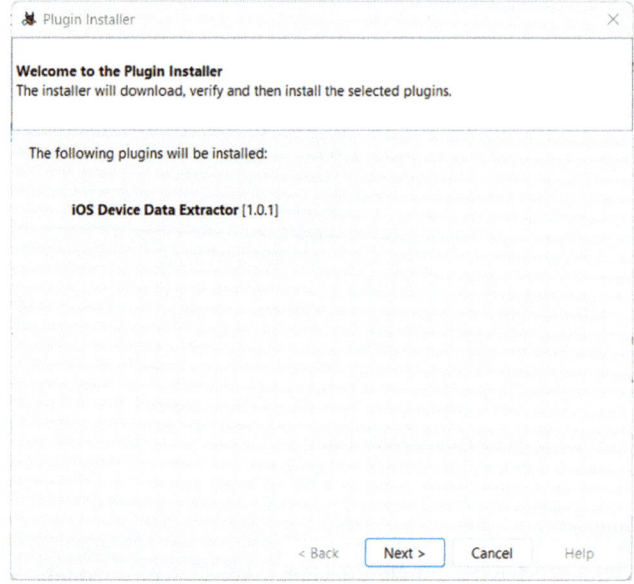

Fig. 5.15 iOS Device Data Extractor: Plugin installer

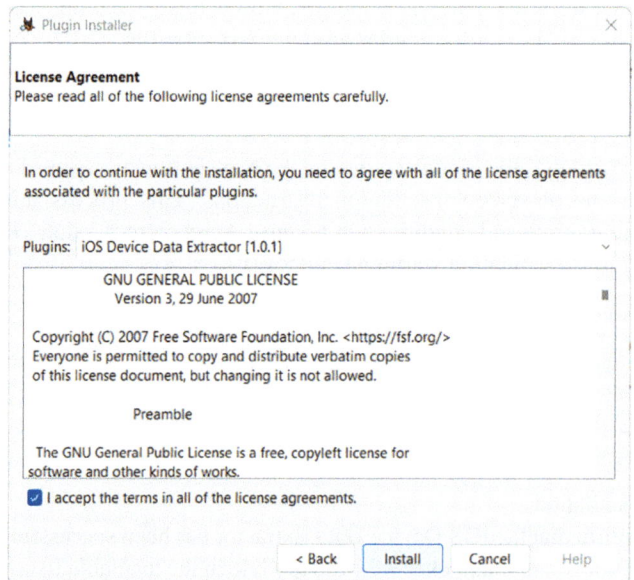

Fig. 5.16 iOS Device Data Extractor: License agreement

5.2 Acquisition of iPhone Drone Controller

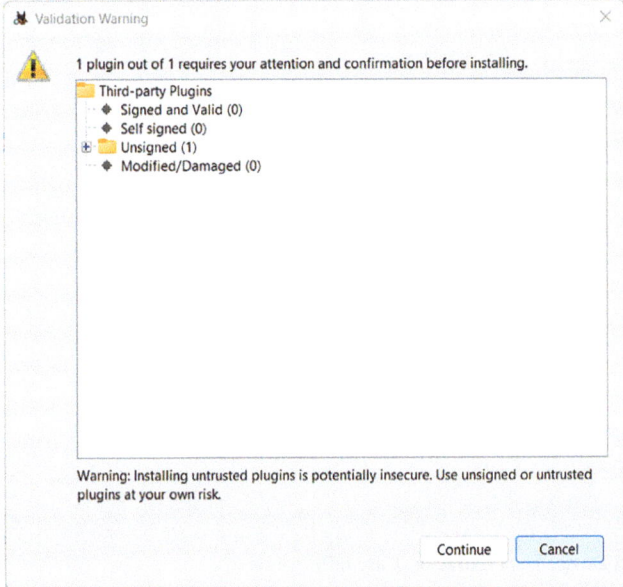

Fig. 5.17 iOS Device Data Extractor: Validation warning

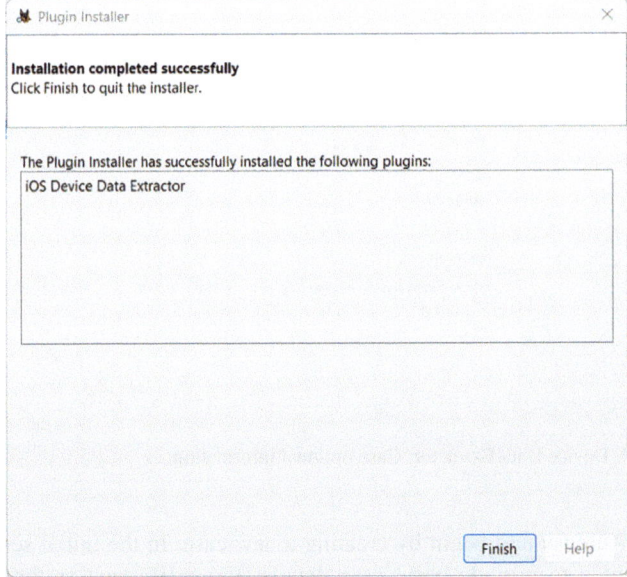

Fig. 5.18 iOS Device Data Extractor: Installation of the plugin is complete

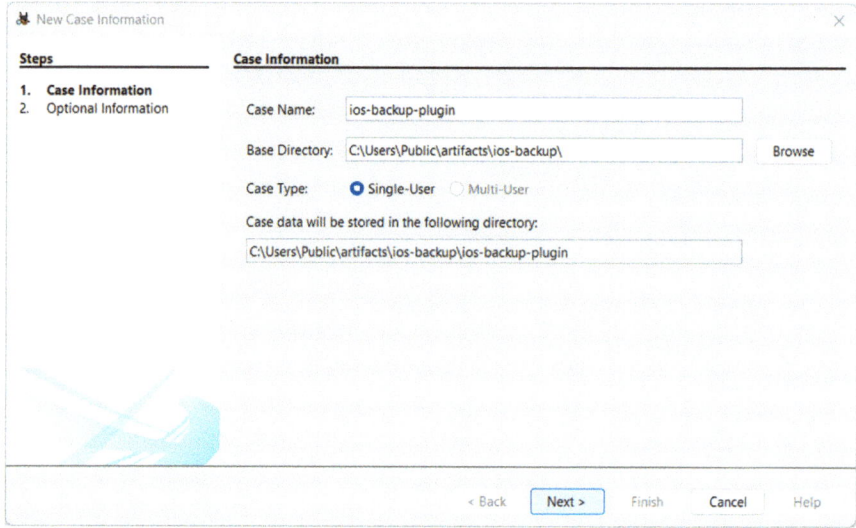

Fig. 5.19 iOS Device Data Extractor: Case information

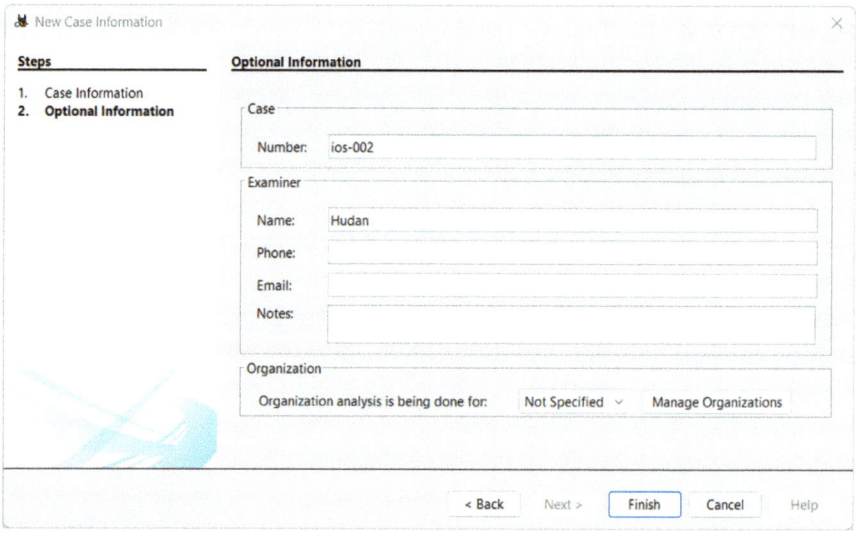

Fig. 5.20 iOS Device Data Extractor: Case optional information

1. Launch Autopsy and begin by creating a new case. In the initial setup, we will be prompted to provide basic case details (Fig. 5.19) such as the case name, description, examiner's name, and other optional information (Fig. 5.20). Enter the relevant details and proceed.

5.2 Acquisition of iPhone Drone Controller

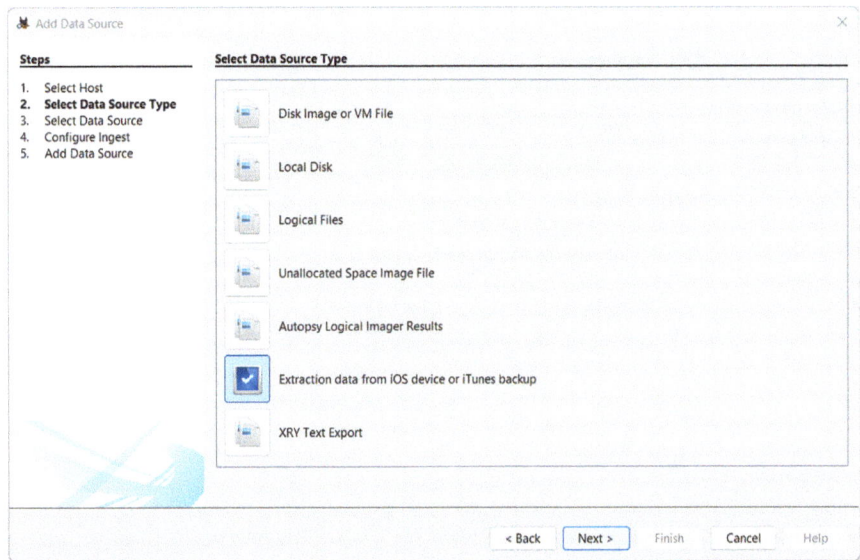

Fig. 5.21 iOS Device Data Extractor: Make sure to select "Extraction data from iOS device or iTunes backup"

2. In the next step, choose the appropriate data source type. From the available options, select "Extraction data from iOS device or iTunes Backup" as shown in Fig. 5.21. It means that the plugin is utilized to process the iOS backup file.
3. Follow the on-screen instructions provided by Autopsy. Users will be asked to specify the location of the backup directory containing the iOS backup data. Once the directory is selected, the iOS Device Data Extractor plugin will begin processing the files. Depending on the size and complexity of the backup, this process may take some time as shown in Fig. 5.22.
4. After the processing is finished, click the Finish button to finalize the operation (Fig. 5.23). At this point, the processed data will be available for analysis within the Autopsy interface.
5. The results generated by the iOS Device Data Extractor plugin can now be reviewed. As shown in Fig. 5.24, the plugin successfully decodes and organizes the backup contents. Instead of hashed file and directory names, the plugin displays meaningful names. Therefore, it is easier to identify and analyze individual files and directories.
6. For example, as illustrated in Fig. 5.24, users can browse specific artifacts extracted from the backup. In the context of drone investigations, this may include accessing and analyzing drone flight log artifacts in a specific directory, such as /AppDomain-com.dji.golite/Documents/FlightRecords. These artifacts can provide insights into flight patterns, timestamps, and other metadata.

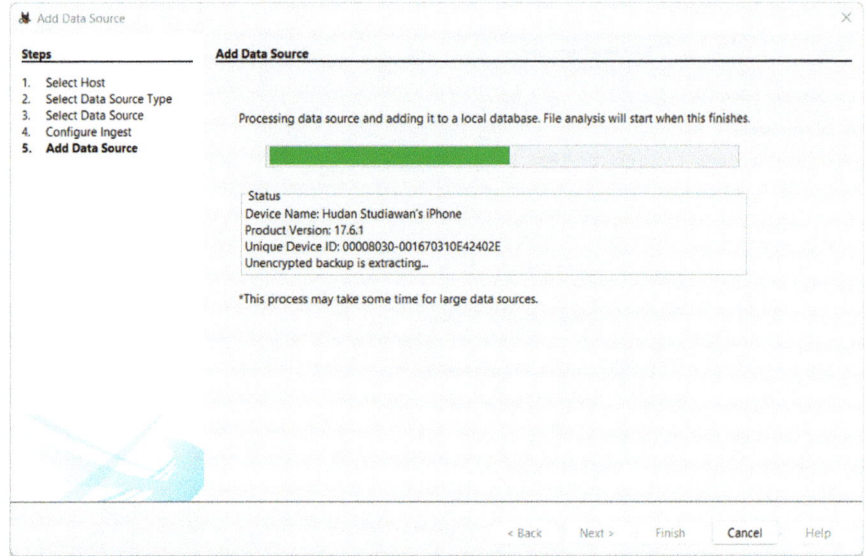

Fig. 5.22 iOS Device Data Extractor: Processing data source

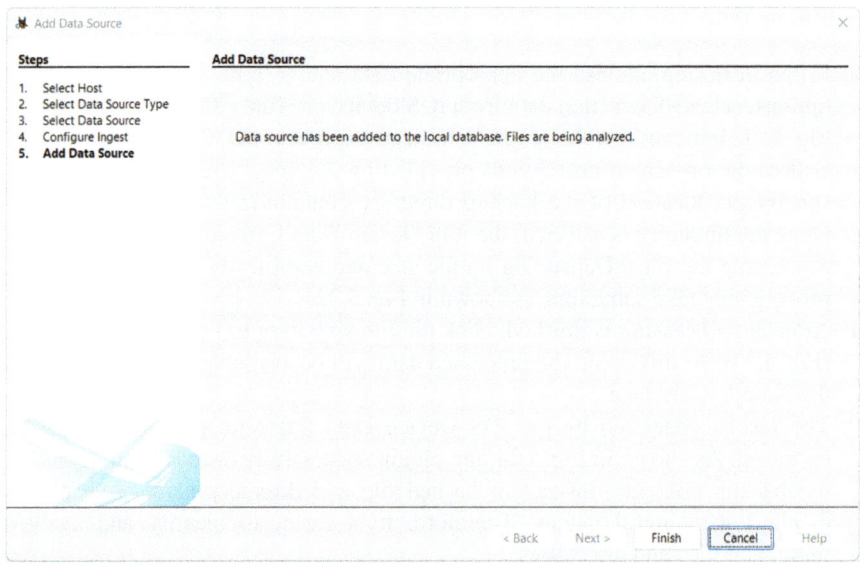

Fig. 5.23 iOS Device Data Extractor: Add data source is successful

5.3 Acquisition of Android Drone Controller

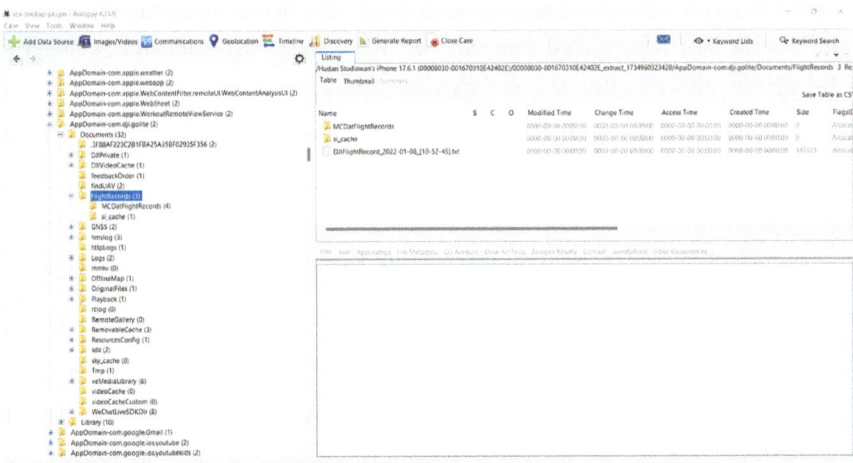

Fig. 5.24 iOS Device Data Extractor: DJI artifacts listing

5.3 Acquisition of Android Drone Controller

In digital forensics, there are two primary types of imaging used to acquire data from devices: physical imaging and logical imaging. Each method serves a different purpose and has its own set of tools and requirements. In this case, we are performing logical imaging, which captures user-level data without directly accessing the entire storage on a bit-by-bit basis.

Logical imaging focuses on extracting data that is accessible through the device's operating system and file system, such as user-level data, application data, and other accessible files stored on the device. This method is often achieved using tools such as Android Debug Bridge (ADB) or other forensic software that interacts with the device's standard file management structure. Logical imaging is generally more straightforward than physical imaging, as it does not require root access to the device. Therefore, it is suitable for cases where accessing user-level files is sufficient. However, it does not capture deleted files, hidden files, or system-level data stored outside user-accessible partitions.

Physical imaging, on the other hand, involves capturing a complete, bit-by-bit copy of the entire device storage, including all data sectors, system files, deleted files, and hidden partitions. This method provides the most comprehensive data acquisition and is often necessary in cases where deeper forensic analysis is required, such as retrieving deleted files or analyzing low-level system data. Physical imaging requires specialized forensic tools, such as Cellebrite UFED, Oxygen Forensic Detective, or MSAB XRY, and may necessitate root access to the Android device to bypass certain security restrictions. Because it involves accessing all sectors of the device's memory, physical imaging can provide a more complete forensic picture but also requires more advanced techniques and equipment.

Creating a forensic image from an Android device involves several steps to make sure that the data is preserved and collected in a manner that maintains its integrity for forensic analysis. To conduct a forensic acquisition of data from an Android device, we need to have the right tools and software in place. Below is a list of tools that will support a smooth and effective forensic examination:

1. Computer with necessary software installed

    Ensure that our computer has all essential forensic and device communication software pre-installed, such as Android File Transfer [2] for communicating with Android devices, and any other specialized forensic imaging or analysis tools needed for the investigation.

2. Data cable for Android device connection

    A compatible data cable is required to securely connect the Android device to the forensic workstation. Use a high-quality cable to prevent data transfer issues and provide a stable connection as it is important for a reliable forensic acquisition.

3. Forensic imaging software

    Select appropriate forensic imaging software to create a complete and verifiable copy of the device's data. Cellebrite UFED is a robust tool for comprehensive data extraction and analysis from Android and iOS devices. On the other hand, Oxygen Forensic Suite is known for its broad support for mobile devices and advanced capabilities in data parsing and analysis. Magnet AXIOM is another powerful option that provides extensive support for data extraction and detailed analysis. Note that aforementioned tools are quite expensive in terms of license pricing. Therefore, we will use free tools, such as Android File Transfer, as it is more affordable for students and researchers.

To enable communication between the Android device and our forensic workstation, we should enable Developer Mode and USB Debugging. This process will allow we to access the device's internal storage and interact with it using forensic acquisition tools.

1. On the Android device, navigate to "Settings" → "About phone". Scroll down to locate the "Build number" entry, which contains version information about the device's software. Tap "Build number" seven times in succession. We may be prompted to enter the device's password or PIN to confirm. Once complete, a message will appear indicating that Developer Mode has been enabled.
2. Return to the main "Settings" menu, and a new option labeled "Developer options" should now be visible. Enter "Developer options" and scroll down to find "USB debugging". Toggle this setting to enable USB debugging, which allows the device to communicate with our computer over USB for advanced functions. A confirmation dialog may appear which asks if we trust the computer that will connect to the device; confirm this to proceed. With USB debugging enabled, we can now use forensic tools to interact with the device and acquire data.

5.3 Acquisition of Android Drone Controller

Furthermore, the following steps explain how to install and use the Android File Transfer tool on macOS for acquisition of Android drone controller.

1. Download and install Android file transfer

   - Start by downloading the Android File Transfer tool. It can be downloaded from its official download page: https://android.p2hp.com/filetransfer/index.html.
   - Alternatively, the installation file can be downloaded directly using the link.[2]
   - Once the download is complete, double-click the .dmg file to open the installer.
   - Drag and drop the Android File Transfer application into the Applications folder on macOS to complete the installation.

2. Connect the Android device

   Make sure that the Android device, which serves as the drone controller and requires forensic analysis, is powered on and ready to connect. Use a compatible USB cable to connect the Android device to the macOS computer.

3. Launching the tool

   - Open the Android File Transfer application.
   - If no Android device is connected or detected, a message will be displayed, as shown in Fig. 5.25. This indicates that the tool cannot establish a connection with an Android device.

4. Troubleshooting connection issues

   If the Android device is not detected, verify the following:

   - The device is properly connected via USB.
   - USB debugging is enabled on the Android device. This can be done by navigating to the Developer Options in the Android settings and enabling USB Debugging.
   - The correct USB mode is selected on the device, typically File Transfer (MTP) mode.
   - On the Android device, allow access to phone data as shown in Fig. 5.26.

   Check the macOS system has the required permissions to access connected devices. If prompted, allow the connection and ensure all necessary drivers or permissions are installed.

5. Accessing and creating the backup

   - Once the Android File Transfer tool detects the connected device, it will display the file directory of the Android device as shown in Fig. 5.27.
   - Navigate through the directory structure to locate important files and data related to the drone controller application. This may include drone flight logs (Fig. 5.28), media (Fig. 5.29), and other application-specific files. For

---

[2] https://dl.google.com/dl/androidjumper/mtp/current/AndroidFileTransfer.dmg.

Fig. 5.25 No Android device found when launching Android File Transfer

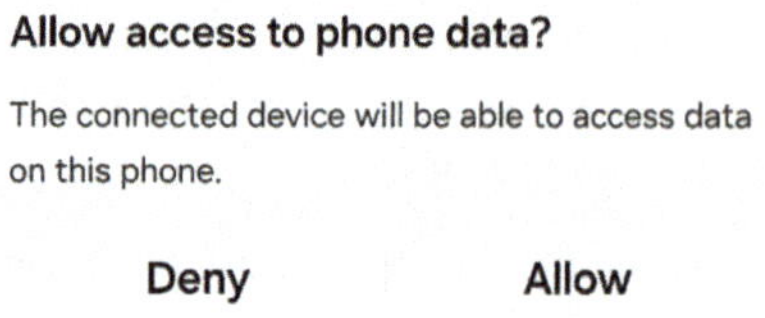

Fig. 5.26 Allow access to phone data before backing up the drone controller data

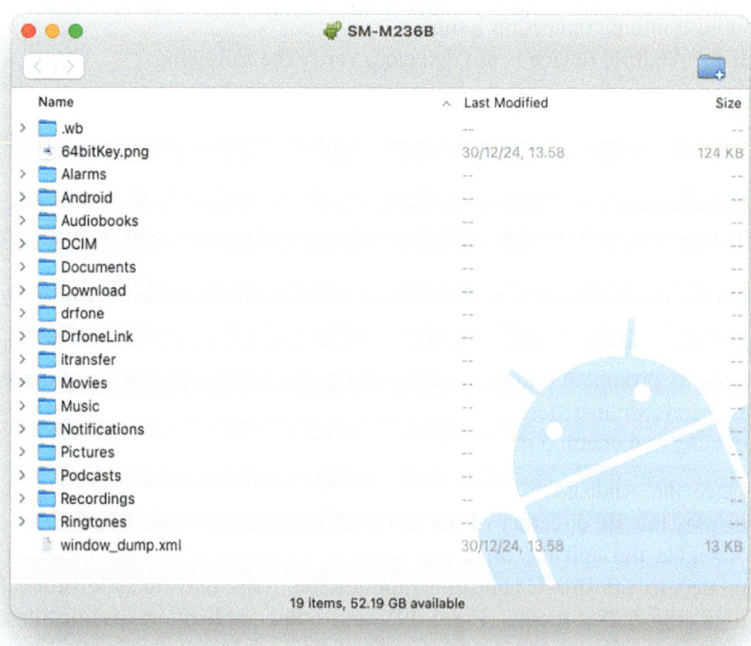

Fig. 5.27 View of directory structures of the Android internal storage

5.3 Acquisition of Android Drone Controller 117

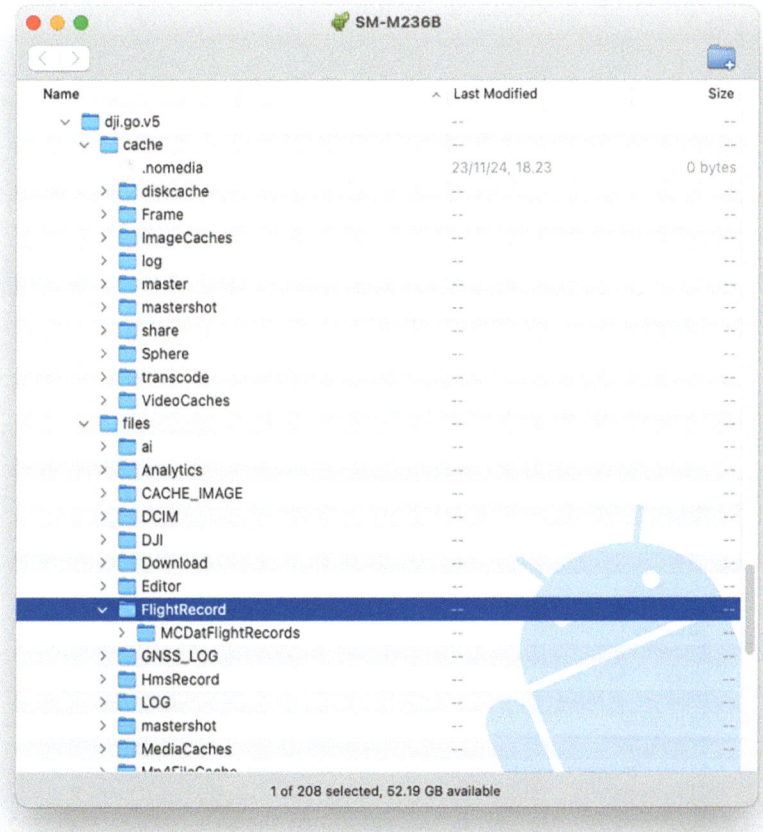

Fig. 5.28 Location of the drone flight logs artifacts

instance, the location of drone flight logs of DJI Fly application version 5 is /Android/data/dji.go.v5/files/FlightRecord, while the location of DJI-related media is /DCIM/DJI Album or /DCIM/DJI Export.
- Copy (drag and drop) the relevant files or the entire directory to the macOS system for further forensic analysis. Make sure that all files are preserved in their original format to maintain evidentiary value.

6. Documenting the process

- Document every step taken during the data extraction process, including details such as the device model, serial number, software versions, and any tools or techniques used.
- Capture screenshots of the process, including any error messages or system prompts, to include in the forensic report.

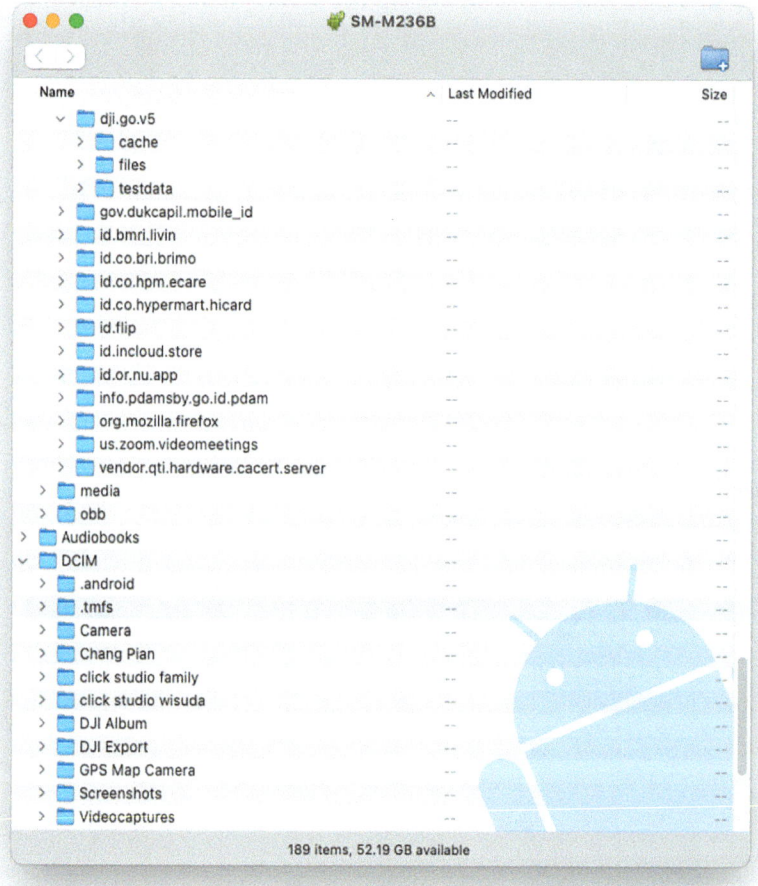

Fig. 5.29 Location of DJI-related media artifacts

After the data backup is complete, verify the completeness and integrity of the copied data. Check the file sizes and contents against the expected values to ensure no data loss occurred during transfer. This verification process is needed in forensic imaging, so we need to maintain detailed documentation of the imaging process, including the date and time of acquisition, the specific commands used, and any noteworthy observations or errors encountered. Documentation is used in forensic procedures as it provides a transparent record of actions. We need to verify the forensic image to confirm that the image is an exact and unaltered copy of the original data on the device. By verifying the image, we maintain the integrity of the evidence for its admissibility in court or further forensic analysis. The verification

5.4 Acquisition of Drone SD Card using FTK Imager

process involves calculating cryptographic hash values for both the original data and the forensic copy, comparing these values to confirm they match.

To verify the integrity of the forensic image, calculate cryptographic hash values (e.g., MD5 or SHA-256) for both the original data on the Android device and the forensic file. Hash values are unique digital fingerprints generated from the data for accurate verification. By calculating and recording these values, we create a point of reference that can be used to confirm the image's integrity over time. Choose secure and widely accepted hashing algorithms such as SHA-256 to guarantee reliable results and reduce the risk of hash collisions (instances where different files produce the same hash).

Once the hash values are calculated for both the original data and the forensic image, compare them to verify they match. Identical hash values indicate that the forensic image is a bit-for-bit duplicate of the original. It confirms that no data was modified, added, or lost during the imaging process. This verification process is important, as any difference between the hash values suggests potential data corruption or alteration, which could compromise the integrity of the evidence. Documenting these hash values and the verification process also provides a transparent audit trail and ensures that the evidence remains legally defensible. To analyze each artifact, readers can check Chap. 6 for each artifact type, Chap. 7 and Chap. 11 for flight logs forensic analysis.

5.4 Acquisition of Drone SD Card using FTK Imager

The acquisition of data from an SD card is one of important steps in forensic investigations involving drones, as these cards often contain evidence such as flight logs, images, videos, and other operational data. In this process, the data must be preserved in its original state while enabling thorough analysis. Using a combination of specialized hardware and software, forensic investigators can create an exact copy of the SD card's contents and safeguard the integrity of the evidence. In this section, the focus is on using FTK Imager, a widely recognized digital forensic tool, to perform the acquisition process [8]. The methodology includes employing an SD card reader and a write blocker to prevent any modifications to the evidence. It is then followed by creating a forensic image of the SD card for subsequent analysis. This approach guarantees compliance with forensic best practices and maintains the evidentiary value of the extracted SD card data.

Figure 5.30 illustrates the setup and components required for a forensic acquisition of data from an SD card. It is a common procedure in digital forensic

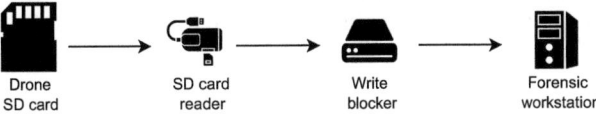

Fig. 5.30 Forensic acquisition workflow for a drone SD card

investigations involving devices in drones. The process begins with identifying the SD card as the source of digital evidence. This card, which is typically removable, stores potentially critical data that needs to be preserved and analyzed without alteration.

To access the data, an SD card reader is employed. This hardware interface allows the SD card to be connected to other devices, such as a forensic workstation. The reader assists in the extraction of data but must be used in conjunction with another tool namely a write blocker. The write blocker will maintain the integrity of the data during the acquisition process. By allowing only read operations, it guarantees that no modifications, intentional or accidental, are made to the SD card. This is a requirement in forensics to uphold the evidentiary value of the digital data. Note that the specification of the write blocker hardware used in this section is explained in Chap. 4.

Finally, the extracted data is processed and analyzed on a forensic workstation. It is a specialized computer equipped with tools and software designed for digital forensics as discussed in Chap. 4. The workstation creates a forensic image of the SD card, which is an exact bit-by-bit copy of the original data. This allows investigators to work on the duplicate while preserving the original data in its unaltered state. To create a forensic image from an SD card using FTK Imager, we can follow these steps.

1. Connect the SD card

    Insert the SD card into a card reader. Connect the card reader to the computer. Use a write blocker to prevent accidental modification of data on the SD card.

2. Open FTK Imager

    Launch FTK Imager on our forensic workstation. Verify that the software detects the SD card.

3. Add evidence source

    In FTK Imager, click "File" → "Create Disk Image" as shown in Fig. 5.31. Select the option Physical Drive (for imaging the entire SD card) as depicted in Fig. 5.32. Click Next to go to the next step.

4. Select drive

    Locate the drone SD card from the list of available drives (Fig. 5.32). Be sure to select the correct drive to avoid imaging the wrong device. Click Finish to load the SD card as an evidence source (Fig. 5.33).

5. Select type of the forensic image

    In the pop-up window, select the type of image format, such as E01 (EnCase Image File) which is preferred for forensic use as it supports metadata and compression. Another format is RAW (dd) where we run a bit-for-bit copy without compression. Click Next after selecting the image type and we will get a configuration window for creating image (Figs. 5.34 and 5.35).

5.4 Acquisition of Drone SD Card using FTK Imager

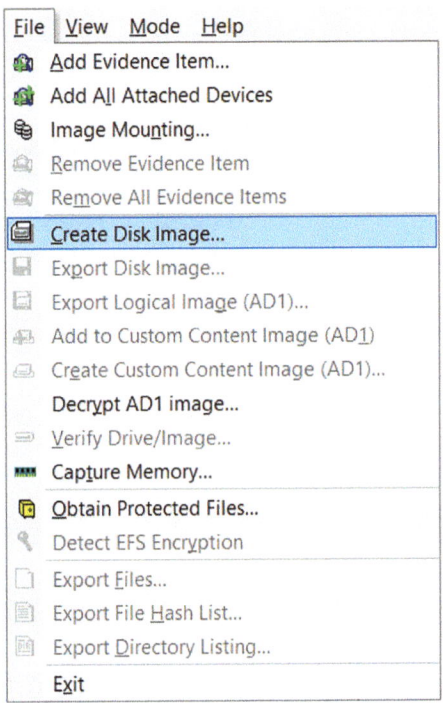

Fig. 5.31 Create disk image in FTK Imager

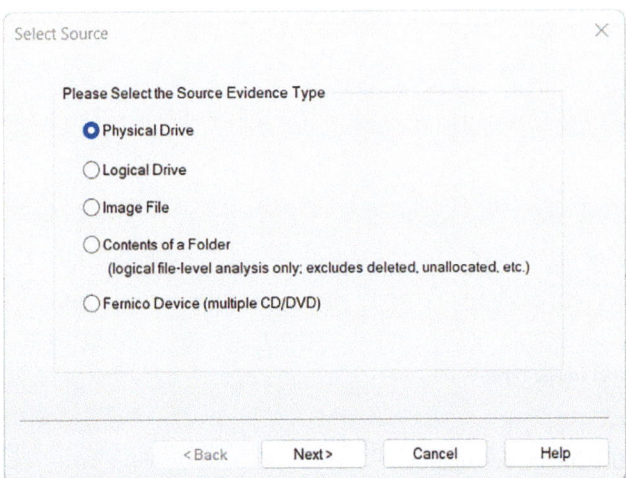

Fig. 5.32 Select source evidence type

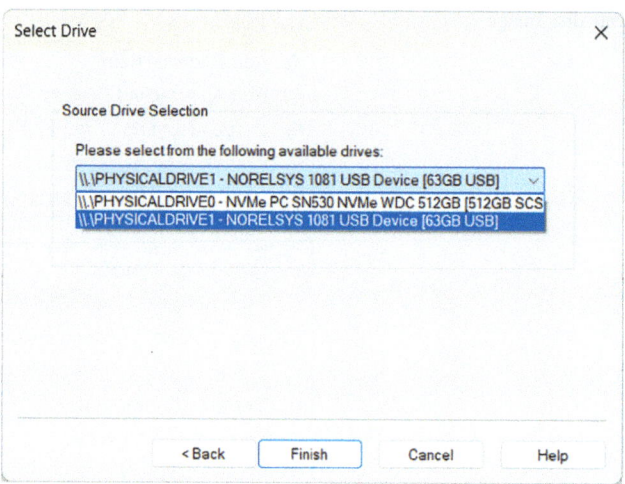

Fig. 5.33 Select source drive

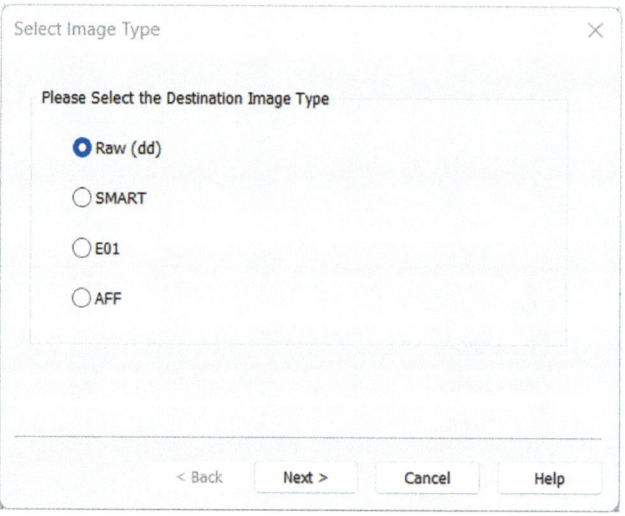

Fig. 5.34 Select image type

6. Configure image destination

    Add image destination with clicking Add button to open the "Image Destination" configuration window.

7. Fill in "Case Information"

    Enter the case identifier in the Case Number field, followed by an appropriate identifier for the evidence item in the Evidence Number field. Provide our name

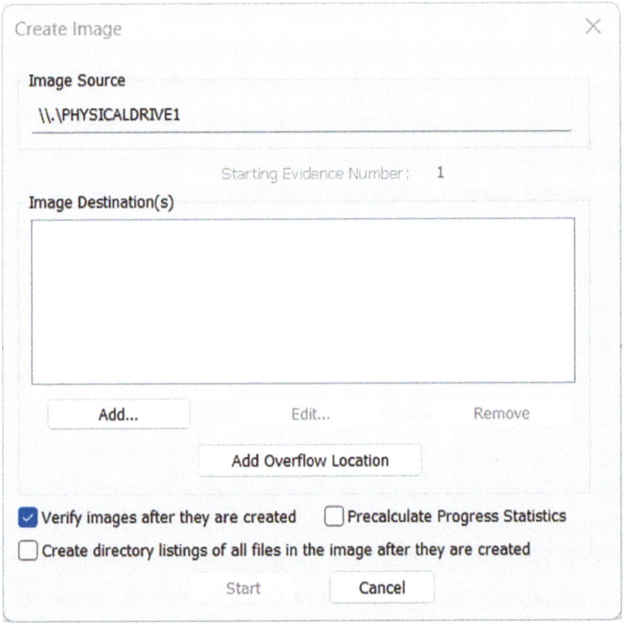

5.4 Acquisition of Drone SD Card using FTK Imager

Fig. 5.35 Configuration of image creation

in the Examiner Name field to document the individual conducting the imaging process. In addition, include a description of the SD card in the Description field; while optional, this is recommended to complete the documentation (Fig. 5.36).

8. Select destination path

   Click Browse to choose a folder where the forensic image will be saved. Set a meaningful file name for the image. Enable compression (if using E01 format). Set segment size if needed (e.g., 2 GB for compatibility). Click Finish to save the configuration (Figs. 5.37 and 5.38).

9. Start imaging

   Review the settings and click Start to begin the imaging process. Monitor the progress bar. FTK Imager will calculate a hash value (MD5/SHA-1) before and after imaging to verify integrity (Fig. 5.39).

10. Verify image integrity

    Once imaging is complete, FTK Imager will display the calculated hash values. Verify that the source hash matches the image hash to confirm a successful, unaltered forensic copy (Figs. 5.40, 5.41, and 5.42).

11. Safely eject the SD card

    Close FTK Imager and safely eject the SD card and write blocker from the acquisition system.

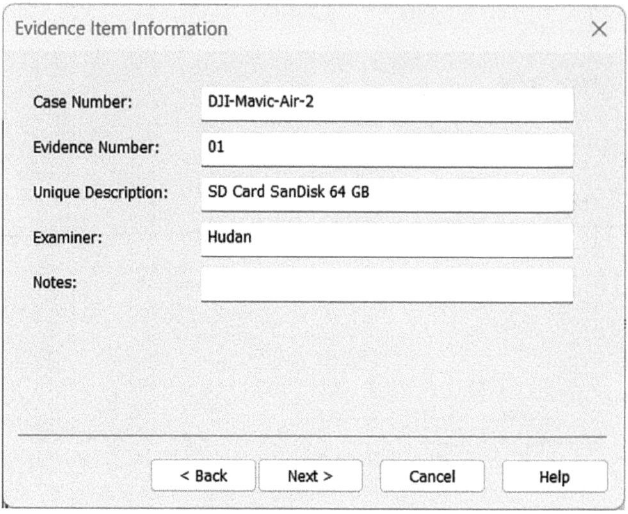

Fig. 5.36 Fill up the evidence item information

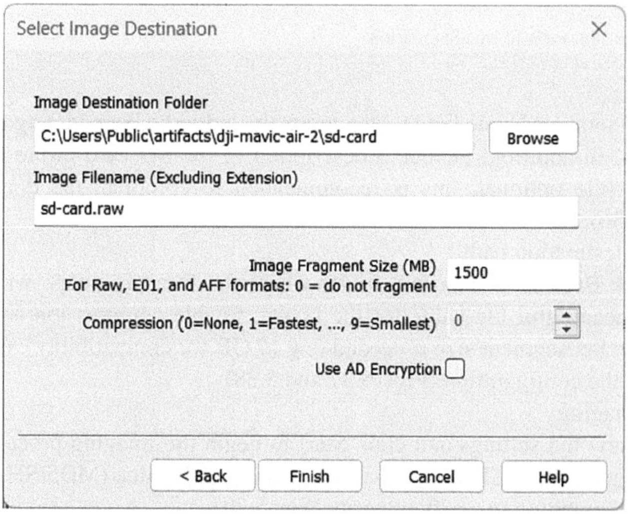

Fig. 5.37 Select image destination

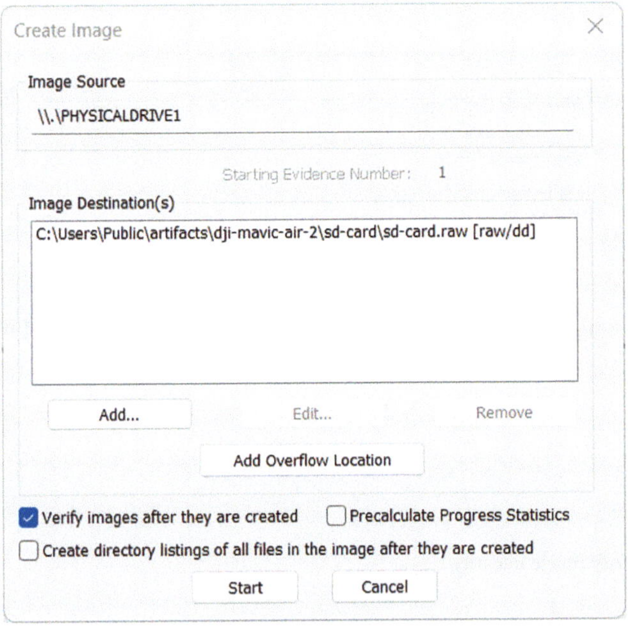

5.4 Acquisition of Drone SD Card using FTK Imager

Fig. 5.38 Configuration of image creation is complete

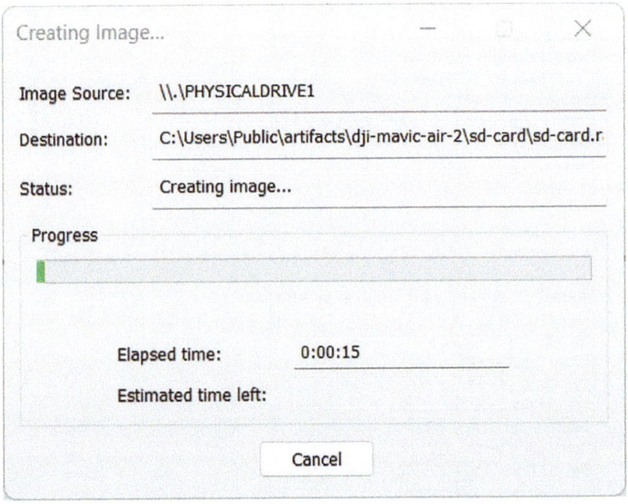

Fig. 5.39 Start imaging the SD card

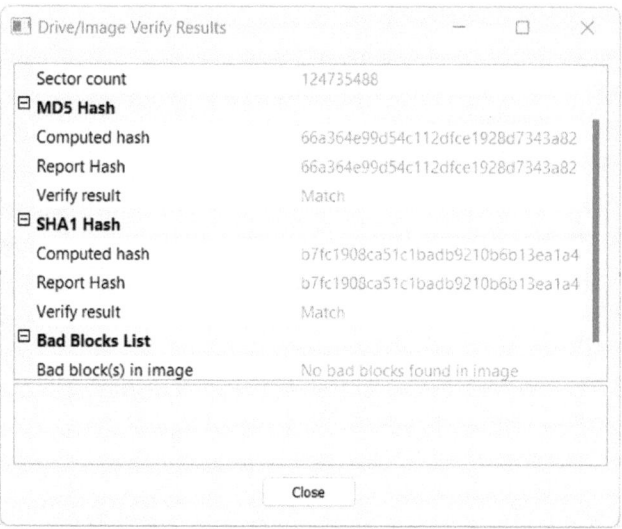

Fig. 5.40 Verify image integrity

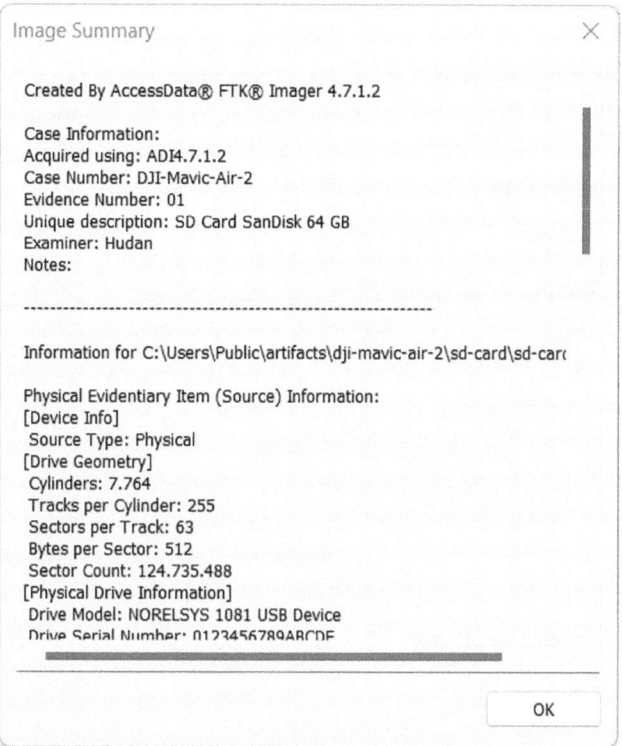

Fig. 5.41 Image summary

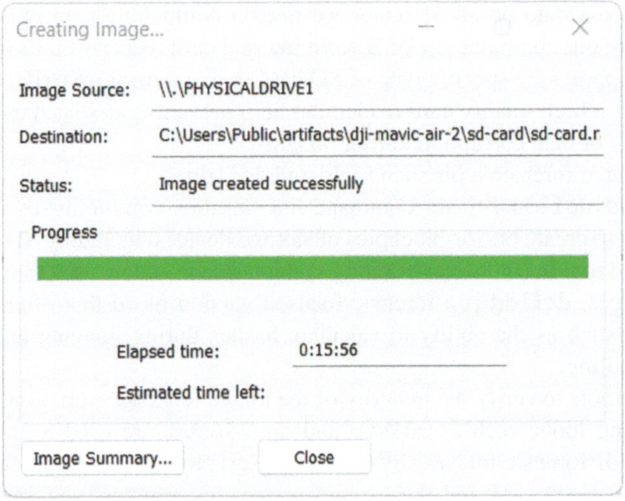

5.5 Acquisition of Drone SD Card with Command Line 127

Fig. 5.42 Image creation is successfully completed

> **Best Practices**

Use a write blocker to protect the original SD card. Keep the forensic image in a secure location with proper chain-of-custody documentation. Work only on the forensic image for analysis to preserve the integrity of the original SD card.

5.5 Acquisition of Drone SD Card with Command Line

Creating a forensic image of an SD card from a drone device involves several steps to preserve the integrity and admissibility of the data. This guide outlines the process using open-source tools. To conduct a forensic investigation and data acquisition from an SD card, we need to prepare the necessary hardware and software tools. Below is a list of the primary tools and resources required for this process.

1. A forensic workstation running a Linux, macOS, or Windows

    For forensic imaging and analysis, a reliable computer equipped with a compatible operating system is essential. Linux distributions are generally preferred in forensics due to their powerful command-line utilities, flexibility, and built-in support for many open-source forensic tools. However, macOS and Windows systems can also be configured to support forensic workflows with the necessary tools and software installed.
2. A card reader to connect the SD card to our computer

To access data on an SD card, we need a compatible card reader that can connect to our computer via USB or an internal card reader slot. Ensure the card reader supports the specific type of SD card (e.g., microSD, SDHC) being used. Choosing a high-quality card reader can help prevent issues such as connection instability or data corruption during imaging.

3. Open-source forensic tools such as dd and dcfldd

    dd and dcfldd [9] are command-line utilities commonly used in digital forensics to create bit-for-bit copies of storage devices, including SD cards. dd is a standard tool in Unix-like operating systems and is widely used for creating raw disk images. dcfldd is a forensic-focused version of dd that offers additional features, such as the ability to calculate hashes during imaging and improved error handling.

4. Hashing tools to verify the integrity of the image (e.g., md5sum, sha256sum)

    Hashing tools, such as md5sum and sha256sum, are used to guarantee the integrity of forensic images. By calculating cryptographic hash values for both the original SD card and the created image, investigators can verify that the image is an exact copy of the original data. md5sum generates MD5 hashes, while sha256sum produces SHA-256 hashes, both of which are commonly used in forensics. Matching hash values confirm that the imaging process was accurate.

Before starting the forensic imaging process, we need to confirm that our system environment is set up correctly and all necessary tools are installed. This preparation involves installing the required forensic tools and verifying that our computer can recognize the connected SD card.

1. Install necessary tools

    If we are using a Linux system, open a terminal and install forensic utilities using a package manager, such as apt (for Debian-based systems such as Ubuntu) or yum (for Red Hat-based systems such as CentOS). Updating our package list before installing ensures we are downloading the latest versions of the tools. In this hands-on approach, we use macOS to perform acquisition of a drone SD card using command line.

    ```
    brew install dcfldd dc3dd
    ```

    The command above installs dcfldd [9] and dc3dd [10] as shown in Fig. 5.43. They are enhanced version of dd specifically designed for forensic imaging, with added functionality for error handling and hashing.

2. Insert the SD card

    Insert the SD card into a card reader, and connect the reader to the computer. Ensure that the connection is stable, as a loose connection can lead to imaging errors or data corruption. Once connected, the SD card should be recognized by the system.

3. Identify the SD card

    To proceed with imaging, we need to identify the device name associated with the SD card. This device name will typically follow a pattern such as /dev/diskX, where "X" represents a specific number assigned by the system.

5.5 Acquisition of Drone SD Card with Command Line

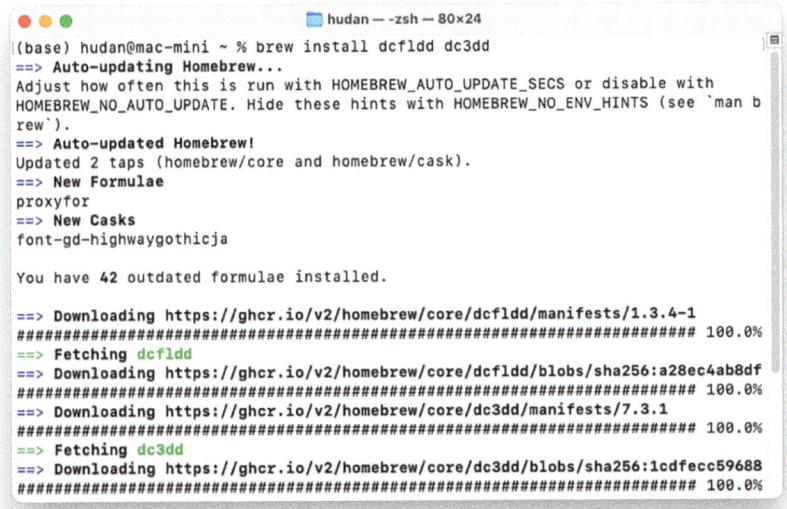

Fig. 5.43 Installation of dcfldd and dc3dd using brew

In macOS, open a terminal and use the diskutil command to list all connected storage devices:

diskutil list

Review the output of diskutil list to identify the SD card. Look for a device entry that matches the storage size of the SD card. Common device names for SD cards are /dev/diskX (for USB-connected readers). Ensure we select the correct device, as using the wrong device name can lead to data loss on other drives. As shown in Fig. 5.44, the SD card name is /dev/disk4.

4. Create a forensic image

Creating a forensic image allows us to capture an exact and bit-by-bit copy of the SD card's data. This image will serve as the preserved evidence for analysis. Therefore, investigators can examine the data without altering the original source. Two widely used utilities for creating forensic images are dd and dcfldd. Both tools are command-line based and offer robust options for copying and converting data, with dcfldd providing additional features specific to forensic needs.

5. Using dd command

dd is a standard Unix utility designed for low-level copying and conversion of data. It reads data from an input source (e.g., an SD card) and writes it to an output destination (e.g., an image file) with user-specified parameters. dd is commonly used in forensics due to its reliability in creating exact copies of storage media.

```
[base] hudan@mac-mini ~ % diskutil list
/dev/disk0 (internal, physical):
   #:                       TYPE NAME                    SIZE       IDENTIFIER
   0:      GUID_partition_scheme                         *251.0 GB   disk0
   1:             Apple_APFS_ISC Container disk1         524.3 MB    disk0s1
   2:                 Apple_APFS Container disk3         245.1 GB    disk0s2
   3:        Apple_APFS_Recovery Container disk2         5.4 GB      disk0s3

/dev/disk3 (synthesized):
   #:                       TYPE NAME                    SIZE       IDENTIFIER
   0:      APFS Container Scheme -                       +245.1 GB   disk3
                                 Physical Store disk0s2
   1:                APFS Volume Macintosh HD            11.2 GB     disk3s1
   2:              APFS Snapshot com.apple.os.update-... 11.2 GB     disk3s1s1
   3:                APFS Volume Preboot                 7.0 GB      disk3s2
   4:                APFS Volume Recovery                1.0 GB      disk3s3
   5:                APFS Volume Data                    209.4 GB    disk3s5
   6:                APFS Volume VM                      24.6 KB     disk3s6

/dev/disk4 (external, physical):
   #:                       TYPE NAME                    SIZE       IDENTIFIER
   0:     FDisk_partition_scheme                         *63.9 GB    disk4
   1:                Windows_NTFS                        63.8 GB     disk4s1
```

Fig. 5.44 List of connected storage devices and their partitions using `diskutil list` command

To create a forensic image using dd, open a terminal and execute the following command:

`sudo dd if=/dev/diskX of=/path/to/save/image.img bs=4M`
`↪ status=progress`

Replace /dev/diskX with the actual device name of the SD card (e.g., /dev/disk4), as identified earlier. Specify the output file location in place of /path/to/save/image.img, where the image file will be saved. Here is a breakdown of the command options:

- `if=/dev/diskX`: Specifies the input file (source), which is the SD card device we want to image.
- `of=/path/to/save/image.img`: Defines the output file (destination) where the image will be stored.
- `bs=4M`: Sets the block size to 4 MB, which can optimize the copying process by speeding up data transfer without compromising accuracy.
- `status=progress`: Displays real-time progress information to monitor the imaging process.

Before creating a disk image on macOS (or any other operating system), it is important to unmount the disk. This step helps protect the data and ensures nothing accidentally gets changed on the disk during the imaging process. As an example, Fig. 5.45 shows the `diskutil unmountDisk` to unmount the disk and dd command to create a disk image from a drone SD card. The device name is

5.5 Acquisition of Drone SD Card with Command Line

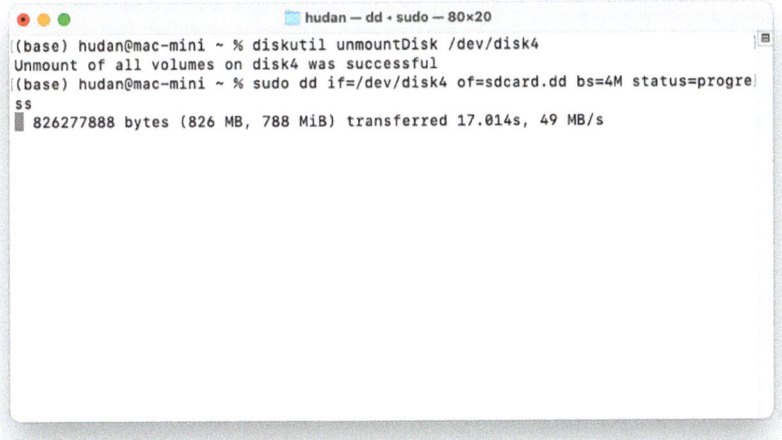

Fig. 5.45 Create a disk image from a SD card using dd command

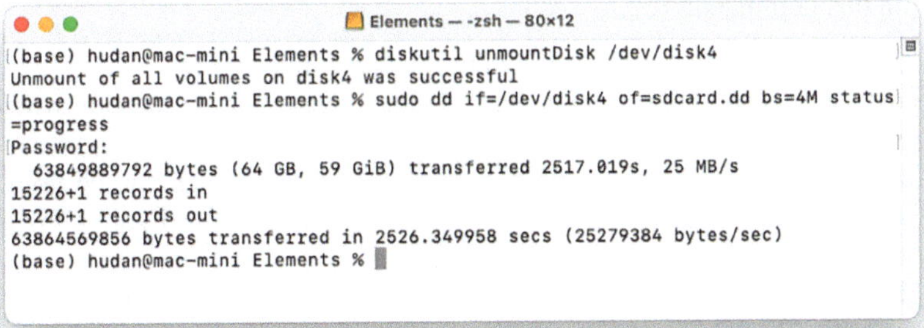

Fig. 5.46 Creating a disk image using dd command is successful

/dev/disk4, the output path is in the current directory, the image file name is sdcard.dd, the block size is 4 MB, and the status option is set to progress to monitor the cloning process. It will take some time to create the disk image. Once it is finished, it will show "X bytes transferred in Y secs (Z bytes/sec)" message as depicted in Fig. 5.46.

6. Using dcfldd command [9]

dcfldd is an enhanced version of dd tailored specifically for forensic imaging. Developed by the U.S. Department of Defense's Computer Forensics Laboratory, dcfldd includes additional features that make it ideal for forensic investigations. These features include hashing (calculating MD5 or SHA hash

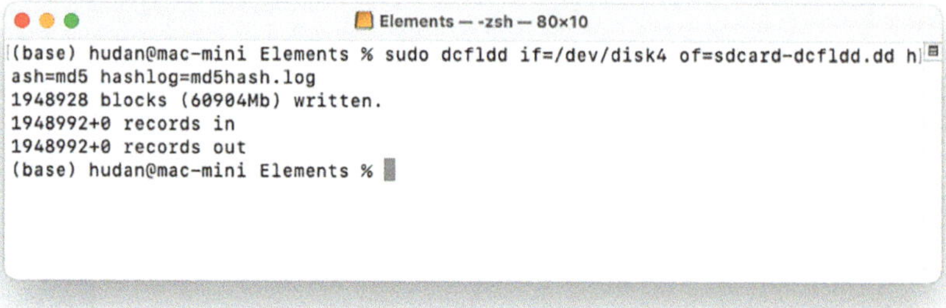

```
(base) hudan@mac-mini Elements % sudo dcfldd if=/dev/disk4 of=sdcard-dcfldd.dd h
ash=md5 hashlog=md5hash.log
1948928 blocks (60904Mb) written.
1948992+0 records in
1948992+0 records out
(base) hudan@mac-mini Elements %
```

Fig. 5.47 Creating a disk image using dcfldd command is successful

values), error logging, and direct output to multiple locations. dcfldd is often preferred in forensic cases due to these capabilities, which help ensure the integrity of the imaging process. To create a forensic image using dcfldd, use the following command:

sudo dcfldd if=/dev/diskX of=/path/to/save/image.img
  ↪ hash=md5 hashlog=/path/to/save/hash.log

In this command:

- if=/dev/diskX: Specifies the SD card as the input source.
- of=/path/to/save/image.img: Defines the output file location where the image file will be saved.
- hash=md5: Calculates an MD5 hash of the image, which can be used later to verify the image's integrity. Alternatively, we can use hash=sha256 for SHA-256 hashing.
- hashlog=/path/to/save/hash.log: Saves the computed hash value to a specified log file for documentation and verification purposes.

Figure 5.47 shows that the dcfldd command has successfully completed the imaging process. It copied the data from the source device /dev/disk4 to the output file sdcard-dcfldd.dd. It creates a forensic image of approximately 60,904 MB (around 59.5 GB) in size. A total of 1,948,992 records were read and written without any errors. Moreover, an MD5 hash was computed during the process and saved to the log file md5hash.log for verification purposes.

7. Create a forensic image using dc3dd command [10]

As an alternative, we can use dc3dd, which is an enhanced version of dd with additional features for forensic imaging.

sudo dc3dd if=/dev/diskX of=/path/to/save/image.img hash=md5
  ↪ log=/path/to/save/image.log

This command also calculates the MD5 hash of the image during the copying process and saves it to a log file. dc3dd offers several advantages over dcfldd, particularly in terms of progress reporting and logging capabilities. It provides

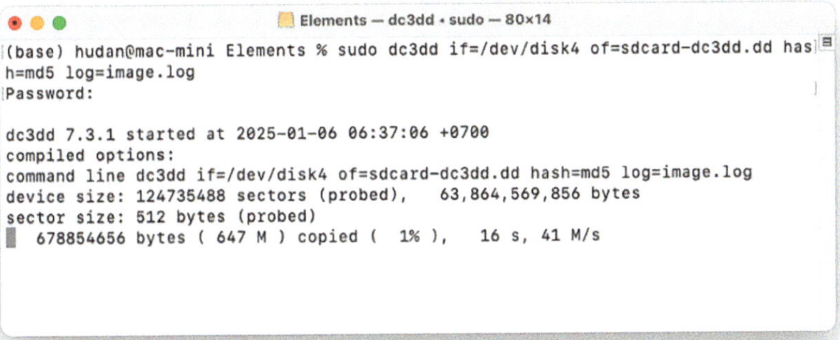

5.5 Acquisition of Drone SD Card with Command Line

```
(base) hudan@mac-mini Elements % sudo dc3dd if=/dev/disk4 of=sdcard-dc3dd.dd has
h=md5 log=image.log
Password:

dc3dd 7.3.1 started at 2025-01-06 06:37:06 +0700
compiled options:
command line dc3dd if=/dev/disk4 of=sdcard-dc3dd.dd hash=md5 log=image.log
device size: 124735488 sectors (probed),    63,864,569,856 bytes
sector size: 512 bytes (probed)
  678854656 bytes ( 647 M ) copied (   1% ),   16 s, 41 M/s
```

Fig. 5.48 Processing a disk image using dc3dd command

real-time progress updates with more granular details. Furthermore, dc3dd displays statistics about input/output performance and errors. One of its features is the detailed logging of the entire imaging process,iyo which includes hash values, progress updates, errors, and performance metric. All information are consolidated into a single log file for easier review and forensic documentation.

Figure 5.48 shows the dc3dd command is currently in progress in creating a forensic disk image of the device /dev/disk4 and saving it to the output file sdcard-dc3dd.dd. The command is also computing an MD5 hash during the imaging process for later verification and logging all details, including progress, errors, and the computed hash, to the file image.log. At this stage, the device size has been identified as approximately 63.8 GB, with a sector size of 512 bytes. So far, 647 MB (1% of the total) has been successfully copied in 16 seconds at a speed of 41 MB/s.

Subsequently, Fig. 5.49 shows the dc3dd command has successfully completed the creation of a forensic disk image of the device /dev/disk4 and saving the output to sdcard-dc3dd.dd. The process copied a total of 63,864,569,856 bytes (approximately 59 GB) across 12,473,548 sectors without encountering any bad sectors. The imaging process took 2559 seconds (around 42.6 minutes) at an average speed of 24 MB/s. d44c23d04332dd722f81f124353ea3f6, an MD5 hash of the data, was generated to verify the integrity of the disk image and confirm it matches the original source data. The operation was completed without errors, as indicated by the matching input and output sector counts, and concluded at 07:19:45 on January 6, 2025.

8. Verifying image integrity

Once the imaging process is complete, compare the hash value of the original device with the hash of the image file to confirm that they match. Identical hash values indicate that the image is an exact duplicate of the original. Using dcfldd for hash calculation directly during imaging simplifies this process and provides an immediate verification point. Both dd and dcfldd are powerful

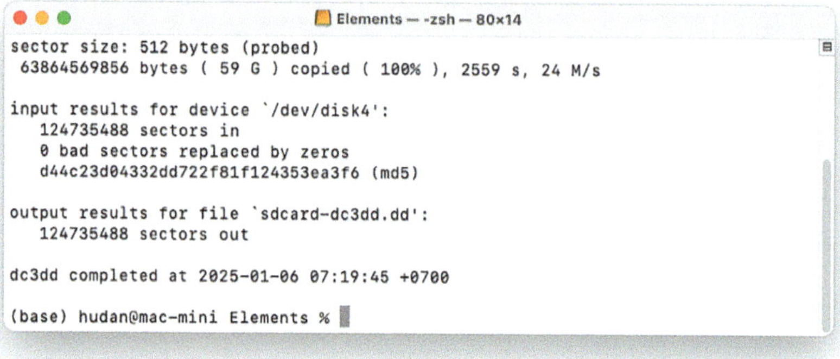

```
sector size: 512 bytes (probed)
 63864569856 bytes ( 59 G ) copied ( 100% ), 2559 s, 24 M/s

input results for device `/dev/disk4':
   124735488 sectors in
   0 bad sectors replaced by zeros
   d44c23d04332dd722f81f124353ea3f6 (md5)

output results for file `sdcard-dc3dd.dd':
   124735488 sectors out

dc3dd completed at 2025-01-06 07:19:45 +0700

(base) hudan@mac-mini Elements %
```

Fig. 5.49 Creating a disk image using dc3dd command is successful

tools for forensic imaging. Choosing between them often depends on the specific requirements of the case, with `dcfldd` providing additional forensic features for secure and documented imaging. After creating the image, verify its integrity by comparing hash values.

`md5sum /path/to/save/image.img`
`sha256sum /path/to/save/image.img`

Compare these values with the hash log generated by `dcfldd` to verify that the image has not been altered. Ensure the SD card is write-protected during the process to avoid accidental modifications.

As an additional considerations, preserving the integrity of evidence is important. Special precautions should be taken to avoid any unintentional changes to the data on the SD card. Even seemingly minor actions, such as mounting the SD card, can lead to automatic writes by the operating system, potentially altering metadata or timestamps. Below are key considerations to ensure that the SD card remains unaltered throughout the forensic process.

Whenever possible, refrain from mounting the SD card entirely to prevent the operating system from performing any automatic background writes. Many operating systems, particularly Windows, may automatically write small amounts of data (such as indexing or thumbnail caches) when a storage device is mounted. These alterations can compromise the integrity of the evidence. By working directly with low-level tools such as `dd` or `dcfldd`, we can access and image the SD card without mounting it and maintain its original state.

If mounting the SD card is unavoidable, we need to make sure that it is mounted in read-only mode to prevent any modifications. Mounting in read-only mode restricts the operating system's ability to write data to the device. On a macOS system, use the following command to mount the SD card in read-only mode:

`sudo diskutil mountDisk readOnly /dev/diskX`

In this command:

- `readOnly` specifies the read-only option, it means that no changes can be made to the SD card.
- `/dev/diskX` should be replaced with the actual device name for our SD card (e.g., `/dev/disk4`).

As we can see that `mountDisk` option is a self-explanatory option in `diskutil` command. Mounting in read-only mode allows investigators to inspect the file system and identify specific files without risking unintentional changes.

> ⚠️ **Attention**
>
> However, remember that even read-only mounting should be minimized in forensics, as it may alter some metadata or device properties upon initial access.

5.6 Summary

This chapter provides an overview of the methods and tools used in forensic data acquisition from drones and other UAVs. It focuses on the steps for collecting and securing data to ensure the evidentiary integrity required in digital forensics. The chapter begins with a discussion on the forensic methods tailored to macOS and Android operating systems. It discusses several tools such as Finder for macOS and forensic write blockers. These tools serve as the foundation for acquiring data while protecting it from unintended modifications.

The chapter details practical and step-by-step instructions for creating forensic images from various devices associated with UAVs, including iPhones, Android devices, and drone SD cards. It covers both logical and physical imaging techniques and empowers forensic practitioners to select the most appropriate method based on the circumstances and requirements of the investigation. Logical imaging, which involves capturing user-accessible data, is contrasted with physical imaging, which creates a complete bit-by-bit copy of the storage device. This dual approach ensures that practitioners are equipped to handle a range of scenarios.

Finally, the chapter places a stress on the application of these data acquisition techniques in digital forensics. It explains the importance of preserving evidence throughout the extraction and analysis phases. Forensic investigators need to keep digital evidence intact and admissible for legal proceedings. By combining theoretical and practical aspects, this chapter serves as a resource for students or forensic professionals and offers them the skills and knowledge required to handle UAV data acquisition.

5.7 Exercises

1. What are some of the key steps involved in data acquisition from drones and UAVs mentioned in this chapter?
2. Which tools are used for data acquisition from Android devices?
3. What are the tools for working with macOS systems?
4. Why is maintaining data integrity important in forensic imaging?
5. What types of devices are covered for drone forensic image acquisition?
6. Explain the role of forensic write blockers in ensuring the integrity of evidence during data acquisition.
7. Discuss the differences between logical and physical imaging methods as outlined in this chapter.
8. Analyze the challenges associated with extracting forensic images from drone SD cards and propose solutions.
9. Evaluate the benefits and limitations of using open-source tools for forensic imaging on drones and UAVs.
10. Propose a standardized workflow for forensic practitioners to acquire and analyze data from drones while maintaining evidence integrity.

References

1. Mac User Guide, *Use the Finder on Mac* (2024). https://support.apple.com/guide/mac-help/organize-your-files-in-the-finder-mchlp2605/mac
2. Android Chinese Network, *Android File Transfer* (2023). https://android.p2hp.com/filetransfer/index.html
3. Open Text, *Quick Reference Quide: Tableau Forensic T35u/T35u-RW SATA/ IDE Bridge* (2021). https://www.opentext.com/file_source/OpenText/en_US/PDF/opentext-t35u-t35u-rw-quick-referenceguide-en.pdf
4. Nikias Bassen and Martin Szulecki, *libimobiledevice: A Cross-platform FOSS library Written in C to Communicate with iOS Devices Natively* (2024). https://libimobiledevice.org/
5. P. Ammon, *Hex Fiend: A fast and clever open source hex editor for macOS* (2024). https://hexfiend.com/
6. R. Peinthor, M. Kleusberg, J. Clift, M. Mgrojo, S.-T. Jeong, S. Furry, *DB Browser for SQLite* (2024). https://sqlitebrowser.org/
7. G. Bieś, E. Bieś, *iOS Device Data Extractor* (2024). https://github.com/ernestbies/iOSDeviceDataExtractor
8. Exterro, *Create Forensic Images with Exterro FTK Imager* (2024). https://www.exterro.com/digital-forensics-software/ftk-imager
9. N. Harbour, J. Eriberto, M. Filho, *dcfldd: Enhanced Version of dd for Forensics and Security* (2024). https://github.com/resurrecting-open-source-projects/dcfldd
10. J. Kornblum, J.C. Lininger, *dc3dd* (2024). https://sourceforge.net/projects/dc3dd/

Chapter 6
Understanding Drone and UAV Forensic Images and Artifacts

Abstract This chapter provides a guide to the process of analyzing forensic images and artifacts through the Autopsy forensic tool. This chapter begins by introducing the reader to the concept of a forensic image in the context of drones and UAVs. Detailed in the chapter is the importance of these digital replicas of physical storage devices for forensic investigations. Following this, the chapter provides a detailed overview of the Autopsy as a leading open-source digital forensics tool. It guides the reader through the installation and configuration of Autopsy and prepares them for hands-on forensic analysis. The core of the chapter focuses on utilizing Autopsy to analyze drone forensic images. It explains how to navigate the Autopsy user interface, how to import forensic images, and how to start the analysis process. We highlight the types of artifact that can be recovered from drone forensic images, such as flight logs, videos, photos, and other data that can provide insight into drone operations and user actions.

6.1 Introduction

The increased use of drones and UAVs has introduced a new frontier for forensic investigations. As these sophisticated devices become increasingly integral to various sectors, including surveillance, agriculture, delivery services, and even recreational use, the potential for their involvement in illicit activities or incidents has increased correspondingly. This chapter is designed to equip the reader with the practical knowledge and skills required to effectively analyze forensic images and artifacts recovered from drones and UAVs.

Forensic images, digital replicas of physical storage devices, preserve an exact copy of storage media for detailed investigation without altering the original evidence. This chapter discusses the significance of these images in the context of drones and UAVs. We will explain how they encapsulate artifacts that can reveal the operations and interactions of these aerial devices within their operational environments.

The main point of this chapter is the exploration of the Autopsy forensic tool [1]. It is an open-source platform renowned for its robust capabilities in digital forensics analysis. We will use Autopsy to analyze drone images and its artifacts. Readers are guided through the steps of installing and configuring Autopsy to set the environment for practical and hands-on forensic investigation. The discussion extends to the practical utilization of Autopsy for the analysis of drone forensic images. We explain the procedural aspects of importing these images into the tool, navigating its user interface, and initiating the analysis process.

In addition, the chapter will elaborate on the identification and examination of various artifacts that can be extracted from drone forensic images. These artifacts, ranging from flight logs and videos to photos, are important in reconstructing the flight paths, operations, and user actions associated with a drone. By providing insights into the types of artifacts recoverable and their potential implications in forensic investigations, this chapter offers the reader with the analytical framework necessary to interpret the narratives hidden within drone and UAV forensic images.

6.2 Related Work

Our recent work provides a comprehensive review of the current state of UAV forensics and reviews digital evidence in drone investigations [2]. It categorizes existing literature into various themes, such as forensic artifacts, frameworks for investigation, and tools available for forensic analysis. The authors identify gaps in current research, particularly in the areas of forensic readiness and the post-investigation phase. The purpose of the paper is to establish a foundation for future research and proposes a new conceptual framework to enhance the effectiveness of UAV forensic investigations. Another study investigates the field of drone forensics as a result of the increasing application of drones and the security risks they pose, particularly when used for malicious activities [3]. It urges the need for forensic analysis of seized drones to recover valuable data that can aid investigations. The authors propose a systematic framework for data recovery and utilize established digital forensic tools to extract both recorded and deleted flight data from drones. The study improves investigator capabilities by providing a structured approach to drone data recovery.

Yousef et al. [4] focus on the emerging field of drone forensics, especially analyzing various models of DJI drones. The increasing use of drones in both the commercial and recreational sectors, along with their potential misuse in criminal activities, should be cited. The paper identifies and extracts digital evidence from drones and their associated components to address challenges due to the proprietary nature of drone technology and the variety of file systems. The study discusses the need for effective forensic methodologies to ensure that valuable data can be recovered and analyzed in a forensically sound manner. Another work provides a comprehensive review of drone forensics and security and focuses on the challenges posed by UAVs [5]. It discusses the need for a structured approach to investigate drone-related incidents as the operational characteristics are different compared to traditional computing devices. The authors analyze existing forensic techniques, categorize various drone artifacts, and discuss the implications of security vulnerabilities in drone systems.

In addition, a detailed investigation of drone forensics is provided in [6] that includes the methodologies and frameworks used to analyze drone-related incidents, malfunctions, and attacks. It describes the fast-moving improvement in drone technology and the corresponding need for effective forensic techniques to address the challenges posed by these developments. The authors discuss various aspects of drone forensics, including evidence collection, the role of machine learning, and the importance of standardized methodologies. The study also brings attention to the complexities involved in analyzing drones, such as in conflict zones, and outlines future research directions to enhance the field. Furthermore, Iqbal et al. [7] provide another comprehensive review of the challenges and methodologies associated with forensic investigations of small-scale digital devices, especially those connected to the Internet of Things (IoT). The paper discusses various types of devices, including smartphones, smart wearables, gaming consoles, smart toys, drones, and the unique forensic challenges each presents. The authors argue that there is a need for specialized tools and frameworks to effectively gather and analyze digital evidence from these devices, as they have diverse operating systems and data storage methods.

A comprehensive study provides a detailed examination of the security challenges and forensic methodologies associated with UAVs [8]. The paper discusses the increasing reliance on drones in various sectors, including military, law enforcement, and commercial applications, raising concerns about their security vulnerabilities. The authors conducted a systematic review of the literature to identify existing threat models, forensic approaches, and persistent security challenges in the field of drone technology. Another work explores the emerging field of drone forensics and the challenges and methodologies associated with UAV investigation [9]. It specifically examines the forensic analysis of the Parrot AR drone 2.0 and explains the importance of retrieving digital evidence from drones to establish ownership and reconstruct flight events. The authors show the need for standardized forensic procedures that can be adapted from traditional digital forensics to address the specific characteristics of drones.

A focused study investigates the increasing use of drones and the potential for their misuse in criminal activities. The study focuses on the DJI RC remote controller, which is widely used to operate DJI drones [10]. The research develops a comprehensive digital forensic methodology for analyzing DJI RC data to identify the pilot of the drone, the specific drone used, the flight path, and the content captured during flights. The study involves collecting and analyzing data from the DJI RC's internal flash memory and external SD card. Reverse engineering is used to decrypt critical information files. Finally, recent research addresses the increasing use of drones in various fields and the subsequent increase in drone-related accidents [11]. It shows the challenges faced in the digital forensic investigation of these accidents and proposes the use of digital twin technology as a solution. The digital twin technology creates a virtual replica of the drone and its environment. The technology allows for detailed simulations and analyses of drone accidents. The paper demonstrates the effectiveness of this approach through a simulation scenario using the Robot Operating System (ROS) and tools such as Gazebo and Rviz.

6.3 Type of Drone and UAV Forensic Images

Drone and UAV investigations rely on the analysis of various storage mediums to extract critical evidence. These forensic images serve as digital replicas of the storage devices associated with drones and their controllers. Three key sources of forensic images in drone investigations are the controller (Android or iPhone), the drone's SD card, and the drone's internal memory. Each source provides unique information on activities and events related to drone operations.

Forensic Image from Controller The controller device, often a smartphone or tablet running Android or iOS, serves as the main component in drone operations. It stores data such as flight logs, user account information, and telemetry data, which are often used for forensic investigations. Creating a controller forensic image involves the use of mobile forensic tools as discussed in Chap. 5. For Android devices, investigators can retrieve data from internal memory, SD card (if available), and cached files. Key artifacts include flight logs stored in proprietary app directories (e.g., DJI GO or DJI Fly) and metadata about flight events. For iPhone devices, encrypted backups and app-specific data can reveal similar details. These forensic images of the controller are used to identify the drone operator and reconstruct flight activities.

Forensic Image from Drone SD Card The SD card used in drones often stores high-resolution photos, videos, and flight logs. Therefore, it becomes an important source of evidence. Forensic imaging of the SD card is straightforward compared to that of mobile devices, as the card can be directly accessed using write-blocking hardware to prevent any data modification. Tools such as FTK Imager can create sector-by-sector copies of the SD card for analysis. From the forensic image, investigators can recover media files that capture the drone's activities, along with timestamps and GPS coordinates embedded in EXIF metadata. In addition, deleted files can often be recovered and provide further evidence of prior usage.

Forensic Image from Drone Internal Memory Many drones also include internal memory that serves as primary storage for flight data. This internal memory can store critical logs, configurations, and occasionally media files, depending on the drone model. Acquiring a forensic image of the internal memory typically requires specialized tools or direct hardware connections to the drone's onboard storage. For instance, tools such as JTAG or Chip-Off techniques may be used for low-level data extraction. The data retrieved from the internal memory often complement the information from the SD card and controller. Therefore, it offers a more comprehensive picture of the drone's operations. It may include additional telemetry data, firmware logs, and operational settings that are not stored elsewhere. These data points are particularly valuable in cases of drone-related accidents or malicious activities, as they provide data about the drone's behavior and potential tampering.

6.4 Autopsy

The Autopsy[1] forensic tool stands at the forefront of digital forensic technology and provides a comprehensive suite of features designed to assist in the analysis of digital devices [1]. Developed by Basis Technology, Autopsy serves as an open-source platform that is powerful and accessible. It fulfills the needs of law enforcement, students, and private sector forensic professionals. Its robust framework supports the investigation of a wide range of digital crimes and incidents, from cyberattacks to unauthorized data access. This tool is one of important resources in the modern digital forensic investigator toolkit.

At its core, Autopsy is built on the Sleuth Kit (TSK), a library of command-line forensic tools that enable us to perform the analysis of disk images. The tool integrates these capabilities into a user-friendly graphical interface that simplifies the process of searching through large volumes of data, recovering deleted files, and examining the file system and file metadata to uncover digital evidence. This seamless integration ensures that even those with limited technical expertise can effectively utilize the tool, while also providing advanced features for experienced investigators.

One of the key strengths of Autopsy is its modular architecture, which supports a wide array of plugins and modules developed by the global forensic community. These extensions augment the capabilities of Autopsy by enabling the examination of specific file types. They facilitate integration with additional forensic tools and automating routine tasks. Whether it is identifying known malicious files with hash databases, extracting geolocation data from images, or analyzing internet browsing history, Autopsy's extensibility ensures it can adapt to the evolving challenges of various forensic artifacts.

Autopsy's case management features are designed to optimize the investigative process. Investigators can create new cases with ease, add and manage data sources, and generate comprehensive reports that detail their findings. The ability of the tool to handle multiple investigators working on a single case simultaneously promotes collaboration and efficiency, especially in complex investigations that require the pooling of expertise and resources. The timeline analysis feature is another useful functionality of Autopsy as we can view events in chronological order. This feature can be used to understand the sequence of actions on a device, help establish user behavior patterns, identify suspicious activities, and pinpoint the timing of specific incidents. By providing a visual representation of data and events, Autopsy aids in constructing a coherent narrative of digital activities.

---

[1] https://www.autopsy.com/.

In summary, the Autopsy forensic tool provides accessibility, functionality, and scalability in digital forensics. Its comprehensive feature set, combined with the support of a vibrant community of developers and forensic experts, ensures that Autopsy remains at the cutting edge of open-source digital investigation tools. In this chapter, we will use Autopsy to perform analysis of drone forensic images and its artifacts.

6.5 Creating a New Case and Importing a Forensic Image to Autopsy

Forensic investigations often begin with the creation of a new case and the importation of forensic images for analysis. This section provides a detailed walkthrough of setting up a digital forensic case in Autopsy. From initializing a new case to configuring ingest modules, each step is provided to ensure the investigation is well organized and the evidence is handled systematically. The hands-on guidance presented here not only provides the setup process, but also highlights best practices for working with forensic datasets, such as those sourced from VTO Labs. Note that the installation steps for Autopsy are provided in Sect. 4.6.

6.5.1 Dataset Preparation

To run hands-on experiments in this section, we need to download the drone dataset from VTO Labs. We focus on DJI Phantom 4 Pro, specifically the forensic image of the internal SD card. The dataset is available for download from the CFReDS website. Below is a step-by-step guide to locate and download the required image for forensic analysis.

1. Access the DJI Phantom 4 Pro dataset
   Navigate to the CFReDS website, which hosts VTO Inc.'s datasets, using the public link.[2]
2. Locate the appropriate directory

   - On the CFReDS webpage, find and open by double clicking the directory named `DJI_Phantom_4_Pro_V2`.
   - Within this directory, select the folder labeled `df_063_DJI_Phantom_4_Pro_V2`.
   - Proceed to open the subdirectory named `2018_June`.
   - Open another subdirectory named `SDCard_internal`.

3. Download the forensic image

---

[2] https://cfreds.nist.gov/all/SteveWatson%2FVTOInc./DroneDataSet.

6.5 Creating a New Case and Importing a Forensic Image to Autopsy

In the SDCard_internal folder, identify and download it to your local directory the file titled DF_063_Internal_SD_Card_Physical_Intact.001.
4. The forensic image is ready to load in Autopsy and we can continue to the next section.

6.5.2 Case Analysis in Autopsy

Creating a new case and adding a data source in Autopsy are initial steps in initiating a digital forensic investigation. When we launch Autopsy, the welcome screen prompts us with several options. To start a new investigation, we will need to create a new case. Here is a detailed guide to navigate these processes:

1. On the Autopsy welcome screen, select the "New Case" button as shown in Fig. 6.1. This action opens a dialogue where we begin the process of establishing a case.
2. Fill "Case Information" (Fig. 6.2):

    - Case Name: Enter a unique and descriptive name for the case. This helps in identifying the case among others in our repository.
    - Base Directory: Specify the location on our computer or network where case data will be stored. It is recommended to use a dedicated large capacity drive because forensic investigations can involve large amounts of data.

3. Fill "Optional Information" (Fig. 6.3):

    - Case Number: If the organization uses case numbers, enter them here for additional tracking and organization.

Fig. 6.1 Autopsy welcome screen and new case selection

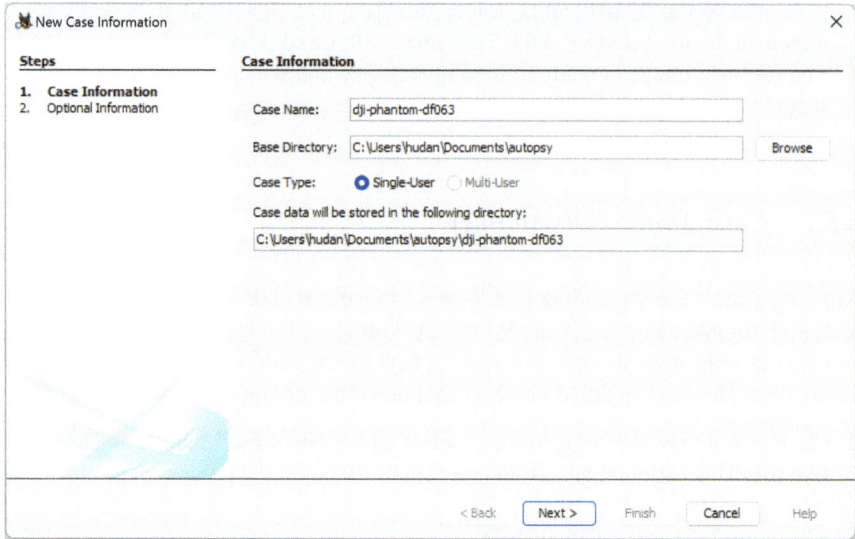

Fig. 6.2 Autopsy case information

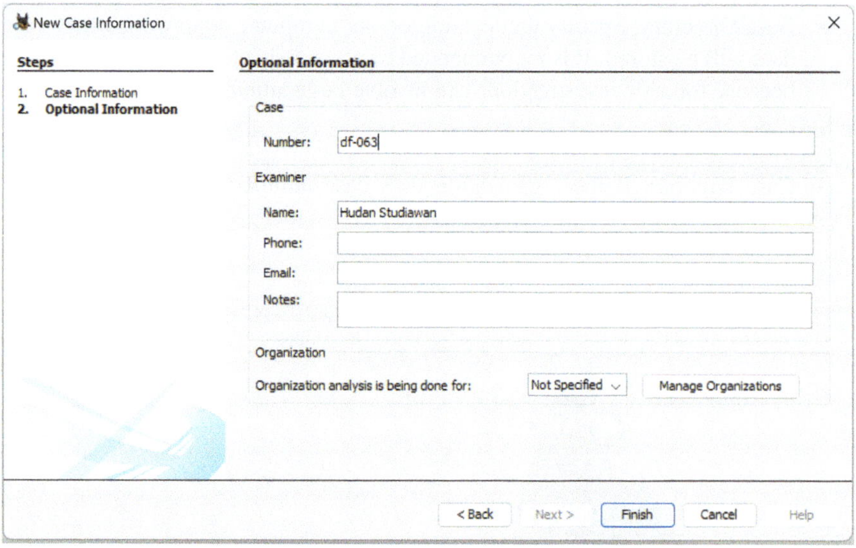

Fig. 6.3 Autopsy optional information

- Examiner: Input the name of the person or team responsible for the investigation. This is for maintaining the chain of custody and for reference in future reviews.

6.5 Creating a New Case and Importing a Forensic Image to Autopsy

- Notes: Although optional, providing a brief description of the case can be helpful for context, especially when revisiting the case or for briefing colleagues.

After filling in the necessary information, click the "Finish" button to create a case. Autopsy will then set up the case directory and prepare the environment for data analysis. Adding a data source is the next step, where we specify the digital evidence we wish to analyze.

1. Select Host

    Hosts in Autopsy are used to organize data sources and other related data. The options presented allow us to determine how we want to associate the new data source with a host as shown in Fig. 6.4.

    - Generate new host name based on data source name: This radio button, when selected, will allow Autopsy to automatically generate a new host name based on the name of the data source we are adding. This is a convenient option if we want to quickly create a host without needing to manually enter a name.
    - Specify new host name: Selecting this option enables us to enter a custom name for a new host. This is useful when we have a specific naming convention or need to maintain consistency with an existing case management system.
    - Use existing host: If the data source we are adding belongs to an already existing host in our case, we would select this option and then select from

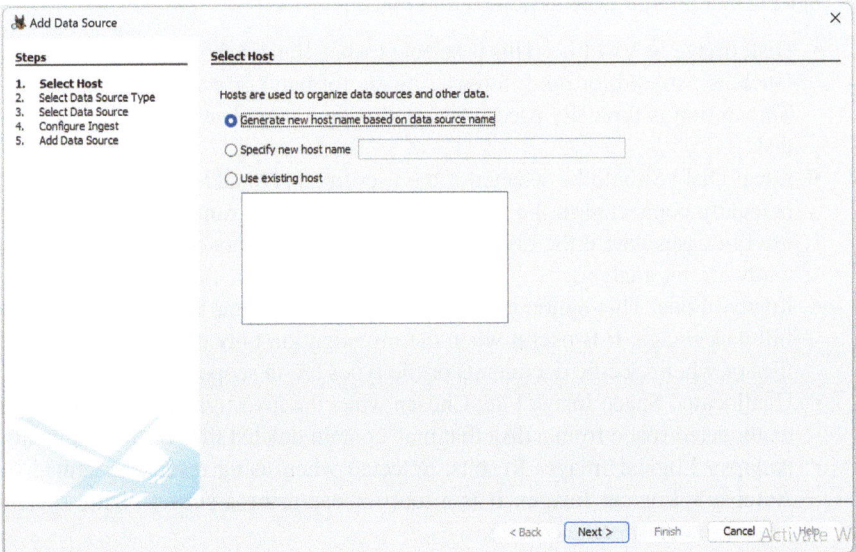

Fig. 6.4 Select host for data source

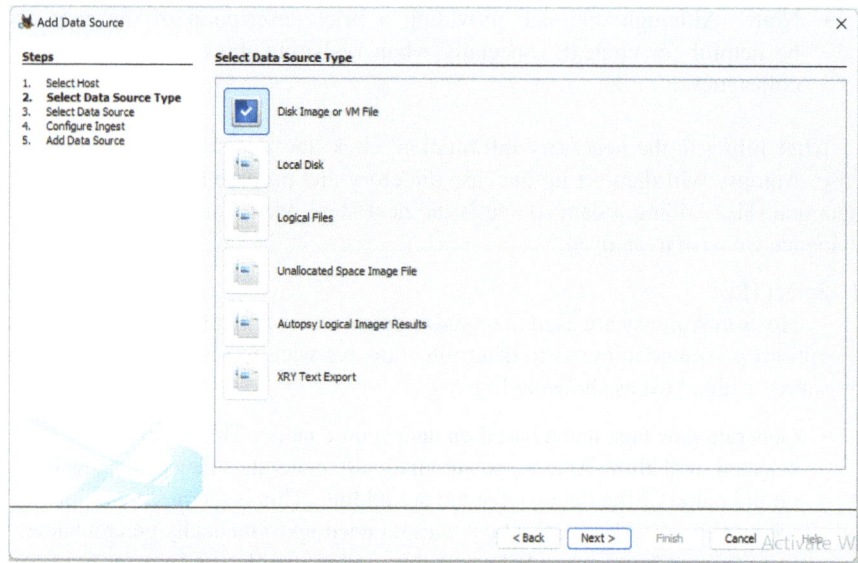

Fig. 6.5 Select data source type

a list of existing hosts. This helps in grouping all data sources related to a single host, which could represent a specific device or individual within the investigation.

2. Select Data Source Type as depicted in Fig. 6.5:
   - Disk Image or VM File: This is selected when the data source is a disk image (such as .iso, .dd, or other forensic image formats) or a virtual machine file. This option is typically used when analyzing a complete copy of a suspect's disk.
   - Local Disk: Should be selected if the user intends to add a physical disk that is locally connected to the system where Autopsy is running. This might be used in cases where the disk has been physically removed from the original hardware for analysis.
   - Logical Files: This option is for adding specific files and folders rather than a full disk image. It is useful when the investigation only concerns a subset of files or when specific documents or file types are in scope.
   - Unallocated Space Image File: Chosen when the investigator has an image of unallocated space from a disk that may contain deleted files or file fragments.
   - Autopsy Logical Imager Results: Selected when using results generated by Autopsy's Logical Imager. It is a tool for capturing a subset of file system data based on specific criteria.
   - XRY Text Export: Used for importing text exports from the XRY forensic tool. It is often applied in mobile device forensics.

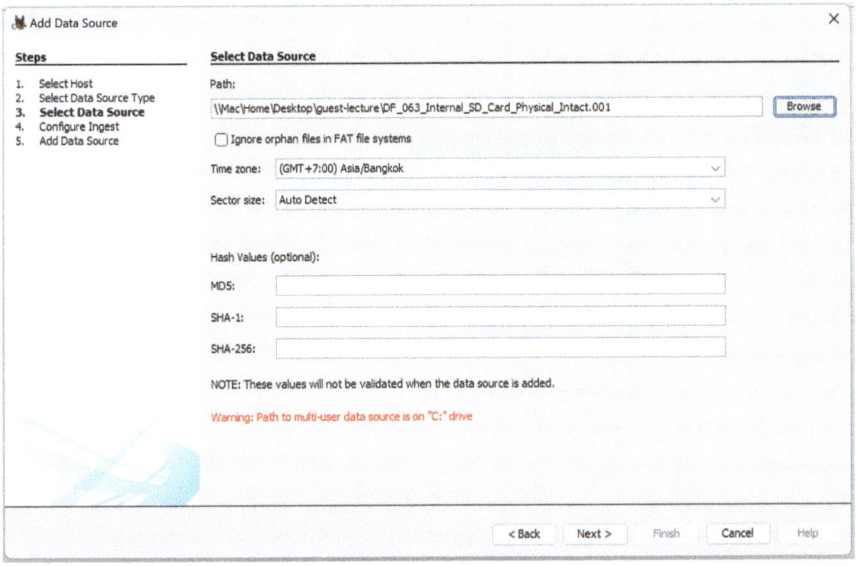

6.5 Creating a New Case and Importing a Forensic Image to Autopsy

Fig. 6.6 Select data source

3. Select Data Source

   Navigate to the location of the disk image, device, or files as shown in Fig. 6.6. For disk images or devices, we may need to specify further details about the time zone and sector size.

4. Configure Ingest:

   - Ingest Modules: Autopsy provides users with various ingest modules to run against the data source. These can include file type identification, keyword search, hash lookup, and more. Choose the modules relevant to our investigation as shown in Fig. 6.7.
   - Settings: For each selected module, configure any necessary settings. This might include specifying keyword lists, hash databases, or other parameters that guide the analysis.

5. After configuring the ingest modules, we can click Finish button as depicted in Fig. 6.8. Autopsy will begin processing the data according to the specified modules. This process can take a significant amount of time, depending on the size of the data source and the selected modules (Fig. 6.9).

Once the data source is added and the ingest process is complete, we can start analyzing the data using Autopsy's various tools and views. We can review files, search for keywords, examine metadata, and much more, all within the context of our newly created case.

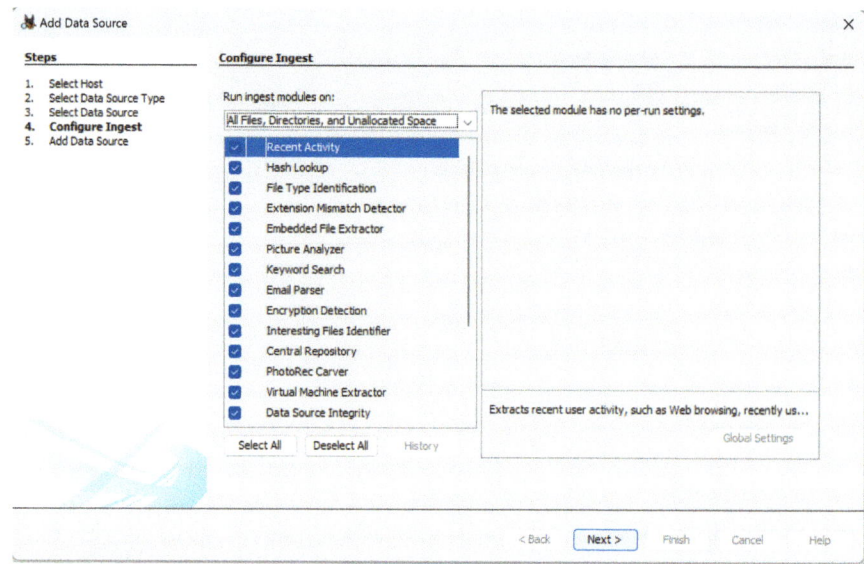

Fig. 6.7 Configure ingest modules

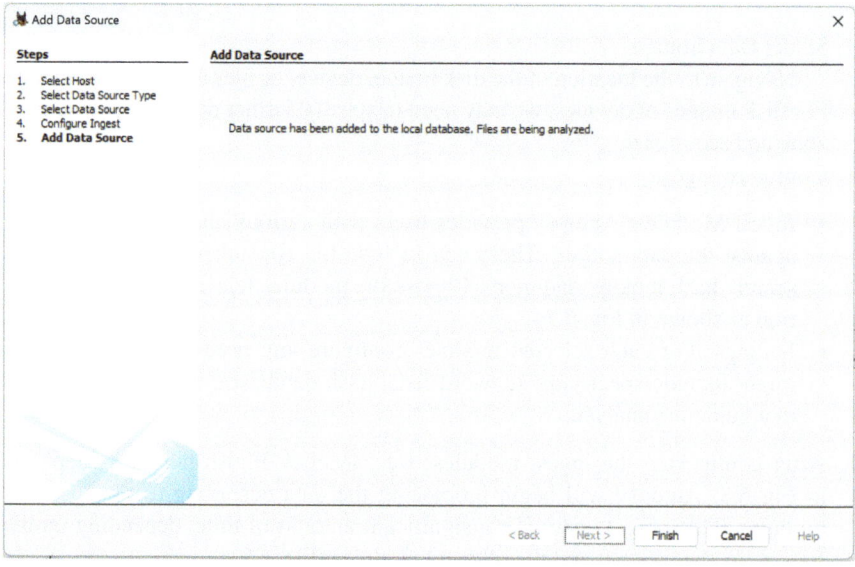

Fig. 6.8 Finish adding data source

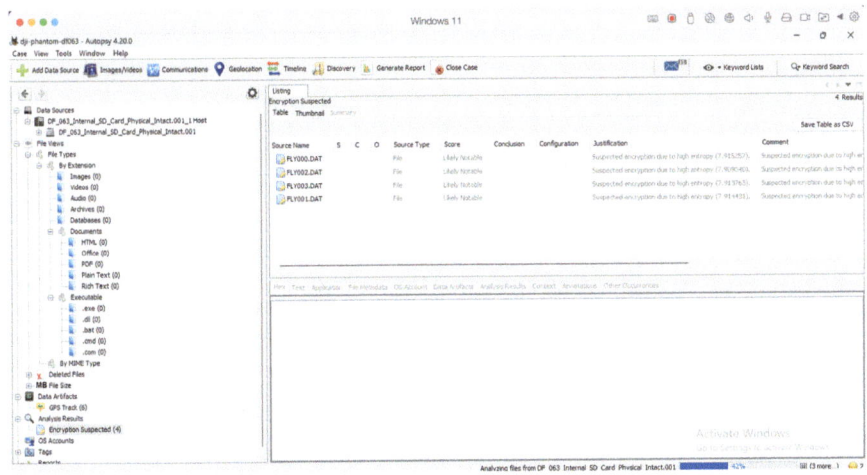

Fig. 6.9 Processing all artifacts takes the time

6.6 Autopsy DJI Drone Analyzer

The DJI Drone Analyzer module in Autopsy supports flight log analysis of files retrieved from DJI drones, specifically by examining data stored on the drone's internal SD card. This module is currently compatible with images from a variety of DJI drone models, including Phantom 3, Phantom 4, Phantom 4 Pro, Inspire 1, Inspire 2, Mavic Pro, and Mavic Air.

Upon locating the DAT files within the SD card image, the module utilizes DatCon to process these files. DatCon decodes the DAT files, extracting detailed telemetry data and other drone-specific information. This processed data provides insight into the drone's flight history, including parameters such as altitude, speed, GPS coordinates, and sensor readings. Therefore, it can help reconstruct events for forensic analysis.

To activate the DJI Drone Analyzer ingest module, check the appropriate box on the Ingest Modules configuration screen to enable its functionality as presented in Fig. 6.10. The analysis results can be accessed in the Results tree, where they appear under the "Extracted Content" section. This location organizes all extracted data for easy review of the processed drone information. GPS data extracted by the module is also accessible in the Geolocation window and provides a visual map view of the drone's recorded locations. In addition, a KML report can be generated that allows for further analysis and visualization of GPS coordinates in mapping software, such as Google Earth.

Using the DJI Drone Analyzer in Autopsy involves using DatCon to analyze drone artifacts, such as flight logs and other metadata. This feature is useful for

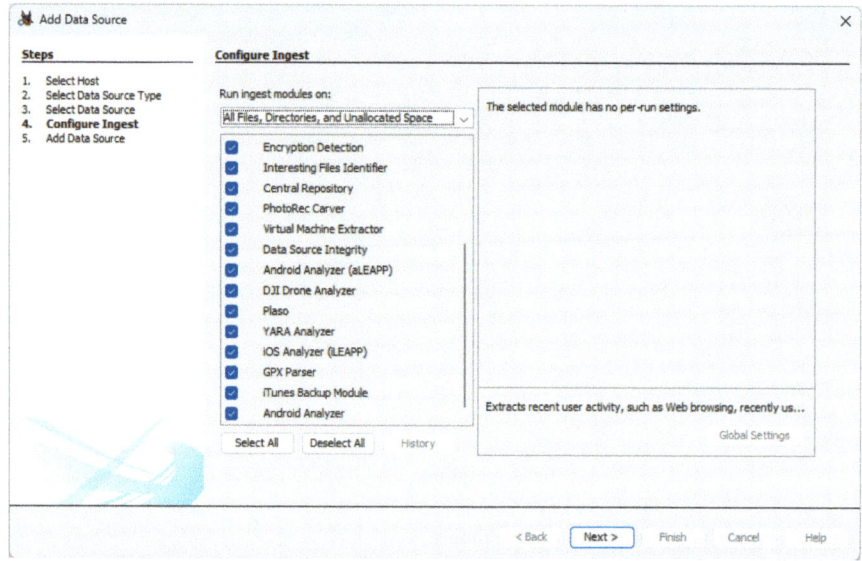

Fig. 6.10 DJI Drone Analyzer in Autopsy

forensic investigations that involve drone data.[3] To use the DJI Drone Analyzer in Autopsy, follow these detailed steps:

1. Open Autopsy and create a new case

    Launch Autopsy on the forensic workstation. If we are starting a new investigation, create a new case by clicking "Create New Case" and provide the necessary details such as the case name, number, and examiner information. However, if we are continuing an existing investigation, we can open the corresponding case file.

2. Add a data source

    Once the case is open, click on the "Add Data Source" button to include the relevant data we want to analyze. Select the appropriate type of data source, such as a local disk, an image file, or logical files. This data source should contain the drone-related files we wish to examine. Follow the on-screen prompts to add the data source and begin the process of mounting it in the case.

3. Enable the DJI Drone Analyzer ingest module

    After adding the data source, the "Configure Ingest Modules" window will appear. In this window, we need to select various analysis modules to run on the data. Make sure to check the box next to "DJI Drone Analyzer" to enable it. This module uses DatCon to parse and analyze drone-specific artifacts such as flight

---

[3] https://sleuthkit.org/autopsy/docs/user-docs/4.21.0/drone_page.html.

6.6 Autopsy DJI Drone Analyzer

data, telemetry, and logs. If needed, we can configure additional options for the Drone Analyzer by clicking the "Settings" button next to the module name.

4. Run the ingest process

   With the DJI Drone Analyzer and any other necessary modules selected, proceed by clicking "Next" and "Finish" to begin the ingest process. Autopsy will now start scanning the added data source, and the DJI Drone Analyzer module will process any drone-related files. Depending on the size of the data, this process may take some time.

5. Review the results

   Once the ingest process is complete, navigate to the "Listing" section in Autopsy. We will find the data extracted by the module. This may include information such as drone flight logs, GPS data, flight paths, and other relevant metadata as displayed in Fig. 6.11. The extracted data are organized in a way that allows us to easily review and analyze the findings.

6. Export or report the findings

   After reviewing the drone data, we may need to export the findings for further analysis as presented in Fig. 6.12 or include them in a report. Autopsy provides options to export specific artifacts or generate a report summarizing the investigation. Use the "Generate Report" feature to create a detailed record of

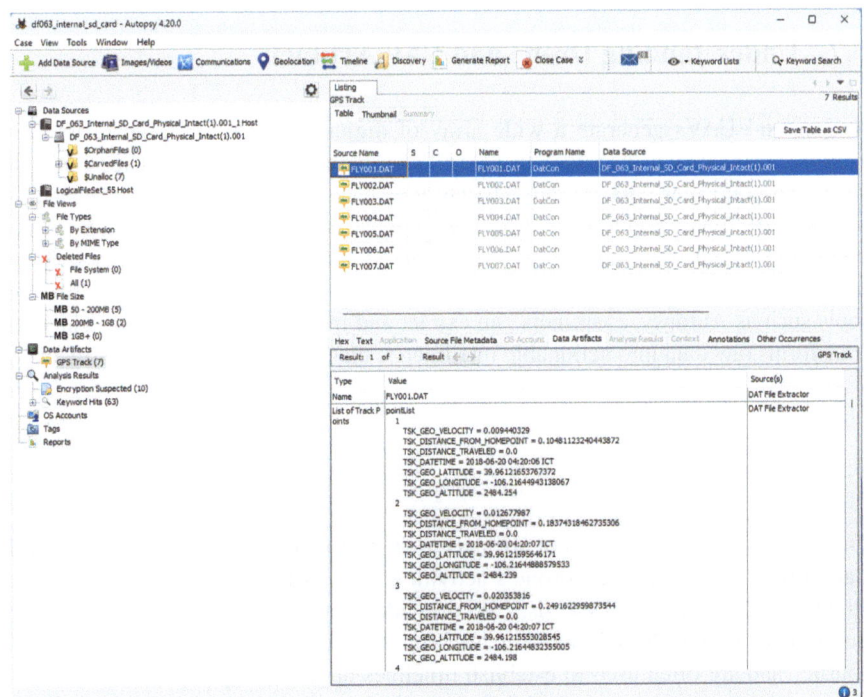

Fig. 6.11 GPS Track is shown when clicking a DAT file

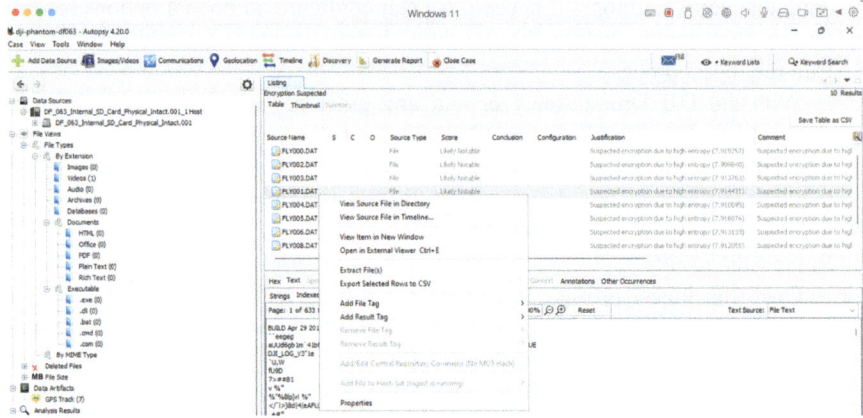

Fig. 6.12 Exporting a flight logs artifact

our findings, which can be shared with other investigators or included in legal documentation.

6.7 Understanding Drone and UAV Artifacts

Drones and UAVs generate a wide array of digital artifacts for forensic investigations. This section discusses these artifacts, with a focus on flight logs, media files, and controller data, to demonstrate how each piece of evidence contributes to a broader understanding of drone operations. By analyzing flight paths, GPS coordinates, and user interactions, investigators can reconstruct drone activities, establish timelines, and even attribute actions to specific individuals. Utilizing tools such as Autopsy, examiners can extract and interpret these artifacts. The tool transforms raw data into actionable intelligence that supports a range of casework.

6.7.1 Flight Logs

One of the most valuable sets of artifacts in drone forensics are flight logs and GPS data. Flight logs record the drone's activities, such as take-off and landing times, flight paths, altitudes, and speeds. GPS data add another layer of detail with precise coordinates of the drone's movements. Together, they can reconstruct a drone's journey and are often used to establish timelines or to locate points of interest that the drone may have interacted with or surveilled.

Figure 6.12 displays the Autopsy tool interface when an investigator exports a flight log artifact. Once we export this artifact, we can use

6.7 Understanding Drone and UAV Artifacts   153

other tools to analyze this flight log. The left pane of the Autopsy tool interface displays the data sources and file types, with the data source labeled DF_063_Internal_SD_Card_Physical_Intact. This source represents a digital forensic image extracted from a drone's SD card. The file structure is organized into categories such as images, videos, audio, and documents. However, in this case, no files are found within these categories. It indicates that the primary focus may be on other file types or data structures within the forensic image.

The main pane lists several files (e.g., FLY000.DAT, FLY001.DAT, etc.) with a .DAT extension. These files are flight log data files from the drone, commonly containing telemetry or event information from drone operations. Each flight log file has a "Likely Notable" score with a justification of "Suspected encryption due to high entropy". This suggests that Autopsy detected high entropy (values above 7.9), which is often associated with encrypted or compressed data. This can indicate that the log files are encrypted or contain complex data structures.

A context menu is open for the FLY001.DAT file, which presents several options for further analysis. These include "View Source File in Directory" and "Timeline", which assist in navigating to the file's location and analyzing its temporal details. "View in External Viewer" allows the file to be opened in another application for deeper examination, while the highlighted "Extract File(s)" option enables the user to export the selected file(s) from Autopsy to an external location for analysis with other tools. Finally, "Export Selected Rows to CSV" offers a way to document the metadata and attributes of these files in a CSV format. Other options include tagging and commenting to facilitate further investigation documentation.

6.7.2 Media: Pictures and Videos

Drones are commonly used to capture aerial imagery. The media files stored in a drone's memory can include photographs and video recordings, which may carry metadata with timestamps and GPS coordinates. Analyzing this content can reveal not just the drone's location at specific times but also what the drone was being used to observe or monitor. This can be especially relevant in cases of privacy violation, surveillance, or even in environmental studies.

The screenshot in Fig. 6.13 shows the analysis of image artifacts from an iOS drone controller using Autopsy. The tool is commonly used to extract and analyze data from digital devices, and in this case, the focus is on identifying and investigating image files likely captured or stored by a drone controller application. On the left side of the interface, the Data Sources Panel lists the extracted logical file systems or datasets under investigation. In this instance, it includes categories such as LogicalFileSet1_Host and By MIME Type, which organize files by their type, including image/png and image/jpeg. This organization allows forensic examiners to quickly navigate to specific artifacts of interest, such as drone-captured images.

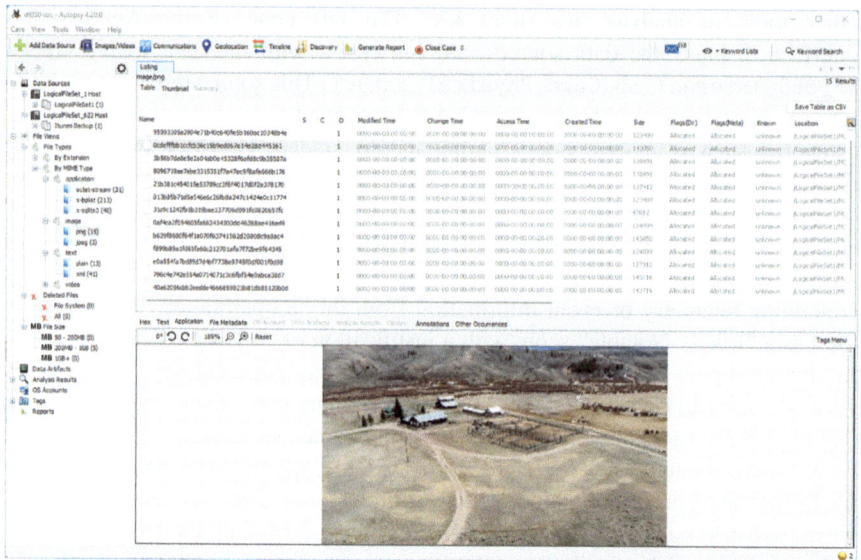

Fig. 6.13 Image artifact from a iOS drone controller

The central image listing panel provides detailed metadata for the filtered image files. The table includes information such as file hashes (checksums to verify file integrity), size, and allocation status (e.g. "Allocated"). However, the timestamp fields, such as "Modified Time" and "Created Time", show null values (0000-00-00 00:00:00), which may indicate missing or incomplete metadata. This could require additional forensic techniques to reconstruct timelines or understand the origin of files. At the bottom of the interface, the preview panel displays a selected image file. In this case, the preview shows an aerial photograph of farmland captured by the drone. This image provides visual evidence to support forensic investigations involving drone activity, such as surveillance, monitoring, or incidents involving unauthorized drone operations. The image's context and metadata can help investigators establish the drone's use and intent.

6.7.3 Controller Artifacts

The controller device, often a smartphone or tablet running on platforms like Android or iOS, can also provide valuable artifacts. These may include the app used to pilot the drone, login information, command inputs, and even screen captures or recordings of the drone's flights. Artifacts from the controller can often provide context to flight logs and media, linking the drone to an individual and establishing user intention.

6.7 Understanding Drone and UAV Artifacts

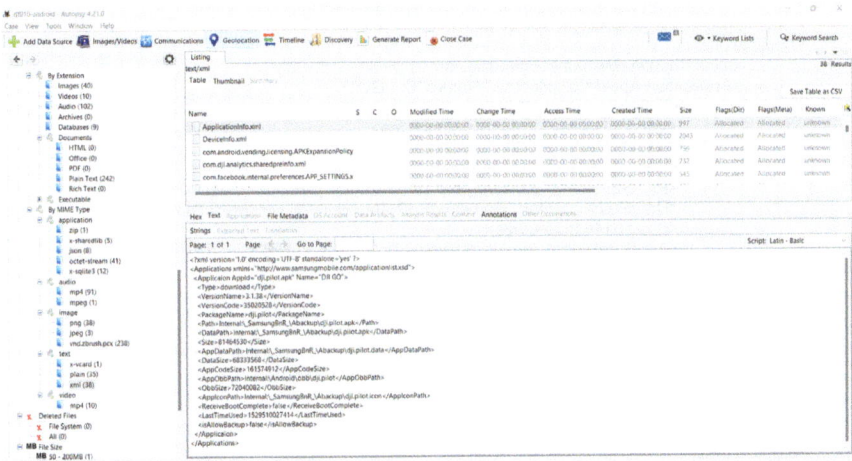

Fig. 6.14 Android controller artifacts: Application information in XML format

Figure 6.14 shows an artifact analysis using Autopsy which focuses on XML files extracted from an Android device that hosts the DJI GO application. This application is widely used for controlling DJI drones and generates metadata stored in XML format. The XML file under examination, `ApplicationInfo.xml`, provides detailed information about the application, such as its version, storage paths, and usage. It offers information on how the application was configured and utilized. The left panel of the interface organizes the extracted files by extension and MIME type, with the selected category being `text/xml`. Several XML files are listed, including `ApplicationInfo.xml` and `DeviceInfo.xml`. These files relate to the functionality and operations of the DJI GO application. The central panel shows the attributes of the file, such as name, size, and allocation status. Interestingly, the timestamps for file modification, access, and creation are all set to 0000-00-00 00:00:00, which suggests missing or zeroed-out metadata. However, the contents of the file itself, displayed in the bottom panel, still reveal forensic details.

The detailed contents of `ApplicationInfo.xml` provide insights into the DJI GO application, as shown in Fig. 6.15. The name and version of the application are explicitly mentioned (DJI GO, version 3.1.38), along with its package name (`dji.pilot.apk`). Storage locations are identified through fields such as DataPath and AppDataPath, which point to specific directories on the device (`SamsungRnR\Abackup\dji.pilot.data`). These paths indicate where application data, such as flight logs or settings, can be stored. Additionally, the XML metadata highlights the sizes of various components, including the application code and OBB (Opaque Binary Blob) files. These details can guide forensic investigators in locating and analyzing drone-related data within the device.

```xml
<?xml version="1.0" encoding="UTF-8" standalone="yes" ?>
<Applications
        xmlns="http://www.samsungmobile.com/applicationlist.xsd">
        <Application AppId="dji.pilot.apk" Name="DJI GO">
                <Type>download</Type>
                <VersionName>3.1.38</VersionName>
                <VersionCode>35020528</VersionCode>
                <PackageName>dji.pilot</PackageName>
                <Path>Internal:\_SamsungBnR_\Abackup\dji.pilot.apk</Path>
                <DataPath>Internal:\_SamsungBnR_\Abackup\dji.pilot.apk</DataPath>
                <Size>81464530</Size>
                <AppDataPath>Internal:\_SamsungBnR_\Abackup\dji.pilot.data</AppDataPath>
                <DataSize>68333568</DataSize>
                <AppCodeSize>161574912</AppCodeSize>
                <AppObbPath>Internal:\Android\obb\dji.pilot</AppObbPath>
                <ObbSize>72040082</ObbSize>
                <AppIconPath>Internal:\_SamsungBnR_\Abackup\dji.pilot.icon</AppIconPath>
                <ReceiveBootComplete>false</ReceiveBootComplete>
                <LastTimeUsed>1529510027414</LastTimeUsed>
                <isAllowBackup>false</isAllowBackup>
        </Application>
</Applications>
```

Fig. 6.15 Extracted application information in XML format

An item of information extracted from the XML file is the `LastTimeUsed` field, which records the last usage time of the application in Unix Epoch format. This timestamp is used to establish timelines in forensic investigations, especially when correlating application activity with drone operations. Furthermore, fields such as `IsAllowBackup` and `ReceiveBootComplete` reveal application settings that provide information on its operational characteristics, such as whether data backups were allowed or if the application was designed to respond to system boot events.

Figure 6.16 depicts another xml artifacts namely `DeviceInfo.xml`. This file contains metadata about the device used as a controller for a DJI drone. We can obtain valuable information about the hardware and software environment. The forensic image is organized by file type in the left panel, with `DeviceInfo.xml` categorized under `text/xml`. The central panel lists file attributes, such as allocation status and size, while timestamps for modification, access, and creation are shown as 0000-00-00 00:00:00.

The contents of `DeviceInfo.xml`, displayed in the bottom panel, include key information about the controller device. The `ModelName` identifies the device as a Samsung Galaxy J7 Perx (SM-J727P), with additional details such as the `PlatformVersion` (Android 7.0). The XML file also lists version codes (`VersionCode` and `USBVersion`) that reflect the hardware and software capabilities of the device. Other features listed such as `SamsungNoteForceRestore`, `KidsModeForceRestore`, and `EncBackupVersion` reveal specific configurations or functionalities enabled on the device, while fields such as `AppInstallAgreement` indicate that terms and conditions were accepted during app installation. The `HiddenVersion` field provides internal versioning information. It is potentially useful for identifying firmware or app updates.

6.7 Understanding Drone and UAV Artifacts

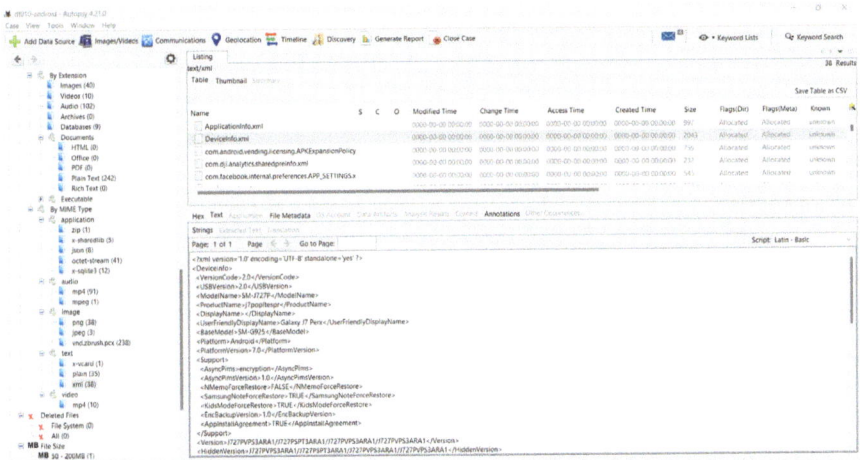

Fig. 6.16 Android controller artifacts: Device information in XML format

Forensically, `DeviceInfo.xml` metadata is important in several ways. First, it helps attribute the device to a specific individual or operator, especially when cross-referenced with user data or app artifacts. Second, it contextualizes the environment in which the DJI drone was operated, offering clues about compatibility and usage. For instance, the Android platform and the details of the version can indicate the features available to the drone controller app. Third, configuration flags such as encryption backup support and restore settings may suggest whether additional data could be recovered from the device or if certain features were intentionally enabled for operational purposes.

Furthermore, Fig. 6.17 illustrates the analysis of an ERROR_POP_LOG file extracted from the DJI Pilot application. This log file, located within the application's LOG subdirectory, contains operational error messages and warnings that occurred during drone operation. The log entries are stored in plain text format and provide timestamps and descriptive messages about errors or warnings that the drone or its controller encountered. We need this artifact as investigators seek to understand operational anomalies or reconstruct flight-related incidents.

The log file, named 20-06-2018, is 796 bytes in size and allocated within the file system. It is part of the logical file set extracted from the device's SD card. However, the modification, access, and creation timestamps are all recorded as 0000-00-00 00:00:00. It indicates that these metadata are missing, wiped, or unavailable due to the extraction method. Despite the lack of traditional timestamp metadata, the content of the log file itself provides precise time-stamped entries, which are needed to create a chronological sequence of events.

The content of the ERROR_POP_LOG file reveals several error messages and warnings. It also includes a guide for the user. For instance, one entry at 08:33:14 advises, "Take off and retry this function". It suggests a failed or aborted takeoff

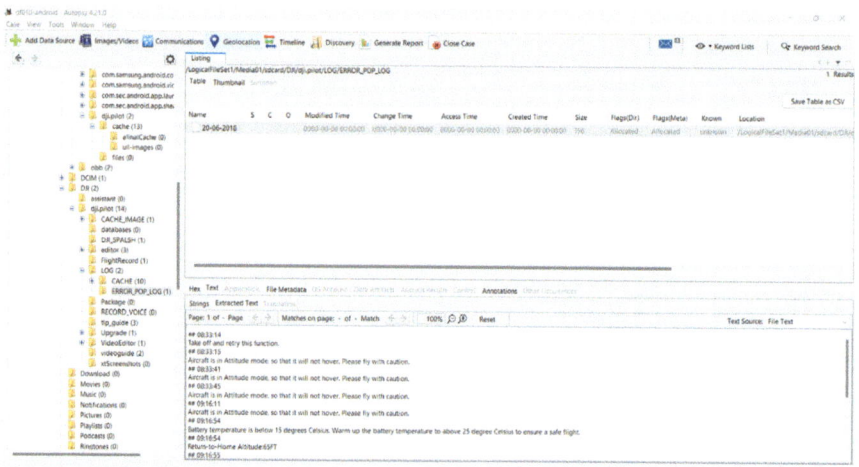

Fig. 6.17 Android controller artifacts: Error pop logs in plain text

attempt. Another entry at 09:16:11 warns, "Aircraft is in Attitude mode, so that it will not hover. Please fly with caution". It indicates a potential issue with the drone's stabilization system. Environmental factors are also noted, such as the log entry at 09:16:54, which states "Battery temperature is below 15° C. Warm up the battery temperature to above 25° C to ensure a safe flight." In addition, the entry at 09:16:55 specifies the "Return-to-Home Altitude: 65FT," and provides operational settings for automatic return.

From a forensic perspective, the ERROR_POP_LOG file is quite important. It provides a detailed view of the operational challenges and errors encountered during a flight session. These logs can help reconstruct a precise timeline of events and link error occurrences to specific actions, such as take-off, hovering, or return-to-home operations. Furthermore, warnings about battery temperature and Attitude mode indicate environmental or technical factors that may have impacted drone performance. Such details are invaluable in understanding whether incidents were caused by user error, environmental conditions, or equipment malfunction.

The log entries also shed light on user behavior and decision making. For example, entries suggesting actions like "Take off and retry this function" can reveal how the user responded to system errors or warnings. If critical warnings, such as low battery or improper flight mode, were ignored, it could indicate negligence. On the other hand, repeated system errors may suggest technical malfunctions that could explain the unintended behavior of drones.

Figure 6.18 illustrates the analysis of JSON artifacts extracted from the DJI Pilot application on an Android device. These artifacts, stored in the directory of /Android/data/dji.pilot/cache/, include several JSON files such as component_upgrade_list.json and component_upgrade_date.json, which contain information about drone firmware updates and associated metadata. These

6.7 Understanding Drone and UAV Artifacts

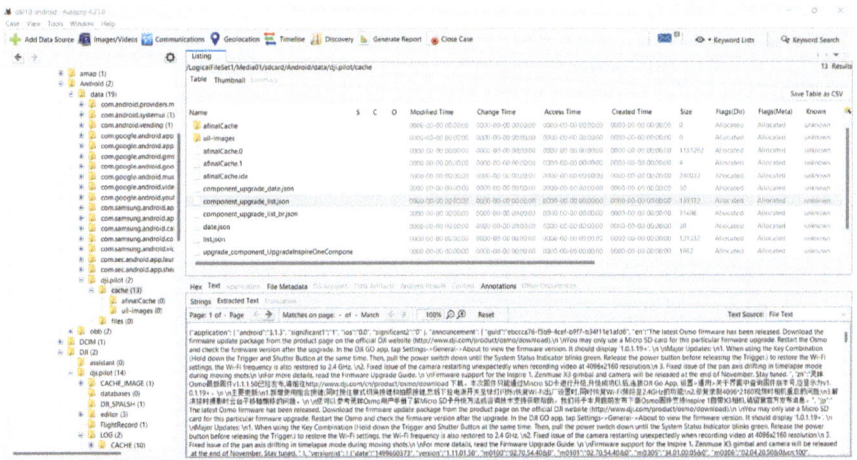

Fig. 6.18 Android controller artifacts: Firmware information in JSON format

files provide a structured record of firmware versions, update history, and technical details. The directory structure reveals that the DJI Pilot app cache folder contains JSON files dedicated to firmware tracking and update management. The attributes of each file, such as the allocation status and size, are displayed in the central pane. The JSON file content, shown in the bottom pane, provides a wealth of information about the firmware details and update history.

The content of the selected JSON file includes fields like application, which indicates the version of the DJI Pilot app (e.g., 3.1.13), and announcement, which details the latest firmware release (1.01.19). The update notes provide insight into what the firmware update addresses, such as resolving Wi-Fi frequency issues, fixing gimbal stabilization, and improving camera functionality. The multilingual instructions in the JSON file demonstrate the global nature of the DJI platform, since the firmware update instructions are provided in multiple languages for a wide user base. The file also contains version metadata for specific drone models and components, including version identifiers for m100, m305, and m306.

From a forensic perspective, these JSON files are quite important. They enable investigators to track the drone's firmware history. One can also determine whether it was up to date at the time of a specific incident. By cross-referencing update release dates with other artifacts, such as flight logs, investigators can create a detailed timeline of maintenance and operational changes. This helps establish whether the drone's firmware played a role in the event, such as a malfunction caused by outdated software or the absence of critical updates. Furthermore, the firmware notes provide insight into the drone's operational capabilities during a specific period. Investigators can assess whether certain features or fixes were available or if the drone was limited by outdated firmware. The presence of detailed update instructions also helps to assess the intent of the user and the compliance with maintenance

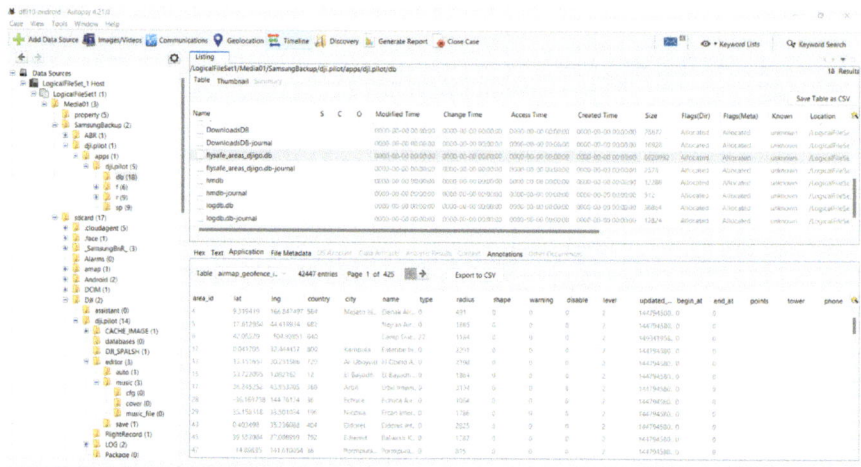

Fig. 6.19 Android controller artifacts: Fly save area in SQLite database

guidelines. If the drone was operated without installing recommended updates, it could indicate negligence or intentional disregard for operational standards.

Figure 6.19 shows an SQLite database file, flysafe_areas_djiGo.db. It is extracted from the DJI Pilot application on an Android device. This database, stored in the /apps/dji.pilot/db/ directory, contains extensive geofencing data related to drone operations. Specifically, it outlines safe and restricted flight zones, including geographic coordinates, zone names, types, and associated restrictions. These artifacts provide information on the regulatory and operational context of a drone's activity.

Active use of this SQLite database for real-time or cached geofencing data is indicated by the accompanying journal file, namely flysafe_areas_djiGo.db-journal. Attributes such as file size (802,092 bytes) and allocation status (Allocated) are displayed, although timestamps for modification, access, and creation are absent (0000-00-00 00:00:00). Despite missing metadata, the database content remains intact and highly informative, containing 42,447 entries that document geofencing zones globally.

The airmap_geofence table, displayed in the bottom panel, provides structured information about geofenced areas. Each entry includes an area ID (area_id), geographic coordinates (latitude and longitude), and descriptive details such as the zone's country, city, and name (e.g., "Mejato Island" or "Elenak Airport"). The table also records the radius of each zone in meters, defining the area of restriction or regulation. Additional columns, such as warning, disable, and level, indicate the operational restrictions or warnings associated with each zone. Metadata fields like updated_at, begin_at, and end_at provide temporal context for when geofences were last updated and their periods of activation.

6.7 Understanding Drone and UAV Artifacts

This database has important forensic implications. First, it helps investigators determine whether a drone's flight path intersected with restricted zones, such as airports or no-fly areas. By mapping the geographic coordinates of these zones, analysts can compare them with GPS data from flight logs to assess compliance with operational rules. Second, the timeline information in the database allows forensic analysts to reconstruct the regulatory environment at the time of a flight. For example, investigators can determine whether a geofence restriction was active when the drone operated in a particular area. Geofencing data also provide information about user behavior and operational capabilities. If warnings and restrictions stored in the database were ignored, it may indicate negligence or intentional violations by the drone operator. In contrast, frequent updates to geofencing information (updated_at) demonstrate how the DJI Pilot app actively informs users about restricted zones. When combined with other forensic artifacts, such as flight logs, error reports, and media files, this database helps build a comprehensive understanding of drone activities and compliance with legal requirements.

Figure 6.20 presents the analysis of drone flight logs stored in binary format, extracted from the DJI Pilot application on an Android drone controller. The flight log, named DJIFlightRecord_2018-06-20_[09-16-53].txt, is located in the /sdcard/DJI/dji.pilot/FlightRecord/ directory. This artifact provides detailed telemetry and operational data recorded during a drone flight. Binary flight logs capture a granular timeline of events, such as flight paths, system parameters, and user input. The file content, displayed in hexadecimal (hex) format in the bottom pane, remains intact and ready for analysis.

The hex view reveals that the file is encoded in binary form, so it is unreadable without specialized tools or decoding scripts. In addition, they may log metadata about the drone's operational state, such as mode changes (e.g., take-off, landing,

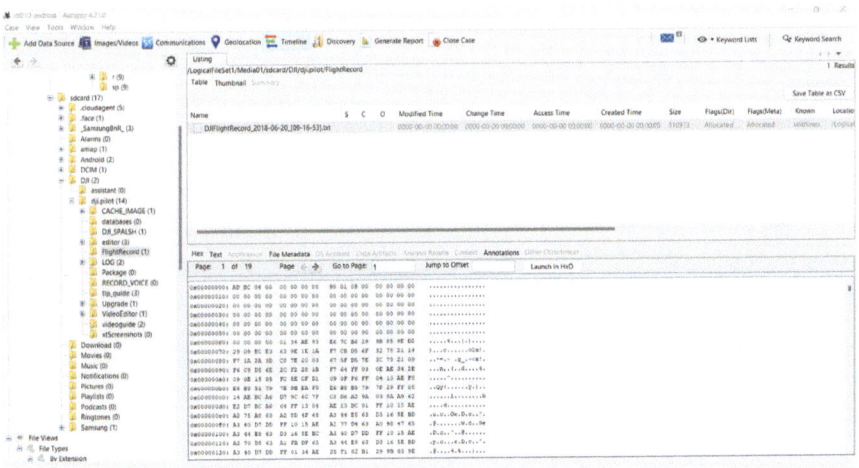

Fig. 6.20 Android controller artifacts: Flight logs in binary text file

hover), errors, and user-triggered commands. These details can be parsed using proprietary DJI software or custom scripts to extract actionable information in a human-readable format, which will be discussed in Chap. 7.

From a forensic point of view, binary flight logs are invaluable to reconstruct drone activity. The logs provide a detailed timeline of drone operations and help investigators map its flight path and analyze its behavior during critical moments. For example, by correlating GPS data with timestamps, analysts can determine where the drone flew, how high it ascended, and whether it entered restricted airspace. The logs can also highlight potential causes of incidents, such as low battery warnings, sudden altitude changes, or signal loss. In addition to telemetry, the logs may contain evidence of user behavior, such as manual inputs for take-off, directional control, or camera operation. This can help establish the intent behind the drone's movements, which is particularly relevant in cases involving unauthorized surveillance or privacy violations.

Another screenshot illustrates video artifacts extracted from an Android controller running the DJI Pilot application as shown in Fig. 6.21. These video files, located in the /LogicalFileSet1/Media01/ directory, are stored in MP4 format and include filenames such as video_guide_land.mp4, osmo_hyperlapse.mp4, and osmo_timelapse.mp4. The videos appear to serve multiple purposes, including guides and user instructions. Note that this directory may also contain media captured by the drone. Video artifacts are highly valuable in forensic investigations as they provide visual evidence of drone operations and context for user activities.

The central pane lists the video files along with their attributes, such as size and allocation status. The sizes of the files range from approximately 300 KB to nearly 1 MB. It indicates they may contain short videos for instructional or demonstration clips. The preview pane at the bottom displays a still image from one of the videos

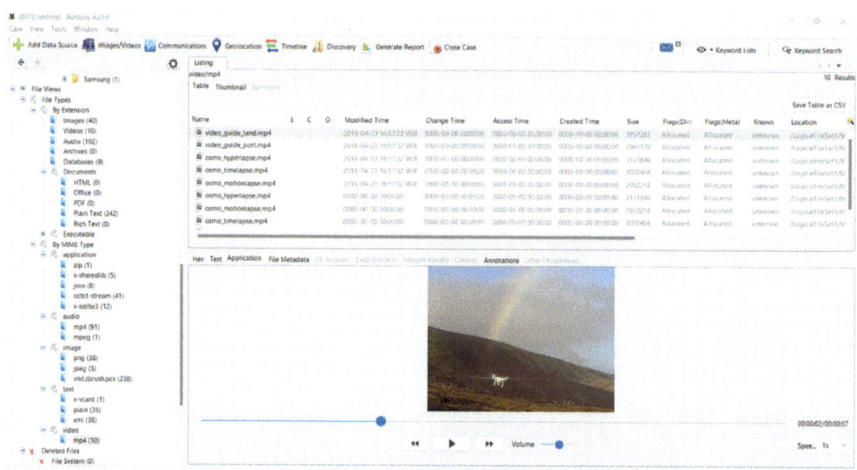

Fig. 6.21 Android controller artifacts: Guide video from the DJI GO application

and shows a drone flying in a scenic outdoor environment with a rainbow in the background. The playback controls in the pane allow investigators to view the video directly within Autopsy.

If the videos were captured by the drone, they could serve as direct evidence of what the drone was observing or surveilling during its flight. This is important in cases of privacy violations or unauthorized surveillance, as the video content could confirm whether the drone was used for illegal or improper purposes. Metadata, such as modification timestamps and file size, helps investigators determine when videos were created or accessed. By correlating these details with other artifacts, such as flight logs, investigators can reconstruct a timeline of drone activities, linking video creation to specific flight events.

The content of the videos can also provide geographic and environmental evidence. For example, the scene displayed in the preview pane of the video, such as landmarks, terrain, or weather conditions, offers clues about the location and time of the drone's operation. This type of analysis can help pinpoint where the drone was used and what it was recording. Finally, videos can be cross-referenced with other forensic evidence, such as flight logs, GPS data, and error reports, to corroborate drone movements and activities. This ensures a comprehensive understanding of the drone's operation and user participation, helping to create a detailed narrative for forensic investigations.

6.8 Summary

The chapter takes the reader through the process of working with drone forensic images. It begins by explaining the different types of forensic images that investigators might come across, such as complete copies of a drone's internal memory. The chapter then introduces the Autopsy tool, which is software used for examining these images, and guides the reader through setting it up.

Subsequently, the chapter shows how to start a new investigation in Autopsy by creating a case and adding drone image files to it. This is where the real investigation work begins, as the reader learns to find and study the digital traces left by drones. These traces, called artifacts, include things such as flight paths and photos taken by the drone, which can tell us a lot about what the drone did. By the end of the chapter, the reader should understand and be confident in using Autopsy to investigate drone-related cases. Finally, the reader can make sense of the technical data to make useful recommendations.

6.9 Exercises

1. What is the primary tool discussed in this chapter for analyzing forensic images and artifacts?

2. What is a forensic image in the context of drones and UAVs?
3. Why are forensic images important for drone and UAV investigations?
4. What can the Autopsy tool do for drone forensics?
5. Name one type of artifact that can be analyzed using the Autopsy tool.
6. Explain the process of configuring the Autopsy tool for forensic analysis as described in this chapter.
7. Discuss the challenges associated with analyzing forensic images from drones and UAVs.
8. Evaluate the effectiveness of the Autopsy tool in handling drone-specific forensic artifacts.
9. How can investigators ensure the integrity of forensic images while importing them into tools like Autopsy?
10. Propose improvements, additional features, or plugins for the Autopsy tool to better support drone and UAV forensic investigations.

References

1. Autopsy, *Autopsy: Digital Forensics* (2024). https://www.autopsy.com/
2. H. Studiawan, G. Grispos, K.-K. Raymond Choo, Unmanned aerial vehicle (UAV) forensics: the good, the bad, and the unaddressed. Comput. Secur. 103340 (2023)
3. S.C. Nayak, B.K. Samanthula, V. Tiwari, Investigating drone data recovery beyond the obvious using digital forensics, in *2023 IEEE 14th Annual Ubiquitous Computing, Electronics & Mobile Communication Conference (UEMCON)* (2023), pp. 0254–0260
4. M. Yousef, F. Iqbal, M. Hussain, Drone forensics: adetailed analysis of emerging DJI models, in *2020 11th International Conference on Information and Communication Systems (ICICS)* (2020), pp. 066–071
5. V. Sihag, et al., Cyber4drone: a systematic review of cyber security and forensics in next-generation drones. Drones **7**(7), 430 (2023)
6. E. Debas, A. Alhuali, M.M. Hafizur Rahman, Forensic examination of drones: a comprehensive study of frameworks, challenges, and machine learning applications, in *IEEE Access* (2024)
7. F. Iqbal, et al., Forensic investigation of small-scale digital devices: a futuristic view. Front. Commun. Netw **4**, 1212743 (2023)
8. A. Adel, T. Jan, Watch the skies: a study on drone attack vectors, forensic approaches, and persisting security challenges. Future Int. **16**(7) (2024)
9. H. Bouafif, et al., Drone forensics: challenges and new insights, in *2018 9th IFIP International Conference on New Technologies, Mobility and Security (NTMS)* (2018), pp. 1–6
10. S. Lee, H. Seo, D. Kim, Digital forensic research for analyzing drone pilot: focusing on DJI remote controller. Sensors **23**(21), 8934 (2023)
11. A. Almusayli, T. Zia, E.-ul-H. Qazi, Drone forensics: an innovative approach to the forensic investigation of drone accidents based on digital twin technology. Technologies **12**(1), 11 (2024)

Chapter 7
Forensic Analysis of Drone and UAV Flight Data

Abstract The chapter explains the process of examining flight data from drones and unmanned aerial vehicles (UAVs) to assist in forensic investigations. This chapter provides an overview of open source forensic tools specifically designed for this task, such as DatCon, CsvView, and DROP (DROne Parser). DatCon is explored for its capability to convert proprietary drone data logs into a more accessible format to support deeper analysis. CsvView is selected for its user-friendly interface that allows investigators to visually inspect flight data captured in CSV files. Lastly, the chapter introduces DROP, an open-source tool to parse drone flight logs. Through hands-on examples and case studies, this chapter guides readers through the practical application of these tools in real-world scenarios. We demonstrate how they can provide information on flight paths, behavior, and potential drone malfunctions.

7.1 Introduction

Flight logs in the context of drones are detailed records that capture a wide array of data points during a drone's flight. They serve as a resource for performance monitoring, troubleshooting, and forensic analysis. These logs meticulously document the drone's flight path and GPS data, including latitude, longitude, altitude, and speed. These data support for an accurate reconstruction of the drone's movements. Each piece of data is associated with precise timestamps and offers a detailed timeline of the flight's sequence of events. Information on the drone's battery status, such as voltage and consumption, is also recorded. They provide information on the drone's operational functionalities and potential issues related to battery performance.

In addition, flight logs record operator control input and details of take-off, landing, and navigation commands. We can use these parameters to understand how the drone maneuvered during its flight. The sensor data from the drone systems onboard contain measuring parameters such as altitude, temperature, and obstacle detection. They further enrich the flight's contextual understanding. Status messages and errors encountered during flight, including GPS signal loss, communication interruptions, or hardware failures, are logged. This collection of data, including telemetry information on the drone's orientation (pitch, roll, yaw), signal strength, and other real-time insights, is important for both hobbyists and professionals.

Flight logs and GPS data can support forensic investigations involving drones. Flight logs and GPS data enable investigators to precisely map out the drone's flight path. This information is used to understand the drone's movements before, during, and after an incident. By analyzing these data, investigators can determine the location of the drone at specific times, its altitude, and speed. These data are important for piecing together the sequence of events.

Moreover, analyzing flight logs can reveal deviations from normal flight patterns, such as sudden drops in altitude, erratic movements, or unexplained accelerations. These anomalies might indicate a malfunction, an operator error, or external interference. It helps to understand the circumstances that led to the incident. Flight logs record the commands sent by the operator and the drone's responses. This information can help investigators assess whether the drone operated as intended or if unexpected behaviors occurred. It can also establish whether the drone was under manual control or following a pre-programmed route. In several cases, it could influence the outcome of the investigation.

GPS data, along with other sensor data recorded in flight logs, can provide information about environmental conditions during the flight, such as weather patterns or obstacles encountered. This information can be used to assess whether external factors played a role in the incident. The precise data contained in flight logs can be used to corroborate or refute witness statements and other forms of evidence. For example, if witnesses claim that a drone was in a particular area at a certain time, flight logs can confirm or deny this assertion based on the recorded GPS data.

Furthermore, flight logs can include information that helps identify the specific drone and potentially the operator or the pilot. Although this may require correlating data from the drone with other sources, such as registration databases or manufacturer records, it can support the attributing responsibility for the drone's actions. By providing an objective and detailed record of the drone's operations, flight logs and GPS data are important artifacts in forensic investigations. They enable a comprehensive analysis that can help reconstruct the facts of a case.

This chapter explains the forensic analysis of drone flight logs and its hands-on exercises. Section 7.2 discusses related research in this research area, while Sect. 7.3 describes the format of drone flight logs. In subsequent sections, we use several free and/or open source tools to perform drone flight log forensics, specifically DatCon (Sect. 7.4), CsvView (Sect. 7.5), DROP (Sect. 7.6), Maraudrone's Map (Sect. 7.7), DRDP (Sect. 7.8), and Digital Drone Forensics Software (Sect. 7.9).

7.2 Related Work

This section reviews related work that examines the forensics of UAVs, especially tools and techniques for flight log analysis. Mantas and Patsakis [1] examine the

7.2 Related Work

unique challenges and gaps in digital forensic investigations involving UAVs. It begins by discussing the increasing need for a standardized drone digital forensics investigation framework to manage the rising use of drones in both legal and malicious contexts. The discussion stresses the complexity of UAV forensics, which is derived from diverse data sources, including flight logs, communication data, and onboard sensors, as well as the lack of adequate tools, standardized procedures, and international collaboration.

Building on the need for tools in UAV forensic investigations, another study introduces DFLER (Drone Flight Log Entity Recognizer), an open source tool designed to extract and analyze human-readable flight log messages [2]. Developed using Python and powered by a fine-tuned BERT-based Named Entity Recognition (NER) model, DFLER focuses on constructing forensic timelines from decrypted flight logs, such as DJI's flight records and error logs, providing detailed forensic reports to aid investigators.

Extending the focus to GPS data analysis, another study explores methodologies to extract, analyze, and visualize GPS data from flight logs of popular commercial drones such as the DJI Phantom 4 Pro, Parrot Bebop 2, and Yuneec Typhoon H [3]. It emphasizes the importance of reconstructing flight paths on satellite maps and introduces the "FlyLog Converter Tool" to simplify flight log analysis. The paper also demonstrates how such tools uncover details such as flight trajectories, altitude variations, and timestamps. Addressing vulnerabilities in drone systems, a separate study focuses on forensic challenges and methods specific to the Parrot Bebop 2 drone and its controller [4]. It highlights the forensic implications of deauthentication and integrity attacks. The authors show that it is important to link a drone to its owner and reconstruct events through collected evidence. The study proposes a drone ontology for modeling context data and a forensic processing framework to support investigators.

Changing attention to the broader forensic challenges posed by the misuse of UAVs, another study investigates the methodologies and frameworks needed to address crimes such as smuggling and surveillance [5]. It explains the value of data stored on drones, such as flight logs and GPS coordinates, while identifying antiforensic techniques and system vulnerabilities that hinder investigations. The study advocates for the integration of forensic tools, legislative improvements, and standardized procedures to combat these challenges effectively. Focusing specifically on micro drones, another work develops a customized forensic framework to analyze the flight logs of DJI, 3DR, and Yuneec drones [6]. The research highlights the growing misuse of drones for illegal activities and introduces a Java-based software tool with a user-friendly GUI to visualize and analyze flight data. The preliminary results of the framework demonstrate its ability to reconstruct flight paths and evaluate compliance with aviation regulations.

Expanding the scope of forensic methodologies, a research proposes a forensic event reconstruction method based on drone log files using directed graph representations [7]. By incorporating sentiment analysis, the approach identifies events of interest, such as anomalies or suspicious activities. Experimental results confirm the method's effectiveness in reconstructing events and highlighting anomalies. In addition to exploring analytical techniques, another paper applies sequence mining methods to analyze drone flight logs for forensic purposes [8]. The paper introduces the transformation of flight logs into sequences of events to enable the application of sequential pattern and rule mining techniques. The findings reveal the potential of sequence rule mining for identifying event relationships, despite challenges in modeling sparse and small datasets.

In terms of cybersecurity and forensic traceability, a comparative study focuses on UAV vulnerabilities to cyberattacks and data integrity issues [9]. By employing a "Purple-Teaming" approach, which combines red and blue team tactics, the study identifies weaknesses in UAV systems and introduces the UAV Kill Chain model to outline critical phases of drone security operations. The research also highlights technical challenges in current forensic tools, such as decryption issues and data consistency. Finally, another study examines the forensic analysis of DJI Mavic Air drones, their associated mobile applications, and remote controllers [10]. It provides a comprehensive methodology for extracting, analyzing, and preserving forensic evidence. They focus on artifacts such as flight logs, GPS coordinates, and user credentials. The study addresses challenges in acquiring data from encrypted and proprietary formats and demonstrates that evidence can be extracted from drone's internal storage and associated devices.

7.3 Format of Drone Flight Logs

Most of the research in drone forensics discusses how to decrypt and extract data from drone flight logs. Before we analyze drone flight logs using various tools, we need to check the structure of these files. Figure 7.1 shows a drone flight log file (FLY003.DAT) opened in Hex Fiend, a hex editor commonly used to view and analyze binary files. The tool presents the file in three main sections: the offset column, the hexadecimal data column, and the ASCII column. The offset column (on the far left) represents the position of each row in the file, measured in bytes, in hexadecimal format. The hexadecimal column (middle section) contains the raw byte values of the file, displayed as pairs of hexadecimal digits. The ASCII column (on the right) attempts to translate the hexadecimal data into human-readable characters where possible, with non-printable characters shown as dots.

.DAT files typically contain telemetry and sensor data, which are used for reconstructing the flight path and understanding the drone's behavior. These logs may include timestamps, GPS coordinates, altitude, speed, inertial measurements, battery information, and user or autopilot control inputs. The ASCII portion of the hex view often provides metadata or readable strings, such as firmware build information or other text fields embedded in the file. For instance, the string BUILD Apr 8 2016 could indicate the firmware version or the date it was compiled.

We describe the structure of drone flight logs based on the work of Clark et al. [11]. The .DAT file format is structured to store a variety of telemetry and operational data recorded during the drone's flight. At the beginning of the file

7.3 Format of Drone Flight Logs

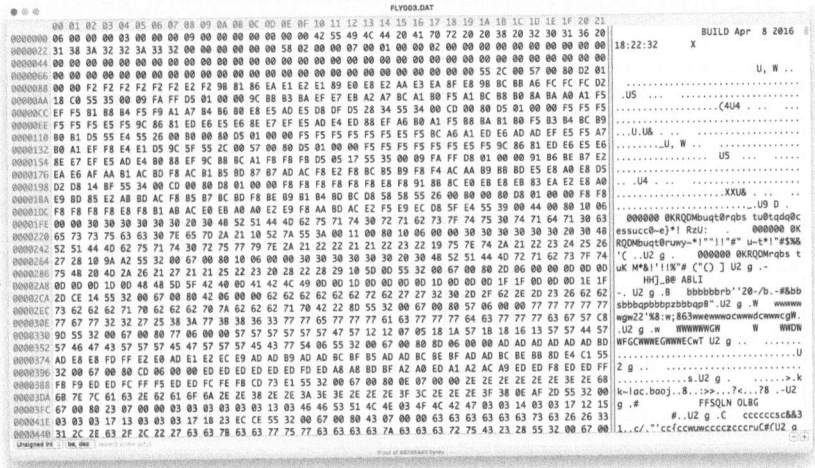

Fig. 7.1 Hex view of a DAT file

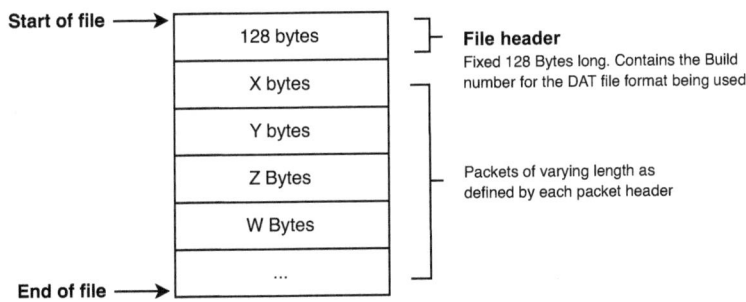

Fig. 7.2 DAT file structure [11]

is a fixed-length header, typically 128 bytes in size, as presented in Fig. 7.2. This header contains metadata, such as the build number of the firmware or software that generated the file. This information provides context about the version and format of the .DAT file and can be useful for decryption and parsing.

Following the header, the file is composed of packets of varying lengths. Each packet begins with a header that specifies its length and may also include additional metadata, such as timestamps or type identifiers. The packets themselves store the telemetry and sensor data captured during flight, including GPS information (e.g., latitude, longitude, altitude, and speed), IMU data (e.g., acceleration, gyroscope readings, and orientation), battery details (e.g., voltage, current, and remaining capacity), and system status messages (e.g., error logs or operational modes such as takeoff, hovering, or landing). The design of the .DAT file allows for variable-length packets to make the format adaptable and space-efficient. This flexibility enables the

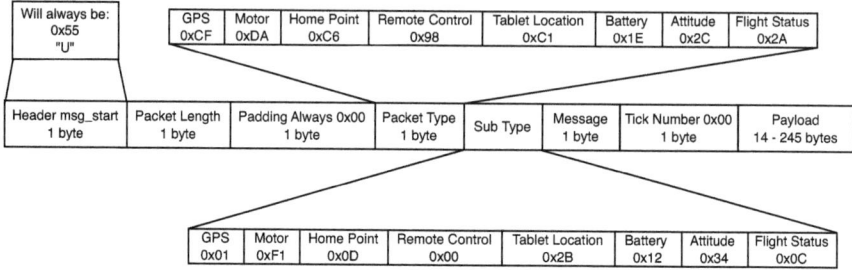

Fig. 7.3 The breakdown of a DAT file structure [11]

file to accommodate different types of data, depending on the drone's capabilities and the recording settings. The file concludes when no more packets are recorded.

The .DAT file is structured to store telemetry and operational data in an organized format. Each file is composed of multiple packets and each packet is further broken down into fields that provide specific pieces of information as displayed in Fig. 7.3. The packet structure begins with a header field (also called msg_start), represented by a single byte with a fixed value of 0×55 (the ASCII character "U"). This header acts as a synchronization marker, indicating the start of a new packet. Following the header is a padding field, another single byte with a constant value of 0×00, which serves as a spacer to maintain consistency and alignment.

The next field, packet type, is a single byte that determines the category of data contained within the packet. The type can indicate telemetry, such as GPS data, motor performance, or battery status. This is followed by the message field, which is also one byte in size and may provide additional metadata or further classify the data. The packet length field, another single byte, specifies the total size of the packet, including the payload. This field allows for efficient navigation between packets during parsing. The packet also includes a tick number field, typically set to 0×00, which might represent a counter or be used for time synchronization.

The most dynamic part of the packet is the payload, which can range in size from 14 to 245 bytes. The payload stores the actual telemetry or operational data and varies depending on the packet type. Subtypes within the payload include information such as GPS coordinates, motor performance data, remote control inputs, battery metrics, and drone attitude (orientation and angular measurements). For instance, GPS packets may log latitude, longitude, altitude, and speed, while battery packets record voltage, current, and capacity.

The .txt binary file is another structured format designed to store detailed telemetry and operational data (Fig. 7.4). Despite its .txt extension, which typically indicates a plain text file, this format is binary, to store datasets generated during a drone's operation. The file begins with a padding section which is filled entirely with 0×00 bytes. The file includes a file length field, which is 4 bytes long and specifies the total size of the file in bytes. This field, denoted as N, provides information for determining the boundaries of the file. Following the file length is a zero-filled section, also filled with 0×00 bytes, which acts as an additional spacing or

7.3 Format of Drone Flight Logs

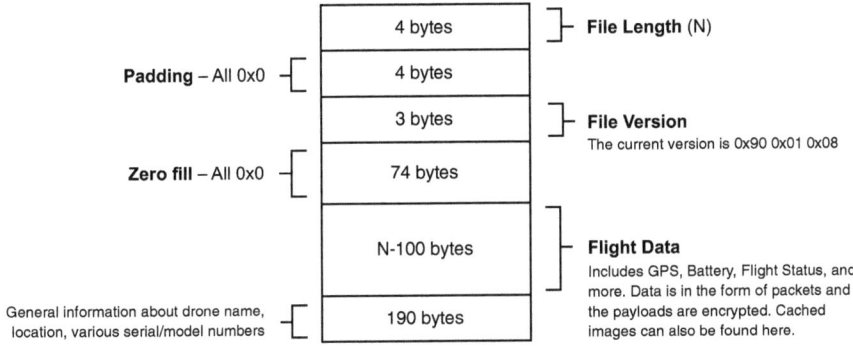

Fig. 7.4 TXT file structure [11]

a placeholder for future metadata. These initial fields serve to organize the file and prepare it for efficient data access.

A file version field, 3 bytes in size, identifies the format version of the .txt binary file. For instance, the version might be represented as 0×90 0×01 0×08. This information is needed for compatibility, as different versions may store data in slightly different formats. Parsing tools must recognize the version to correctly interpret the data structure and access the stored information. The file also contains a general information section, typically 74 bytes long, which includes metadata about the drone, such as its name, location, and serial or model numbers. This section provides context for the flight log so that analysts can identify the specific drone and its configuration during the logged session.

The core of the .txt binary file is the flight data section, which spans N - 100 bytes. This section is composed of packets that store telemetry and operational data. Each packet may include GPS coordinates, battery metrics, flight status information, and other performance indicators such as motor data and attitude. In particular, the payloads in these packets are encrypted to protect sensitive information. Decrypting these data requires manufacturer-provided tools or reverse engineering. In addition to telemetry data, the file may include cached images, which are stored alongside the flight data. These images could be used for visual recording, navigation, or additional analysis.

The .txt binary file structure for drone flight logs is designed to store and organize telemetry and operational data. Each file comprises multiple packets, each packet broken down into distinct fields that enable parsing. As shown in Fig. 7.5, the structure begins with a packet-type field, which is 1 byte long. This field identifies the kind of data the packet contains, such as GPS information, battery metrics, flight status, or other telemetry. Following the packet type is the payload length field, another 1 byte. This field specifies the size of the payload. We can use this information to parse the number of bytes to read in the next section. By clearly defining the payload length, the format supports variable-length packets.

The payload forms the core of the packet that contains the actual telemetry and operational data. The size of the payload varies, as determined by the payload length

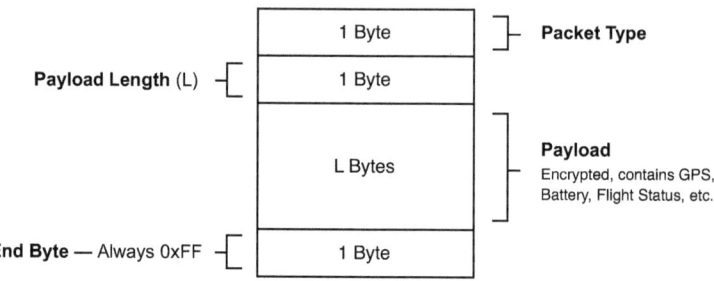

Fig. 7.5 TXT file structure breakdown [11]

field. Note that these data are also encrypted. The payload may include details such as GPS coordinates, altitude, speed, battery status, and system messages. Each packet ends with an end byte, set to a constant value of 0xFF. This consistent marker signals the end of a packet, which can help maintain synchronization during parsing. Even if a portion of the file is corrupted, the clear boundaries provided by the end byte allow analysts to recover and process the remaining data.

7.4 DatCon: A Tool for Drone Flight Log Analysis

DatCon[1] is a software application specifically designed to read and parse DAT files. These files were produced by drones, most notably those of DJI, a leading drone manufacturer. DAT files are proprietary in nature and contain detailed flight data logs that include GPS positioning, altitude, speed, battery status, control inputs, and much more. However, the proprietary format of these files often makes them inaccessible for analysis without specialized tools. DatCon addresses this challenge by converting the DAT files into more user-friendly formats, specifically CSV (comma-separated values); thus supporting straightforward analysis using widely utilized software such as Excel or advanced data analysis systems.

The main strength of DatCon lies in its ability to extract a range of data from the drone's DAT files. Investigators can obtain in-depth insight into the drone's flight behaviors and operational parameters. This detailed extraction is invaluable for forensic investigations, where it helps reconstruct flight paths, analyze operational anomalies, and understand the conditions that led to an incident. Based on its official documentation, DatCon reads DAT files from Phantom 3, Phantom 4, Phantom 4 Pro, Inspire 1, and Mavic Pro and produces output files that can then be used by CsvView, Excel, Dashware, and Google Earth. Despite the complexity of the data it processes, DatCon is designed with a user-friendly interface. Therefore, it is

---

[1] https://datfile.net/DatCon/intro.html.

7.4 DatCon: A Tool for Drone Flight Log Analysis

accessible to people without extensive technical knowledge of drone operations or data analysis.

The DatCon utility has several practical applications. In accident investigations, it can be a tool for determining causative factors by providing a detailed account of the drone's flight path and operational status prior to an incident. Security professionals also benefit from DatCon when investigating unauthorized or suspicious drone activities, as it allows them to track the drone's flight path and behaviors accurately.

To install DatCon on a Windows system, follow the detailed steps below:

1. Download DatCon

    Visit the DatCon Downloads page.[2] Choose the appropriate version of DatCon that suits our needs and click on the download link to save the file to our computer.

2. Unzip the downloaded file

    The downloaded file will be in a compressed .zip format. Extract the contents of the .zip file into a temporary directory on our system. We can do this by right-clicking the file and selecting "Extract All" or using any compatible file extraction software.

3. Install DatCon

    In the extracted files, locate the `DatConSetup` executable file. Double-click on `DatConSetup` to begin the installation process. Follow the on-screen instructions to complete the installation. We may be prompted to choose an installation directory and accept the license agreement.

4. Start and use DatCon

    After the installation is complete, we can launch DatCon by double-clicking its desktop icon or by finding it in the Start Menu. If we are new to DatCon, it is recommended to familiarize with the program by accessing the user manual. The manual is available via the Help menu within DatCon and provide detailed instructions and useful tips on how to use the software effectively.

By following these steps, we will have DatCon installed and ready for use on our Windows computer. To install DatCon on a macOS system, follow the detailed steps below:

1. Download and unzip the file

    Visit the DatCon Downloads (see Footnote 2) page and download the appropriate version of DatCon for macOS. The downloaded file will be in compressed .zip format. Extract the contents of the .zip file to a directory on our computer.

2. Run DatCon using the terminal

    Open the Terminal application and navigate to the directory where we extracted the DatCon files. Run the following command:

    ```
    java -jar DatCon.X.Y.Z.jar
    ```

---

[2] https://datfile.net/DatCon/downloads.html.

Replace X.Y.Z with the actual version number of the DatCon file we downloaded, such as 4.3.0. This command will start the DatCon application.
3. Verify Java installation

   Ensure that we have a 64-bit version of Java installed on our system. DatCon requires 64-bit Java to function correctly. We can verify if Java is installed by running:

   ```
   java -version
   ```

   If Java is not installed or if we need to update to a 64-bit version, visit the Eclipse Temurin download page[3] to install the latest version. Note that Temurin is one of the Java Development Kit (JDK) alternatives.
4. Create a shortcut (optional step)

   To make launching DatCon easier, we can create a shortcut or alias for the command. This will allow us to start DatCon from any directory without having to navigate to the installation folder each time. We can add an alias to our shell configuration file (e.g., .bashrc or .zshrc) as follows:

   ```
   alias datcon='java -jar /path/to/DatCon.X.Y.Z.jar'
   ```

   Replace /path/to/DatCon.X.Y.Z.jar with the full path to the extracted DatCon file. After adding the alias, we can start DatCon by simply typing datcon in the Terminal.

By following these steps, we will have DatCon installed and ready for use on our macOS system. Figure 7.6 displays the DatCon tool (version 4.3.0), which is used to analyze and convert drone flight log files, specifically .DAT files. This tool enables the transformation of raw flight data into more accessible formats such as CSV, KML, and TXT. It also assists in a detailed investigation of the drone's flight path and behavior.

In the DatCon interface as shown in Fig. 7.6, the ".DAT file" section shows the file path to the raw drone flight log namely FLY002.DAT. It is located in the directory /Users/hudan/git/DROP/sample-data/DAT-Files/. The "Output Dir" specifies where the converted files will be saved, which is set to the same directory. The "Time Axis" section offers control over the time range for data conversion, based on specific drone events. These events include "Recording Start", "Motor Start", "Flight Start", "Motor Stop", "Recording Stop", and "GPS Lock". By selecting an event, the user can focus the data extraction on particular moments in the flight. The fields for "Lower" and "Upper" indicate the time range in seconds, while "TickNo" shows the tick numbers that correspond to these times.

In the "CSV" section, users can convert .DAT data to the CSV format, with the option to set a sample rate, such as 30 Hz. This section provides the file name for the CSV output (FLY002.csv), which can be viewed after conversion in spreadsheet software such as Excel (Fig. 7.7). The "Log Files" section enables the generation of text-based log files, including an "Event Log File", a "Config Log

---

[3] https://adoptium.net/temurin/releases/.

7.4 DatCon: A Tool for Drone Flight Log Analysis

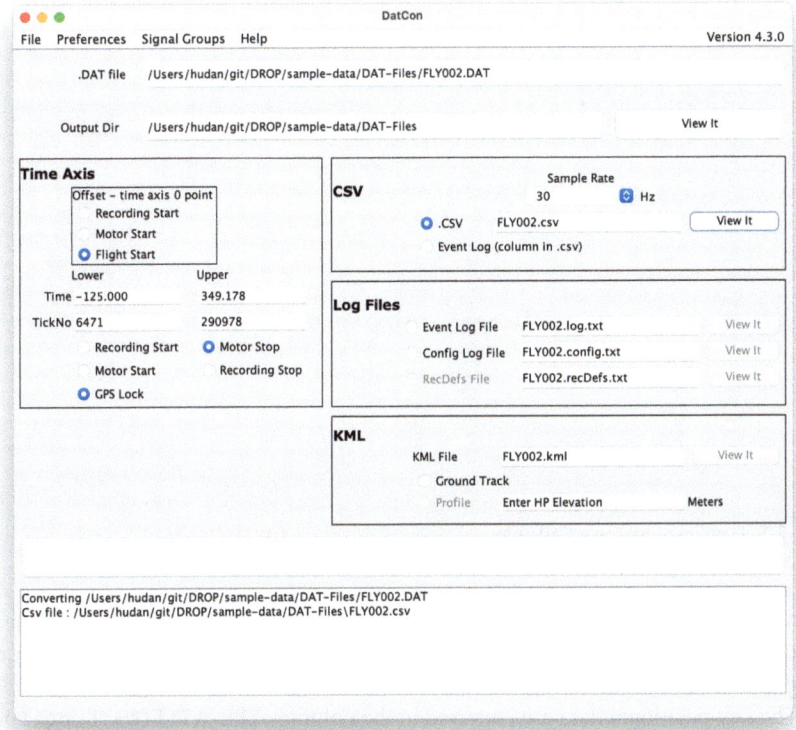

Fig. 7.6 DatCon reads and parses FLY0002.DAT

File", and a "RecDefs File", which contains definitions for the .DAT data structure. Each log file serves a different purpose: event logs capture specific occurrences, configuration logs store configuration details, and the "RecDefs" file defines data record structures.

The "KML" section allows users to create KML files, a format compatible with platforms such as Google Earth, to visualize the flight path of the drone. Options include "Ground Track" and "Profile", with the ability to specify the HP (Home Position) elevation if needed. At the bottom of the interface, a status message indicates the conversion progress, showing the paths for the source .DAT file and the output CSV file.

Figure 7.7 presents the DatCon CSV output file. This CSV file represents data extracted from a drone flight log, generated by DatCon. Each row corresponds to a specific time point or event during the flight, and each column provides detailed measurements of the drone's performance and navigation. The `Clock:Tick#` column serves as a sequential identifier and represents the tick count of the drone's data logger. It acts as an internal counter that tracks events recorded in the log.

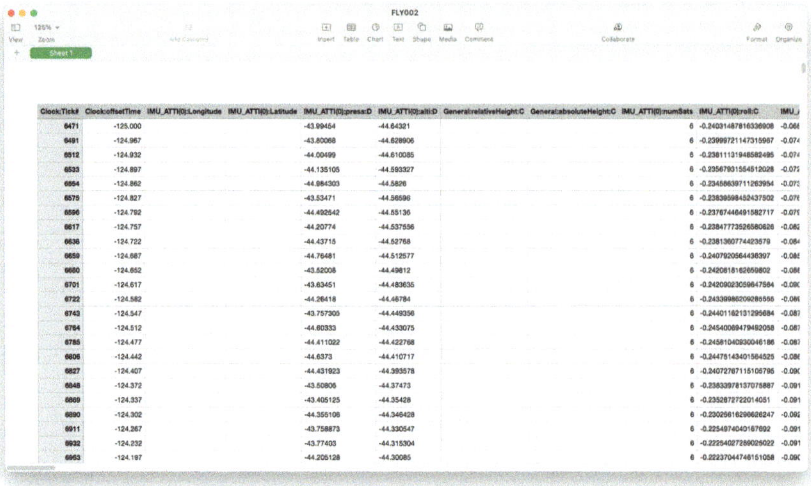

Clock:Tick#	Clock:offsetTime	IMU_ATTI(0):Longitude	IMU_ATTI(0):Latitude	IMU_ATTI(0):press:D	IMU_ATTI(0):alti:D	General:relativeHeight:C	General:absoluteHeight:C	IMU_ATTI(0):numSats	IMU_ATTI(0):roll:C	IMU_
6471	-125.000			-43.99454	-44.64321			6	-0.24031487816336908	-0.06
6491	-124.967			-43.80068	-44.628906			6	-0.23999721147315967	-0.074
6512	-124.932			-44.00499	-44.610085			6	-0.23811131948582495	-0.074
6533	-124.897			-44.135105	-44.593327			6	-0.23567931554512028	-0.072
6554	-124.862			-44.984303	-44.5826			6	-0.23458639711263954	-0.073
6575	-124.827			-43.53471	-44.56596			6	-0.23839598452437502	-0.076
6596	-124.792			-44.492542	-44.55136			6	-0.23767446491582717	-0.079
6617	-124.757			-44.20774	-44.537556			6	-0.23847773526580626	-0.082
6638	-124.722			-44.43715	-44.52768			6	-0.2381360774423579	-0.084
6659	-124.687			-44.76481	-44.512577			6	-0.2407920564436397	-0.085
6680	-124.652			-43.52008	-44.49812			6	-0.2420818162659802	-0.086
6701	-124.617			-43.63451	-44.483635			6	-0.24209023059647564	-0.090
6722	-124.582			-44.26418	-44.46784			6	-0.24339986209285555	-0.089
6743	-124.547			-43.757305	-44.449356			6	-0.24401162131295684	-0.087
6764	-124.512			-44.60333	-44.433075			6	-0.24540069479492058	-0.087
6785	-124.477			-44.411022	-44.422768			6	-0.24581040930046186	-0.087
6806	-124.442			-44.6373	-44.410717			6	-0.24476143401564525	-0.086
6827	-124.407			-44.431923	-44.393578			6	-0.24072767115105795	-0.090
6848	-124.372			-43.50806	-44.37473			6	-0.23833978137075887	-0.091
6869	-124.337			-43.405125	-44.35428			6	-0.2352872722014051	-0.091
6890	-124.302			-44.355106	-44.346428			6	-0.23025616296626247	-0.092
6911	-124.267			-43.758873	-44.330547			6	-0.2254974040167692	-0.091
6932	-124.232			-43.77403	-44.315304			6	-0.22254027289025022	-0.091
6953	-124.197			-44.205128	-44.30085			6	-0.22237044746151058	-0.090

Fig. 7.7 DatCon output in a CSV file

Complementing this, the `Clock:offsetTime` column provides the relative time (in seconds) from a reference point, such as the start of the flight. This parameter can help to align events chronologically.

The key positional data are captured in the columns `IMU_ATTI(0):Longitude`, `IMU_ATTI(0):Latitude`, and `IMU_ATTI(0):Alt:D`. These columns record the drone's geographic location and altitude during the flight. The longitude and latitude reflect the position of the drone on Earth, while altitude (`Alt:D`) specifies its height. Subsequently, the `General:relativeHeight:C` column measures the drone's height relative to its take-off point, which is crucial for analyzing its vertical movements. The `General:absoluteHeight:C` column records the absolute altitude above sea level.

The drone's connection to GPS satellites is tracked in the `IMU_ATTI(0):num Sats` column, which records the number of satellites connected at any given moment. A higher satellite count typically indicates a stronger GPS signal. Finally, drone orientation and stability are detailed in the `IMU_ATTI(0):roll:C`, `pitch:C`, and `yaw:C` columns. These represent the drone's rotation around its three axes: roll (side-to-side tilting), pitch (front-to-back tilting), and yaw (rotational heading). These measurements provide information on how the drone maintained balance and executed movements during flight.

7.5 CsvView: Viewer of DatCon Results

CsvView[4] emerged as a solution to the challenges of visualizing and interpreting the CSV files produced by DatCon. While DatCon can transforms DAT files into CSV formats, which are easier to manage and analyze using standard spreadsheet software, the volume and complexity of the data captured in these CSV files can still pose challenges. Users found that despite conversion, making sense of the detailed flight data and extracting actionable insights required a more intuitive approach to data visualization and analysis.

Recognizing this gap, CsvView was developed to offer a user-friendly interface specifically designed for the detailed examination of drone flight data encoded in CSV files. This software enables users to view and interpret complex flight information through a more accessible and visually intuitive platform. CsvView provides visualization of flight paths, altitude profiles, speed variations, and other data points through graphical representations. Therefore, CsvView makes it easier for users to understand the drone's behavior and operational parameters at a glance. By offering tools for plotting, analyzing, and comparing flight data, CsvView allows for a deeper dive into drone flight. The software's ability to visually map out flight paths and patterns also aids in identifying anomalies or deviations from expected behavior for forensic analyses and investigative contexts.

In essence, CsvView addresses the need for an easier interaction with drone flight data. It complements DatCon by improving the accessibility and interpretability of the data it converts. This synergy between DatCon and CsvView provides a toolkit for anyone looking to analyze the specifics of drone flight data. To install CsvView on a Windows system, we follow the detailed steps below:

1. Download the installer

    Visit the CsvView downloads page.[5] Select the appropriate version of CsvView for Windows and click the download link to save the installer file (.exe) to the forensic workstation.

2. Run the installer

    Locate the downloaded .exe file in our Downloads folder or the directory where we saved it. Double-click on the installer file to start the installation process. Follow the on-screen instructions to proceed. We will be prompted to choose an installation directory, agree to the software license terms, and confirm any additional settings or components that we want to install.

3. Complete the installation

    Once we have followed all the installation steps, the process will finish and CsvView will be installed on our computer. The installer may also prompt we to add CsvView to our Start Menu or create a desktop shortcut for easier access.

---

[4] https://datfile.net/CsvView/intro.html.

[5] https://datfile.net/CsvView/downloads.html.

4. Launch CsvView

After the installation is complete, we can start CsvView by clicking on its icon in the Start Menu or by using the desktop shortcut if we choose to create one. Upon launching, CsvView will be ready for use and we can begin importing and analyzing CSV files.

5. Optional step: Check for updates

It is advisable to check for updates periodically to check that we are using the latest version of CsvView with the most recent features and bug fixes. We can check for updates directly from within the software by accessing the Help menu or visiting the CsvView Downloads page for manual updates.

By following these steps, we will have CsvView installed and ready to use on our Windows computer. On the other hand, to install CsvView on a macOS system, we can follow the detailed steps below:

1. Download the .zip file

Visit the CsvView Downloads page and download the latest .zip file for macOS. This file contains the installer package for CsvView. At the time of writing, the latest version is 4.3.0.

2. Open the .zip file

After the download is complete, locate the .zip file in our Downloads folder or the directory where it was saved.

```
cd ~/Downloads/CsvViewMac.4.3.0/
```

3. Run CsvView

To run CsvView, type this command in the terminal:

```
java -jar CsvView.4.3.0.jar
```

4. Optional step: Add CsvView to the Dock

For easier access, we may want to add CsvView to our Dock. To do this, locate CsvView in the Applications folder, click and hold the icon, then drag it to the Dock. This will create a shortcut that we can use to quickly launch the application in the future.

Figure 7.8 displays the CsvView tool, which is used to examine, visualize, and interpret CSV files generated by the DatCon tool. This feature helps reconstruct the flight behavior of the drone and identify key metrics related to its performance. At the top of the interface, the .DAT, .csv, or .tsv field shows the path of the file to the CSV file processed from the original .DAT data, located at /Users/hudan/git/DROP/sample-data/DAT-Files/FLY002.csv. The Source Type field is marked as DATGENERIC, which denotes a generic data format originating from the .DAT file. The tool suggests a standardized structure for this type of drone data.

The "SigPlayers" (Signal Players) section provides a list of selectable data categories, such as "Motor Speeds," "LeftFront Motor," "MagMod/Compass Error," and "navHealth". These options allow the investigator to focus on specific aspects of the drone's telemetry and performance, such as motor speeds or navigation health.

7.5 CsvView: Viewer of DatCon Results

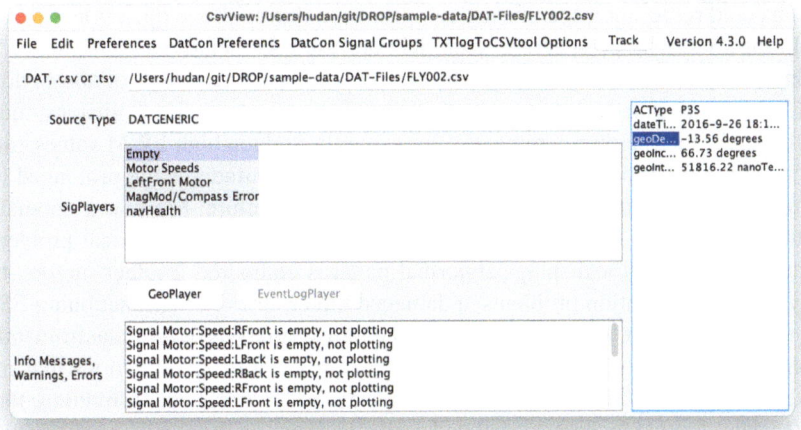

Fig. 7.8 CsvView opens the parsing result FLY002.csv

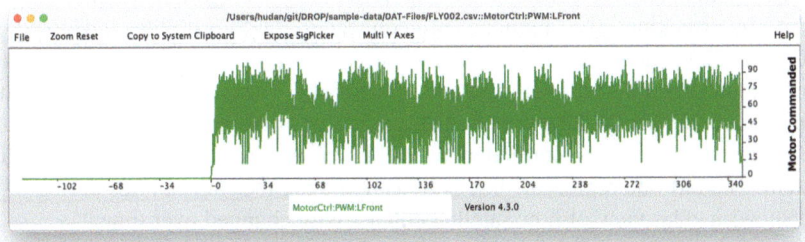

Fig. 7.9 Left front motor visualization in CsvView

There is also an "Empty" option, which indicates that in some cases no data are currently available for display.

The visualization presented in Fig. 7.9 is a plot of the left front motor commanded PWM signal (pulse width modulation) from a drone's flight log. It represents how the drone's electronic control system commanded the left front motor throughout the flight. The horizontal axis indicates the time of the flight progression in terms of sequential log data. The vertical axis represents the motor's commanded PWM values and reflects the amount of electrical signal sent to control the motor speed. PWM values are used to regulate motor speed, where higher values indicate increased motor power output. The graph reveals fluctuations in the motor power demand as the drone executes various flight maneuvers, such as altitude changes, rotations (yaw, roll, and pitch), or stability corrections. The jagged pattern in the green plot line is typical of drones actively balancing their flight. It is a response to environmental conditions, such as turbulence, sudden control inputs, or internal adjustments from the flight controller to maintain a smooth flight path. Each fluctuation on the graph could represent a specific action or reaction during drone operation.

From a forensic perspective, such data are used for behavioral analysis of the drone during an incident. Investigators can pinpoint irregular spikes or abrupt drops in PWM values that could correspond to significant events such as collisions, motor stalls, or abrupt operator commands. For example, sudden high PWM values could indicate that the drone struggled to gain or maintain altitude, while prolonged low values may signal motor disengagement or a power failure. Moreover, anomalies in the PWM graph can also point to potential mechanical or electrical problems. If the left front motor displays abnormal patterns compared to other motors, this may suggest calibration problems, a damaged motor, or even malfunctioning ESCs (Electronic Speed Controllers). This comparative analysis with data from other motors is needed to identify systemic failures or pinpoint malfunctions that may have contributed to flight anomalies or crashes. For incident reconstruction, these PWM data should ideally be correlated with other logs, such as GPS, accelerometer, or gyroscope data, to obtain a full picture of the drone's movements and operating conditions. By integrating these data, forensic investigators can establish a detailed timeline of flight behavior and determine the root causes of incidents.

The visualization represents in Fig. 7.10 the "Navigation Health" of the drone by tracking the number of connected GPS satellites, denoted as IMU_ATTI(0):numSats. This metric is used for assessing the drone's ability to maintain accurate positional data and stable flight. The x-axis represents the time or the sequence of log entries, while the y-axis shows the number of GPS satellites used, which is typically correlated with the accuracy of the drone's navigation system. The red line represents fluctuations in the number of satellites during flight and shows a clear view of how satellite connectivity changed over time.

From a forensic perspective, these data can support understanding of the drone's positional accuracy. A higher number of satellites indicates reliable GPS positioning, while drops in satellite count can suggest potential navigation challenges. For instance, sudden declines in connectivity could lead to reliance on inertial navigation, which is less accurate and might result in deviations from the planned

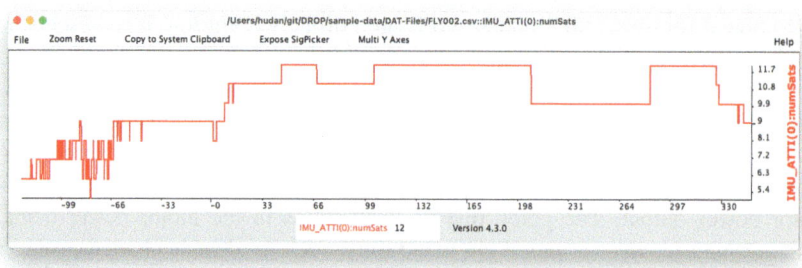

Fig. 7.10 Navigation health visualization in CsvView

flight path. This can help explain incidents such as erratic movements, unintended trajectory changes, or crashes. Visualization also helps identify potential environmental factors that affect navigation. Drops in satellite connectivity might indicate that the drone was flying in GPS-denied zones, such as urban environments with tall buildings, dense foliage, or areas with significant electromagnetic interference.

Furthermore, erratic changes in the number of connected satellites may point to hardware or firmware issues. Problems with the GPS module, antenna, or software could lead to unreliable or intermittent satellite connections. Such issues might warrant further investigation of the drone's components to determine whether they played a role in any anomalous behavior. These data can be used to verify compliance with flight regulations. Low satellite connectivity could result in the drone straying from pre-defined flight paths or geofenced areas. For forensic investigators, this information is crucial for determining whether the drone operated within its intended boundaries or violated any operational restrictions.

In the middle section, the "GeoPlayer" button offers additional visualization or playback options for geographic data and event logs, as shown in Fig. 7.11. Using this feature, investigators is able to review and understand the drone's flight path and events logged during its operation. On the other hand, "EventLogPlayer" button appears, but it is inactive in this screenshot.

At the bottom of the interface of Fig. 7.8, the "Info Messages, Warnings, Errors" section displays messages related to the data being analyzed. Here, it notes that several motor speed signals (for example, RFront, LFront, LBack, and RBack) are empty and therefore not plotted. This information may indicate missing or unavailable data for these motors, which could be used to investigate flight stability problems or mechanical failures during flight.

On the right side of Fig. 7.8, the data panel provides specific details such as drone type (ACType: P3S) and several geolocation metrics. These metrics include the date and time of log entry (2016-9-26 18:1 ...), geoDeclination ($-13.56°$), geoInclination ($66.73°$), and geomagnetic intensity (51816.22 nanoTesla). These geolocation data points are valuable for forensic analysis, as they help pinpoint the drone's location, orientation, and environmental conditions during the flight.

7.6 DROP: DRone Open source Parser

DROP,[6] which stands for Drone Open Source Parser, is a tool designed to parse and analyze data from drone flight logs [11]. As detailed in the paper published in the Digital Investigation journal, DROP aims to address the challenges associated with the proprietary nature of drone data logs, which can be a barrier to forensic investigations. The proprietary formats used by many drone manufacturers limit the

---

[6] https://github.com/BiTLab-BaggiliTruthLab/DROP.

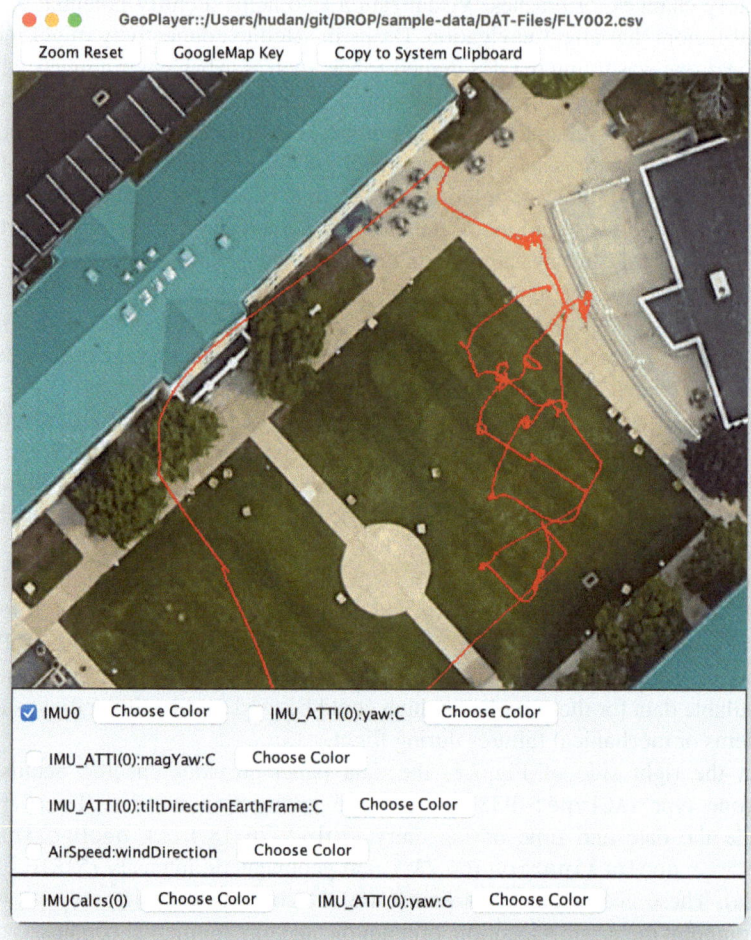

Fig. 7.11 GeoPlayer in CsvView

ability of investigators to access and analyze flight data, including reconstruction of drone flight paths, understanding drone behavior, and investigating incidents involving drones.

The development of DROP as an open source tool allows investigators, researchers, and students to access a resource for the drone data analysis process. By providing the means to parse drone flight logs, DROP facilitates the extraction of valuable data such as GPS coordinates, flight times, speeds, altitudes, and operator inputs, among others. One of the key benefits of DROP is its open-source nature, which encourages collaboration among developers, forensic investigators,

7.6 DROP: DRone Open source Parser

and the wider community. As an open-source tool, DROP is an accessible option for individuals and organizations who may not have the resources to invest in proprietary software.

To install and use the DROP tool from the GitHub repository, we can follow these steps:

1. Clone the repository and go to the cloned directory

   ```
   git clone https://github.com/BiTLab-BaggiliTruthLab/DROP.git
   cd DROP
   ```

2. Create Anaconda virtual environment and activate it

   ```
   conda create --name drop python=3.10
   conda activate drop
   ```

3. Install dependencies
   Ensure we have Python installed (version 3.6 or later) as performed in the previous step. Install the required dependencies:

   ```
   pip install -r requirements.txt
   ```

4. Run the DROP tool
   Use the provided DROP.py script to parse DJI drone log files:

   ```
   python DROP.py --input /path/to/our/logfile.DAT --output
     ↪ /path/to/output/directory
   ```

   Replace the directory /path/to/our/logfile.DAT with the path to our .DAT file and /path/to/output/directory with the directory where we want to save the output CSV file.

5. For detailed usage and additional options, refer to the repository's README file or run the following command as shown in Fig. 7.12:

   ```
   python DROP.py --help
   ```

Figure 7.13 shows the output from running the DROP tool on a drone flight log file, specifically the file FLY003.DAT. The purpose of this tool is to analyze flight data for forensic examination. The analysis began on 2024-11-09 at 17:31:11, using the command python DROP.py sample-data/DAT-Files/FLY003.DAT, which specifies the DROP script and the input file being analyzed. The tool processes the flight log file located at sample-data/DAT-Files/FLY003.DAT, and the results are saved in an output file named FLY003-Output.csv. The size of the input file is approximately 6.979 MB.

Before and after processing the file, hash digests (MD5, SHA1, and SHA512) are calculated to verify that the data remains unaltered throughout the analysis. The initial MD5 hash is e70428f...16046ea, the SHA1 hash is 3aa260df...dccb33a, and the SHA512 hash is 06a0d6a2...85fc8616. These values match the corresponding "After" hashes, which confirms the integrity of the data during processing. During the analysis of the DAT file, several packet length errors were detected. At byte

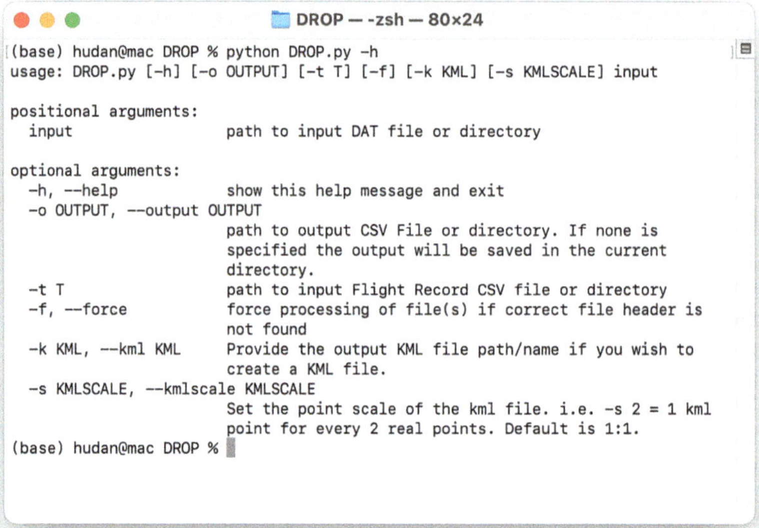

```
... DROP — -zsh — 80×24
(base) hudan@mac DROP % python DROP.py -h
usage: DROP.py [-h] [-o OUTPUT] [-t T] [-f] [-k KML] [-s KMLSCALE] input

positional arguments:
  input                 path to input DAT file or directory

optional arguments:
  -h, --help            show this help message and exit
  -o OUTPUT, --output OUTPUT
                        path to output CSV File or directory. If none is
                        specified the output will be saved in the current
                        directory.
  -t T                  path to input Flight Record CSV file or directory
  -f, --force           force processing of file(s) if correct file header is
                        not found
  -k KML, --kml KML     Provide the output KML file path/name if you wish to
                        create a KML file.
  -s KMLSCALE, --kmlscale KMLSCALE
                        Set the point scale of the kml file. i.e. -s 2 = 1 kml
                        point for every 2 real points. Default is 1:1.
(base) hudan@mac DROP %
```

Fig. 7.12 DROP help command output

13,127,534, the tool expected the start of a packet (0×55) but found a different byte ("u"). Another error occurred in byte 67,432,401, where it expected 0×55 but found "v". Similarly, in byte 69,791,579, it expected 0×55 but encountered "p". These errors could suggest corrupted data, missing packets, or potential issues with how the data was recorded.

The analysis was completed at 17:31:37, with a processing time of only 26.8 seconds. The tool displayed "Processing complete," and it indicates that DROP finished successfully without any fatal errors. This output provides forensic analysts with an initial verification of data integrity and highlights anomalies, such as packet errors, that may require further investigation. Figure 7.14 displays a spreadsheet output file generated by the DROP tool. It contains data parsed from a drone flight log file named FLY003.DAT. The output includes several columns representing different types of data recorded by the drone during its flight.

The first section of the spreadsheet contains timing information, including messageid, which assigns a unique identifier to each message or data packet in the log. The offsetTime and logDateTime columns record time offsets or timestamps for each message, possibly in seconds or milliseconds. The time (millisecond) column offers a high-resolution timestamp in milliseconds for each entry. The location and GPS data are represented in columns such as latitude, longitude, satnum, and gpsHealth. These columns should contain GPS coordinates and details about the number of satellites and the quality of the GPS signal. However, in

7.6 DROP: DRone Open source Parser

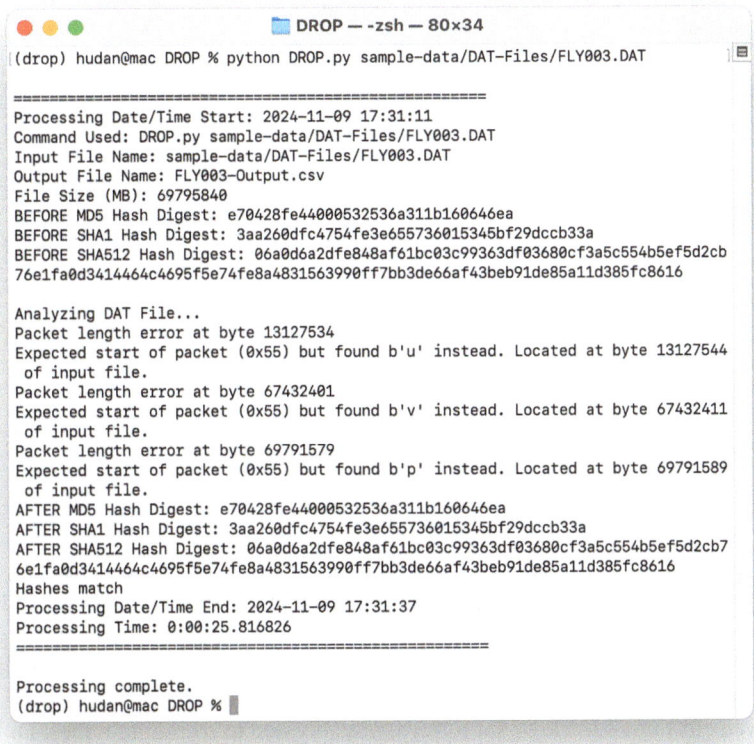

Fig. 7.13 DROP parses a flight log namely FLY003.DAT

this output, these values are all set to zero which indicate missing data or problems in GPS recording.

The spreadsheet also includes altitude and barometric information in `altitude` and `baroAlt` columns, which would typically represent the height of the drone based on GPS and barometric sensors. The acceleration and gyroscope data are recorded in columns labeled `accelX`, `accelY`, `accelZ` (acceleration along the X, Y, Z axes) and `gyroX`, `gyroY`, `gyroZ` (gyroscope readings for rotation along each axis). In this output, these values are all zero, which suggests either a lack of recorded data or an issue with capturing this sensor information during the flight.

Furthermore, the spreadsheet contains `errorX`, `errorY`, `errorZ` columns, which represent calibration or error metrics for each axis, and `magX`, `magY`, `magZ`, and `magMod` columns, which represent magnetic field measurements along each axis and a calculated magnitude. Unlike other data, the magnetic field readings are populated, which indicates that these sensor data were successfully captured. Finally, velocity data are shown in `velN` and `velE` columns, which represent the North and East

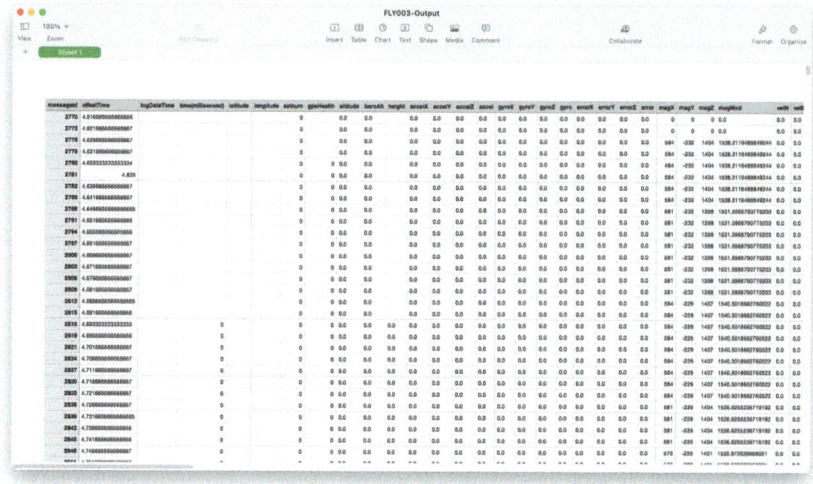

Fig. 7.14 The output of DROP when parsing FLY003.DAT

components of the drone's velocity based on GPS or other sensors. These fields also show zero, revealing that no velocity data are recorded. Overall, the output suggests that while the tool parsed the columns successfully, the log file contained limited or incomplete data, with most sensor readings (GPS, altitude, acceleration, gyroscope, and velocity) unavailable or set to zero, possibly due to recording issues. Only the magnetic field data appear fully captured.

7.7 Visualization of Flight Logs with Maraudrone's Map

Another research [12] proposes an enhanced fork of the DROP, with added features to parse DATv3 files and generate JSON output files.[7] The JSON files produced can then be visualized on a map using Maraudrone's Map for easy analysis of flight data. To run the tool, use the following command:

`python DROP.py -j <DAT_FILE>`

The option `-j` indicates that we want to format the output in JSON. Replace `<DAT_FILE>` with the name of the DAT file we wish to parse. This command will create a JSON output compatible with Maraudrone's Map. Using visualization, we can view and analyze the flight path of the drone and related telemetry visually.

On the other hand, one of the main drawbacks of command-line tools is their lack of built-in visualization capabilities. Without a graphical representation, analyzing

---

[7] https://github.com/dumbledrone/DROP.

7.7 Visualization of Flight Logs with Maraudrone's Map

flight data solely from the command line or a CSV file can be challenging and less intuitive. Interpreting raw numbers, coordinates, or other flight metrics without visualization can make it harder to identify trends or anomalies in the data.

To address this limitation, we present a visualization of the results from the DROP tool using a complementary tool named Maraudrone's Map [12]. Maraudrone's Map is a tool designed to visualize drone flight data and aims at easier understanding and analysis. The tool provides an interactive map that displays the drone's flight path, key data points, and other visual metrics that improve comprehension of the flight data at a glance. The application visualizes DJI drone flight paths overlayed on a map using OpenStreetMap data. Users can examine telemetry data such as latitude, longitude, altitude, speed, and orientation in an interactive interface. A dedicated flight information panel displays details such as drone location, velocity, and flight duration. In addition, the controller status panel visualizes the user joystick inputs, including pitch, roll, yaw, and throttle values. Therefore, investigators can identify the operator's commands and drone response during flight.

One of the distinguishing features of the tool is its ability to detect and highlight anomalies in flight log data. Sudden changes in speed, altitude, or loss of control can be referred for further investigation. The interactive playback controls allow users to navigate the flight logs at various speeds to analyze the flight behavior of the drone in detail. Investigators can also import or export log files to facilitate evidence collection and reporting. Maraudrone's Map is available as an open-source project on the GitHub repository.[8] Using Maraudrone's Map alongside DROP, users can benefit from a comprehensive approach that combines the command-line parsing utilities of DROP with the intuitive and visual insights provided by Maraudrone's Map. The step-by-step installation guide for Maraudrone's Map is provided as follows.

1. Download and install Node.js

   First, download and install Node.js from the official website. Node.js is needed because it includes npm (Node Package Manager), which allows us to install the packages needed for Maraudrone's Map. After downloading, follow the installation instructions based on our operating system (e.g., Windows, macOS, or Linux). To verify the installation, open terminal or command prompt and type:

   ```
   node -v
   npm -v
   ```

   We should see the versions of Node.js and npm printed in the terminal to confirm that the installation was successful.

2. Install Angular CLI

   Maraudrone's Map is built using Angular, which requires the Angular Command Line Interface (CLI) for development tasks. To install Angular CLI

---

[8] https://github.com/dumbledrone/MaraudronesMap.

globally, open the Node.js terminal and run:

```
npm install -g @angular/cli
```

This command installs Angular CLI across our system so that we can use Angular-specific commands such as `ng serve` or `ng build`.

3. Clone the project repository

```
git clone https://github.com/dumbledrone/MaraudronesMap.git
```

4. Install project dependencies

Navigate to the Maraudrone's Map project directory. Once in the project directory, install all the necessary project modules and dependencies specified in the `package.json` file by running:

```
npm install
```

This command downloads and installs all required packages locally within the project. These include Angular packages, libraries, and other tools necessary for running Maraudrone's Map.

5. Run the development server

With the dependencies installed, we can start a local development server by running:

```
NODE_OPTIONS=--openssl-legacy-provider ng serve
```

This command compiles the project and launches a development server at the default address. `-openssl-legacy-provider` is a flag passed to Node.js to instruct it to use the legacy OpenSSL cryptography provider. This is required if the project or its dependencies use cryptographic algorithms that are no longer supported in the latest versions of OpenSSL (which newer Node.js versions use by default). This is often necessary when working with older libraries that rely on deprecated cryptographic methods. We should see a message indicating that the server is running, typically at `http://localhost:4200/`. Open the browser and go to `http://localhost:4200/` to view the Maraudrone's Map application.

Figure 7.15 shows the terminal output where the tool is executed locally using the Angular development server. The Angular CLI has successfully compiled the application and generated the necessary files for the browser. These files include `vendor.js` (5.21 MB), which contains third-party libraries and dependencies, `main.js` (269.95 kB) for the main logic and components of the application, and `polyfills.js` (128.51 kB) for browser compatibility. Furthermore, the `styles.css` file (91.46 kB) includes the application styles, and `runtime.js` (6.63 kB) handles the loading and execution of JavaScript modules. The total size of the generated application bundle is 5.69 MB, which represents all files necessary for the application to run.

The terminal output confirms that the Angular Live Development Server is active and serving the application on `localhost:4200`. This means that the Maraudrone's Map tool can now be accessed in a web browser by navigating to the provided

7.7 Visualization of Flight Logs with Maraudrone's Map 189

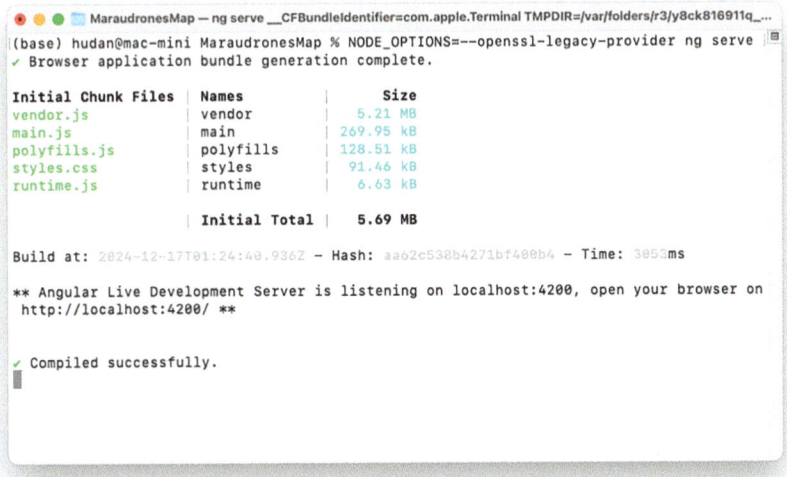

Fig. 7.15 Terminal window when running ng serve

URL, where users can interact with the interface to analyze DJI drone flight logs. The server provides a local environment for development and the build process took approximately 3053 milliseconds (about 3 seconds). A unique hash (aa62c538b4271bf400b4) identifies the current version of the compiled application. The ISO 8601 format-based timestamp (2024-12-17T01:24:40.936Z) shows when the build was generated. Finally, the confirmation message "✓Compiled successfully" indicates that the application was built without issues and is ready to use. This output demonstrates that the tool is now fully operational in a local environment. By navigating to `http://localhost:4200`, users can use the application's features for forensic analysis and visualization of DJI drone telemetry data as depicted in Fig. 7.16.

Figure 7.17 demonstrates the side panel in the Maraudrone's Map tool which serves as a centralized hub for analyzing and visualizing drone telemetry data. It provides essential flight information, telemetry details, controller input, and anomaly detection to support forensic investigations of DJI drone logs. The "Flight Info" section presents basic metadata about the flight, including the drone model (e.g., Phantom 4 Advanced), date and time of operation, total flight duration, and elapsed time in seconds. This information helps to establish the flight's timeline and context.

The "General Info" section provides detailed telemetry data such as the drone's current geographic coordinates (longitude and latitude), altitude, and traveled distance. It also displays the speed of the drone, both horizontal and vertical, along with the percentage of battery and the temperature of the component. We need these details to understand the operational behavior, the trajectory of the movement,

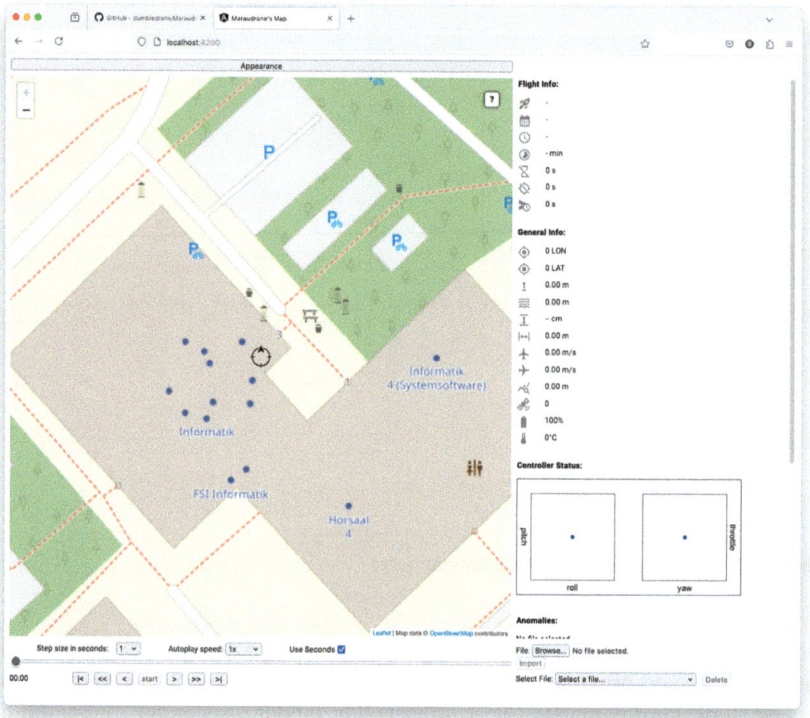

Fig. 7.16 Home page of Maraudrone's Map

and the environmental conditions during flight. Investigators can use this section to analyze flight patterns and identify unusual altitude or velocity changes. The "Controller Status" section visualizes the operator's control input during the flight, including pitch, roll, yaw, and throttle. These parameters reflect how the drone was controlled in terms of tilt (forward, backward, or sideways), rotation, and lift. By observing these joystick movements, investigators can determine whether drone responses matched operator inputs and detect irregular manual control behavior.

The "Anomalies" section highlights inconsistencies or irregularities in the flight data, such as gaps in timestamps (e.g. missing or delayed entries) or unexpected changes in control inputs like rotor speeds. This section is particularly useful for identifying corrupted logs, flight interruptions, or manipulated data, which are common issues in forensic investigations. By pinpointing anomalies, investigators can better understand events that may have caused flight irregularities or failures. Finally, the tool includes file management capabilities, such as importing flight log files in JSON format for analysis and export processed data for reporting. In general, the side panel integrates flight metadata, telemetry visualization, control input, and anomaly detection into one interactive interface.

7.8 DRDP: An Alternative for DatCon and DROP

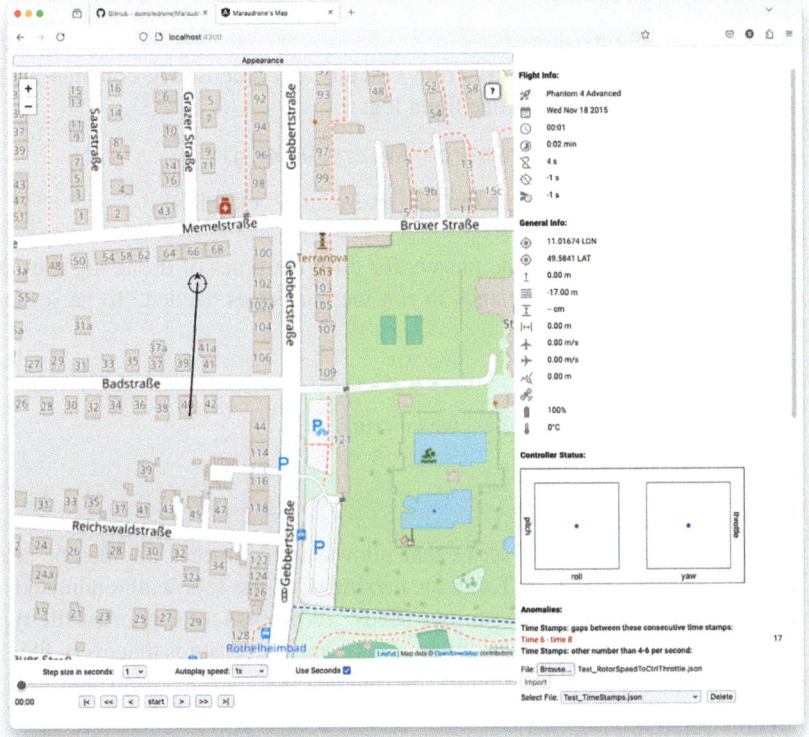

Fig. 7.17 Maraudrone's Map of Test_RotorSpeedToCtrlThrottle.json

7.8 DRDP: An Alternative for DatCon and DROP

This section discusses a step-by-step guide for installing and using the DRDP tool, a drone flight log parser [13]. DRDP is an alternative for DatCon and DROP tool. Based on our experiments, DROP is unable to successfully parse binary TXT drone flight logs. To install and run DRDP, we can follow the steps below.

1. Create an Anaconda environment
   First, ensure that we have Anaconda installed. Open terminal or command prompt and run the following command to create a new Conda environment named drdp with Python version 3.11:

```
conda create --name drdp python=3.11
```

2. Activate the environment
   Once the environment is created, activate it by running this command. This ensures that all installations and commands are executed within the drdp environment.

```
conda activate drdp
```

3. Clone the DRDP repository

    Clone the DRDP repository from GitHub. This repository contains all the necessary files for parsing drone flight logs:

```
git clone https://github.com/EveZzy/DRDP.git
```

After cloning, navigate into the DRDP directory: `cd DRDP`.

4. Run the parsers

    DRDP comes with specific parsers for different types of drone log files. We can use the following commands to run the parsers as needed. To parse .DAT files, run:

```
python dat_parser.py
```

To parse .TXT flight log files, we can run:

```
python txt_parser.py
```

5. Run the main script

    DRDP includes a main script that allows us to process all files or select specific file types (DAT or TXT) for parsing. Use the following commands to run the main script according to our needs. To process all supported file types within the directory, use the -a option. This command ensures that every file that the script supports is included in the parsing process:

```
python main.py -a
```

If the focus is exclusively on .DAT files, the -d option allows investigators to limit parsing to these files only. This option is particularly useful when we want to exclude other file types and target-specific data:

```
python main.py -d
```

Similarly, the -t option enables the script to process only .TXT files. This is ideal for cases where binary text-based flight data needs to be parsed without including other file formats:

```
python main.py -t
```

The results of this parsing process are shown in Tables 7.1 and 7.2 for DAT and TXT files, respectively.

7.9 Digital Drone Forensics Software

In this section, we discuss the Digital Drone Forensics Software tool, which is a specialized application for drone forensics, designed to analyze flight log

Table 7.1 Sample output of DRDP for a DAT file

Time	Longitude	Latitude
2020-07-10 12:31:09	121.4127124	31.1766991
2020-07-10 12:31:09	121.4127132	31.1767014
2020-07-10 12:31:09	121.4127145	31.1767029
2020-07-10 12:31:10	121.4127167	31.1767049
2020-07-10 12:31:10	121.4127170	31.1767063
2020-07-10 12:31:10	121.4127167	31.1767063
2020-07-10 12:31:10	121.4127170	31.1767063
2020-07-10 12:31:11	121.4127171	31.1767063
2020-07-10 12:31:11	121.4127167	31.1767064
2020-07-10 12:31:11	121.4127168	31.1767059
2020-07-10 12:31:11	121.4127164	31.1767055
2020-07-10 12:31:11	121.4127159	31.1767052
2020-07-10 12:31:12	121.4127153	31.1767051
2020-07-10 12:31:12	121.4127153	31.1767050
2020-07-10 12:31:12	121.4127153	31.1767050
2020-07-10 12:31:12	121.4127151	31.1767051
2020-07-10 12:31:14	121.4127120	31.1767051
2020-07-10 12:31:14	121.4127120	31.1767049
2020-07-10 12:31:14	121.4127117	31.1767048
2020-07-10 12:31:14	121.4127116	31.1767047
...	...	...

files from DJI drones.[9] Its primary objective is to assist forensic investigators in extracting, organizing, and visualizing metadata and telemetry recorded during drone operations [14]. This data can provide information during investigations, particularly in legal cases, regulatory audits, or accident analysis involving drones.

7.9.1 Environment Settings for Digital Drone Forensics Software

For this tool, we utilize Eclipse IDE version 2024-09 (4.33) and it can be easily downloaded from its official website.[10] Eclipse serves as the primary platform for developing, debugging, and running Java-based application code. Moreover, we employ Java, specifically OpenJDK version 17.0.8.1 (released on 2023-08-24), which provides a reliable rntime environment for executing Java-based applications.

---

[9] https://github.com/ankitrlps/DroneForensicsSoftware.
[10] https://eclipseide.org/.

Table 7.2 Sample output of DRDP for a TXT file

Update time	Longitude	Latitude	Flight distance	Flight altitude
2020-07-10 19:31:19.653000	121.41270997085715	31.17670558106035	0	0.00000000000000000
2020-07-10 19:31:19.750000	121.41270991227476	31.176705627780670	0	0.00000000000000000
2020-07-10 19:31:19.852000	121.41270992615043	31.176705647701223	0	0.0025785835459828377
2020-07-10 19:31:19.956000	121.41270994141857	31.176705665179668	0	0.005004924256354570
2020-07-10 19:31:20.057000	121.41270999874992	31.176705682708900	0	0.010796995833516210
2020-07-10 19:31:20.160000	121.41271003152663	31.176705713323646	0	0.015413485467433930
2020-07-10 19:31:20.261000	121.41271009258789	31.176705723055125	0	0.021322507411241530
2020-07-10 19:31:20.364000	121.41271012688783	31.176705734474233	0	0.024823987856507300
2020-07-10 19:31:20.466000	121.41271020485124	31.176705750696335	0	0.032457273453474045
2020-07-10 19:31:20.569000	121.41271026510580	31.176705779610170	0	0.039029683917760850
2020-07-10 19:31:20.673000	121.41271033812900	31.176705817124840	0	0.047132965177297590
2020-07-10 19:31:20.775000	121.41271038384703	31.176705861463095	0	0.053707472980022430
2020-07-10 19:31:20.877000	121.41271039951192	31.176705884162280	0	0.056638635694980620
2020-07-10 19:31:20.979000	121.41271039741355	31.176705902371390	0	0.058673214167356490
2020-07-10 19:31:21.082000	121.41271039117476	31.176705901538950	0	0.059273920953273770
2020-07-10 19:31:21.183000	121.41271038865582	31.176705904776956	0	0.059706427156925200
2020-07-10 19:31:21.285000	121.41271038349595	31.176705916425725	0	0.061091609299182890
2020-07-10 19:31:21.388000	121.41271038125862	31.176705939483476	0	0.063664332032203670
2020-07-10 19:31:21.491000	121.41271037697553	31.176705975053935	0	0.067640520632267000
2020-07-10 19:31:21.595000	121.41271037542350	31.176706013346184	0	0.071900986135005950
⋮	⋮	⋮	⋮	⋮

7.9 Digital Drone Forensics Software

OpenJDK 17 provides compatibility with the libraries and frameworks required for the running of the software.

To run the Digital Drone Forensics Software, we also rely on the 3DS Model Importer. This component is used to import 3D models into the application and visualizes a complete 3D representation of the flight. This visualization provides an intuitive and detailed perspective for forensic analysis. The 3DS Model Importer can be downloaded from the official source.[11] To set up the 3DS Model Importer in Eclipse, follow these detailed steps to ensure proper integration into the project environment:

1. Clone the repository

    Begin by cloning the repository that contains the project files from its official GitHub repository. Open a terminal or command prompt, and use the following command to clone the repository:

    ```
    git clone
        ↪https://github.com/ankitrlps/DroneForensicsSoftware.git
    ```

    Once the cloning process is complete, open the cloned project in Eclipse by selecting File → Open Projects from File System and navigating to the directory where the repository was cloned.

2. Access the project build path

    In the Project Explorer view, locate the project we have just opened. Right-click on the project name to access the context menu.

3. Configure the build path

    From the context menu, follow these steps:

    - Select Build Path from the options.
    - Choose Configure Build Path to open the Build Path configuration window.

4. Add external JAR files

    - In the Build Path configuration window, click on the Libraries tab.
    - Click the Add External JARs button to open a file selection dialog.
    - Navigate to the location of the 3DS Model Importer JAR file that downloaded previously (e.g., from the link: http://www.interactivemesh.org/models/download/JFX3DModelImporters_EA_2014-02-09.zip).
    - Select the JAR file and click Open to add it to the project as shown in Fig. 7.18.

5. Apply changes and close

    - After adding the JAR file, click Apply and Close to save the changes and exit the Build Path configuration window.
    - Eclipse will automatically update the project build path and include the 3DS Model Importer in the class path.

---

[11] http://www.interactivemesh.org/models/download/JFX3DModelImporters_EA_2014-02-09.zip.

Fig. 7.18 Add an external JAR file for 3DS Model Importer

To set up and use JavaFX with the project in Eclipse, follow these detailed steps:

1. Download JavaFX

   Start by downloading the latest version of JavaFX, specifically version 21.0.5, from the official website.[12]

2. Extract the JavaFX package

   Once the download is complete, extract the contents of the JavaFX package into a directory on the local machine. Take note of the directory path where the package is extracted as it will be required later.

3. Access the Run Configurations

   In Eclipse, navigate to the project in the Project Explorer. Right-click on the project name and follow these steps:

   - Select Run As from the context menu.
   - Choose Run Configurations to open the Run Configurations window.

4. Configure VM Arguments

   In the Run Configurations window, follow these steps:

   - Select the tab labeled Arguments.
   - In the VM Arguments section, add the following configuration, making sure to replace the module path with the directory where we extracted the JavaFX package:

   ```
   --module-path
       ↪"/Users/hudan/Downloads/javafx-sdk-23.0.1/lib"
   ```

---

[12] https://gluonhq.com/products/javafx/.

7.9 Digital Drone Forensics Software

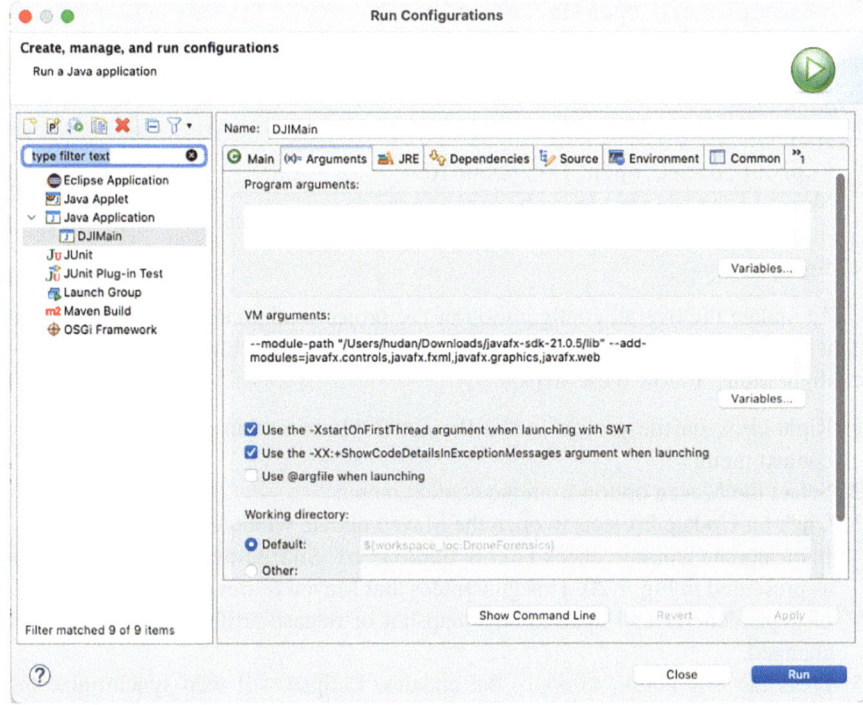

Fig. 7.19 Configuration of JavaFX

```
--add-modules=javafx.controls,javafx.fxml,
javafx.graphics,javafx.web
```

Adjust the path as needed to point to the lib folder of the locally extracted JavaFX directory as shown in Fig. 7.19.

5. Apply and Run

After entering the VM arguments, click Apply to save the changes. Finally, click Close to close the dialog.

Furthermore, to ensure proper configuration of the Maven project, add the following settings to the pom.xml file. These configurations will allow the project to integrate necessary dependencies, plugins, and other build settings required for the application. Below is an example of the required settings:

```
<dependency>
    <groupId>org.openjfx</groupId>
    <artifactId>javafx-controls</artifactId>
    <version>23.0.1</version>
</dependency>
<dependency>
```

```xml
    <groupId>org.openjfx</groupId>
    <artifactId>javafx-fxml</artifactId>
    <version>23.0.1</version>
</dependency>
<dependency>
    <groupId>org.openjfx</groupId>
    <artifactId>javafx-web</artifactId>
    <version>23.0.1</version>
</dependency>
```

To update the overall configuration of the project and guarantee that all dependencies and project settings are correctly synchronized with the current Maven configurations, follow these steps:

1. Right-click on the project in the Project Explorer within Eclipse to open the context menu.
2. Select the Maven option from the context menu.
3. Click on Update Projects to open the Maven update window.
4. In the update window, check `Force Update of Snapshots/Releases` option as presented in Fig. 7.20. This guarantees that Maven retrieves the latest versions of dependencies and updates any snapshot or release artifacts that might have changed.
5. Press the OK button to apply the updates. Eclipse will then synchronize the project with Maven and refresh the configuration.

By following these steps, the project will be updated to reflect the latest changes in the Maven settings and dependencies.

7.9.2 Dataset Requirements for DJI Phantom 4

To work with the DJI Phantom 4 flight data, we will need to download the DJI Phantom 4 dataset provided by VTO Inc. This dataset contains flight log data for the DJI Phantom 4 drone, which can be accessed through the CFReDS website. The following steps outline the process for finding, downloading and extracting the required flight logs for further forensic analysis.

1. Download the DJI Phantom 4 dataset
   Visit the CFReDS website, which hosts datasets from VTO Inc., by following this link: https://cfreds.nist.gov/all/SteveWatson%2FVTOInc./DroneDataSet.
2. Navigate to the correct directory

   - On the CFReDS page, locate and open the `DJI_Phantom_4` directory.
   - Within that directory, select `df004_DJI_Phantom_4`.
   - Next, open the folder labeled `2018_June`.

7.9 Digital Drone Forensics Software

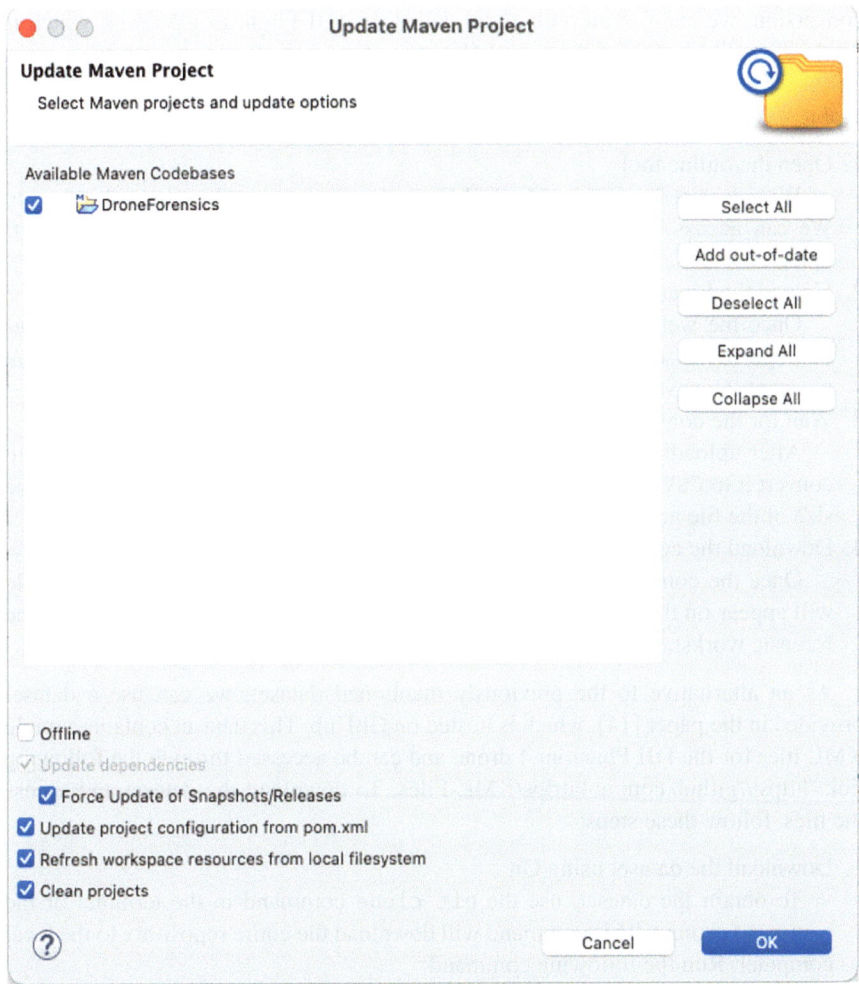

Fig. 7.20 Update maven project for digital drone forensics

3. Download the flight logs
   In the 2018_June folder, look for a file named flight_logs.zip and download it to the forensic workstation.
4. Extract the flight log data
   Once the download is complete, extract the contents of the flight_logs.zip file using a file extraction tool such as WinRAR, 7-Zip, or the default unzip utility on the operating system. This will provide access to the flight log data.

This software used to process drone flight data can only handle files in CSV or KML formats. However, flight log data from DJI drones are often stored in a binary .TXT format, which needs to be converted into a CSV file. To perform this conversion, we can use an online tool called the DJI Flight Log Viewer, which is freely accessible at the following link: https://www.phantomhelp.com/logviewer/upload/. The following steps outline the process to convert a binary .TXT file into a usable CSV file:

1. Open the online tool

    We can start by opening the DJI Flight Log Viewer tool in the web browser. We can access it through this link: https://www.phantomhelp.com/logviewer/upload/.

2. Upload the binary .TXT file

    Once the webpage loads, locate and click on the Upload Log button. This will open a file selection dialog box. Browse the computer, find the .TXT file that we want to convert, and upload it to the tool.

3. Wait for the conversion process

    After uploading the file, the tool will begin processing the binary .TXT file to convert it to CSV format. This process may take a few minutes, depending on the size of the file and the server load.

4. Download the converted CSV file

    Once the conversion is complete, a download link for the resulting CSV file will appear on the webpage. Click the link to download the converted file to the forensic workstation.

As an alternative to the previously mentioned dataset, we can use a dataset provided in the paper [14], which is hosted on GitHub. This dataset contains sample KML files for the DJI Phantom 4 drone and can be accessed through the following link: https://github.com/ankitrlps/KML-Files. To download the dataset and access the files, follow these steps:

1. Download the dataset using Git

    To obtain the dataset, use the `git clone` command in the terminal or the command prompt. This command will download the entire repository to the local computer. Run the following command:

    `git clone https://github.com/ankitrlps/KML-Files.git`

2. Navigate to the correct directory

    Once the cloning process is complete, navigate to the downloaded repository folder. We can do this by using the `cd` command on the terminal. For example:

    `cd KML-Files`

3. Locate the DJI Phantom 4 sample data Inside the `KML-Files` directory, we will find sample data files in KML format that correspond to DJI Phantom 4 drone flight logs in KML or CSV format.

7.9 Digital Drone Forensics Software

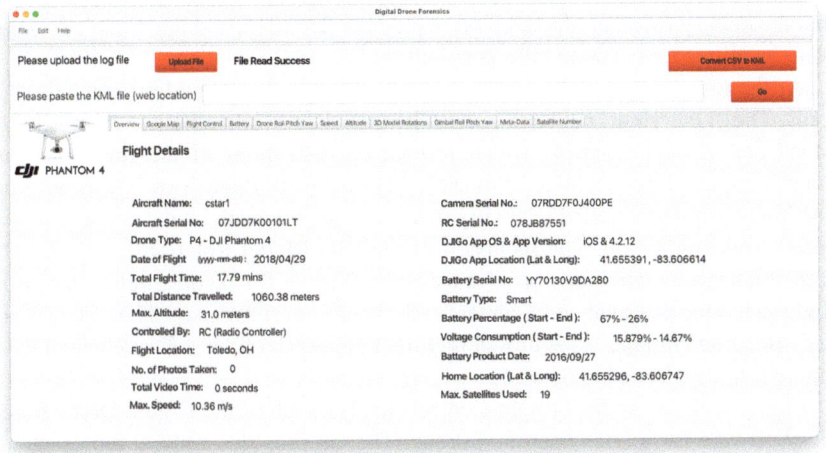

Fig. 7.21 Digital drone forensics software: overview

7.9.3 Running Digital Drone Forensics Software

Figure 7.21 presents a Digital Drone Forensics Software interface, a tool designed to process and analyze drone flight logs. Using this tool, forensic investigators can extract metadata from drone operations, particularly for DJI Phantom 4 drones, and assist in understanding flight behavior and verifying technical details. The software allows users to upload log files and convert extracted data into KML (Keyhole Markup Language) format, which can then be visualized in mapping tools such as Google Earth.

The main section of the interface is dedicated to "Flight Details", which are divided into two parts. On the left, general flight parameters are displayed, including aircraft name, serial number, and type, confirming the drone as a DJI Phantom 4. Details such as the date of flight (April 29, 2018), the total duration of the flight (17.79 minutes), and the distance traveled (1060.38 meters) provide information on the operating performance of the drone. The maximum altitude reached during the flight was 31 meters, and the drone was controlled by a radio controller (RC) in Toledo, Ohio. Moreover, the data shows that no photos or videos were taken, while the drone achieved a maximum speed of 10.36 meters per second.

On the right side, the software focuses on technical metadata extracted from flight logs. This includes hardware details such as the camera serial number, remote controller serial number, and the DJI Go app version (iOS 4.2.12) used during the flight. The latitude and longitude coordinates provide the precise flight location (41.655391, −83.606614) and the home return location (41.655296, −83.606747) to allow geospatial tracking. Battery performance is also analyzed, showing the serial number of the battery, the initial and final percentages of the battery (67–

26%), and the voltage consumption (15.879–14.67%). The metadata also reveals that the battery was manufactured on September 27, 2016, and the drone utilized up to 19 satellites for accurate GPS positioning.

At the top of the interface, several tabs provide access to additional analytical tools. The "Overview" tab summarizes the metadata, while others such as "Google Map" and "Flight Control" which provide visualizations of the flight path and control input. Tabs such as "Battery", "Drone Roll Pitch Yaw", "Speed", and "Altitude" suggest detailed telemetry data analysis, focusing on the drone's motion, speed, and orientation. The inclusion of the "3D Model Rotations" and "Gimbal Roll Pitch Yaw" tabs hints at advanced visual tools for reconstructing the drone flight behavior and analyzing camera stabilization performance. The "Meta-Data" and "Satellite Number" tabs further highlight the depth of the software in extracting and presenting drone log details.

Figure 7.22 shows the "Flight Control" analysis tab of the Digital Drone Forensics Software for a DJI Phantom 4 drone. The tab provides a visual representation of the drone's flight states and shows key phases such as take-off, GPS-controlled flight, and landing. The graph tracks drone flight behavior over time and displays specific actions that occurred during the operation. The vertical axis on the graph represents flight states or control modes, which are numerically coded and explained in the legend on the right-hand side. For instance, '10' represents an Assisted Take Off, '6' corresponds to Flying (GPS), '11' indicates an Auto Take Off, and '12' denotes Auto Landing. The horizontal axis represents timestamps, which probably correspond to specific events logged during the flight.

From the graph, the drone begins its operation at a higher state ('11'—Auto Take Off), followed immediately by entering the state '6' (Flying under GPS control). The graph remains relatively stable for a significant portion of the flight duration, as

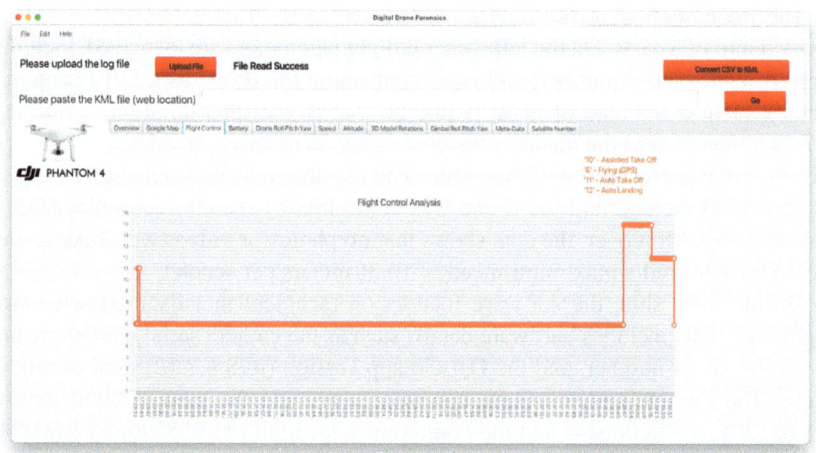

Fig. 7.22 Digital drone forensics software: flight control

7.9 Digital Drone Forensics Software

indicated by the flat horizontal line, suggesting a consistent GPS-controlled flight. Toward the end of the timeline, the graph jumps to the state '12' (Auto Landing), showing a clear change where the drone begins the automated landing process. After reaching this state, the graph remains flat, indicating the end of the flight.

This analysis is used for drone forensics as it helps investigators reconstruct the exact sequence of flight events. By visualizing the transitions between different flight states, investigators can identify anomalies, such as unexpected mode changes, forced landings, or disruptions in GPS connectivity. The clear labeling of states and timestamps ensures that the timeline of events can be cross-referenced with other metadata, such as altitude, speed, and location.

Figure 7.23 illustrates the "Battery" tab of the Digital Drone Forensics Software, presenting a graph titled "Battery Over Time." The graph provides a visual representation of the drone's battery percentage throughout the flight duration, plotted against timestamps recorded at regular intervals. Forensic investigators can track the gradual decline in battery power as the drone operates and obtain information on battery usage and overall flight efficiency. The Y-axis of the graph represents the battery percentage, which starts at approximately 67% and steadily decreases to around 26% at the end of the flight. The X-axis shows the timestamps in the format "hh:mm:ss:ms," to mark specific moments during the drone's operation. The smooth and continuous decline of the battery percentage indicates that the drone's power consumption was consistent throughout the flight, with no sudden drops or anomalies.

This tab is used to understand the battery performance and identify any irregularities in power usage. For instance, sudden drops in battery percentage could indicate power failures, environmental factors, or issues with the drone's hardware. By analyzing this graph, the investigators can also confirm the duration of the flight,

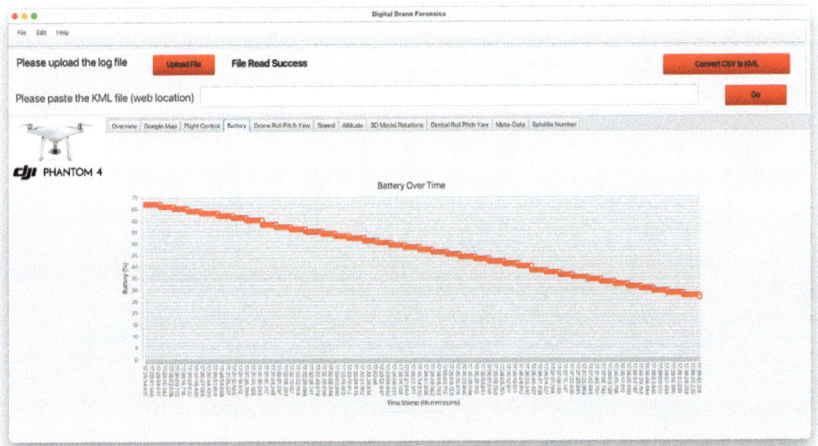

Fig. 7.23 Digital drone forensics software: battery

the rate of battery depletion, and whether the drone operated within safe power limits. The consistent linear decline seen here reflects normal battery consumption during a stable flight.

In forensic investigations, battery analysis can help determine the timeline of events, validate flight durations, and identify whether the drone returned to its home location due to low battery power. It also serves as supporting evidence when reconstructing the flight path and ensures that the drone's energy usage aligns with its recorded activities. For example, if an incident occurred during flight, investigators could examine the battery levels at that specific time to understand the drone's operational status.

Figure 7.24 represents the "Drone Roll Pitch Yaw" tab of the Digital Drone Forensics Software. It displays a time-series analysis of the drone's orientation dynamics during its flight. The graph plots three key metrics–Roll, Pitch, and Yaw–over time, providing a detailed view of the drone's motion and stabilization behavior. These metrics represent how the drone maintained its orientation and balance during flight. The Y-axis measures the values of Roll, Pitch, and Yaw, ranging approximately from −200 to 200 units, which likely correspond to angular degrees or standardized measurements of orientation. The X-axis represents the timestamps in the format hh:mm:ss:ms, allowing investigators to track changes in orientation at specific moments during flight. Each orientation parameter is color-coded for clarity: Roll (orange), Pitch (yellow), and Yaw (green).

The graph reveals that the Yaw (green line) shows significant and frequent variations throughout the flight indicating directional changes and rotations along the vertical axis. These spikes in yaw often suggest deliberate adjustments or corrections made to the drone's heading, such as turns or changes in direction. The Pitch (yellow) and Roll (orange) remain more stable near the centerline for most

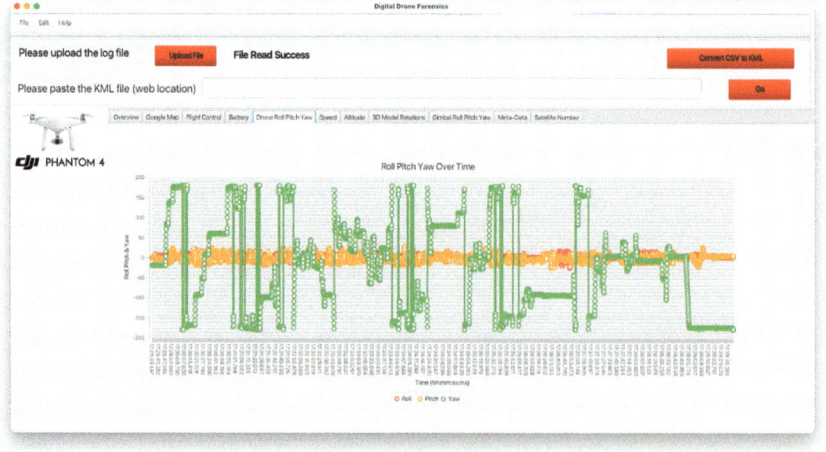

Fig. 7.24 Digital drone forensics software: drone roll, pitch, and yaw

7.9 Digital Drone Forensics Software

of the flight, with occasional fluctuations. Roll represents the tilt along the drone's longitudinal axis (side-to-side motion), while pitch reflects the tilt along the lateral axis (forward-backward motion). Both parameters show periodic changes, which are likely adjustments made to stabilize the drone during flight maneuvers or counteract environmental disturbances such as wind.

The chaotic sections of the graph with abrupt spikes and changes in values may indicate moments when the drone underwent significant motion, such as sudden turns, changes in altitude, or corrections to its position. For example, the sharp vertical spikes in the Yaw values suggest rapid rotations or heading adjustments. Similarly, small but consistent deviations in Roll and Pitch could reflect attempts to stabilize the drone during its flight path.

From a forensic perspective, the analysis of Roll, Pitch, and Yaw data is used to examine drone flight behavior and control stability. It allows investigators to reconstruct flight events, detect unusual maneuvers, and identify instances of instability that may have led to crashes or erratic behavior. By correlating these data with other metrics, such as battery levels, altitude, and speed, investigators can build a complete picture of the drone's operation and any anomalies that occurred.

Figure 7.25 presents the Speed tab of the Digital Drone Forensics Software, which visually analyzes the drone's velocity over time. The graph titled "Speed Over Time" plots the drone's speed in meters per second (m/s) against the corresponding timestamps recorded during the flight. Using this time-series representation, we can understand the drone's movement dynamics, performance, and potential anomalies throughout the operation. The Y-axis represents the speed of the drone, with values ranging from 0 to 11 m/s. The X-axis displays the timestamps in a detailed format (hh:mm:ss:ms), which indicates the exact moments the speed data were recorded.

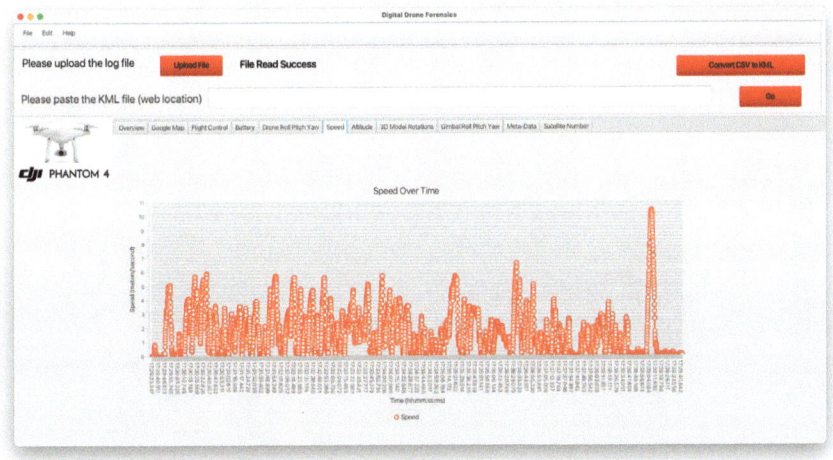

Fig. 7.25 Digital drone forensics software: speed

The data points are plotted as orange circles, forming a dense pattern that reveals fluctuations in the drone's speed throughout the flight.

From the graph, it is clear that the drone speed varied during the operation. The initial speed starts low, around 1 m/s, suggesting a gradual take-off phase. Shortly after, the speed increases and fluctuates between 2 and 6 m/s, indicating active movement under GPS control or during flight maneuvers. There are periodic peaks and troughs throughout the graph, which likely correspond to the drone accelerating, decelerating, or making adjustments in its path. Near the end of the timeline, a sharp spike occurs, where the speed rapidly reaches its maximum value of approximately 10 m/s, followed by a sudden drop to nearly 0 m/s. This sharp change likely represents the final descent and automatic landing phase of the drone.

The "Speed Over Time" graph is valuable in forensic investigations, as it helps reconstruct the flight behavior of the drone. By analyzing the speed fluctuations, investigators can identify periods of consistent travel, rapid acceleration, or abrupt stops. Such insights are critical for determining whether the drone performed specific actions, such as evading obstacles, following a predefined flight path, or responding to external factors such as wind or user input. In addition, sudden changes in speed could indicate anomalies, such as hardware issues, environmental interference, or collisions.

Figure 7.26 from the "Altitude" tab of the Digital Drone Forensics Software presents a graph titled "Altitude Over Time". It visualizes the drone's vertical movement during the flight and shows changes in altitude over the recorded timestamps. Based on this data, we can understand the drone's flight trajectory, including its take-off, hovering, descent, and landing phases. The Y-axis of the graph represents the altitude in meters, ranging from 0 to 32 meters, while the X-axis provides the timestamps (hh:mm:ss:ms) for each recorded data point. The

Fig. 7.26 Digital drone forensics software: altitude

7.9 Digital Drone Forensics Software

graph's orange circles mark the altitude readings at specific times throughout the flight, forming a step-like pattern that clearly highlights distinct phases of vertical movement.

From the graph, the drone initially starts at ground level (0 meters) and begins a steady ascent, reaching an altitude of approximately 22 meters. This phase likely represents the take-off and climb portion of the flight. After achieving this height, the drone appears to hover at a consistent altitude for some time, as indicated by the relatively flat portion of the curve. This stability suggests that the drone maintained a constant elevation, probably under GPS control.

Following the hovering phase, the graph shows a descent to a lower altitude of around 7 meters, where the drone hovers again for a period before further lowering to approximately 5 meters. This behavior could indicate maneuvering at a lower height, possibly for inspection or controlled positioning. The drone's altitude remains stable at this lower level for a significant duration, as seen by the extended flat portion of the curve. Toward the end of the flight, the graph reveals a sharp and rapid ascent to a peak altitude of approximately 30 meters, followed immediately by a steep descent back to ground level (0 meters). This sudden vertical movement likely corresponds to a final maneuver before the auto-landing phase and marks the conclusion of the flight.

The "Altitude Over Time" graph supports forensic investigators in reconstructing the drone's vertical trajectory and identifying critical phases such as take-off, hovering, and landing. Any unusual altitude changes, such as sudden drops or rapid climbs, can be analyzed to determine whether the drone encountered issues such as user input errors, environmental interference, or mechanical failures. By correlating the altitude data with other telemetry, such as speed, roll-pitch-yaw dynamics, and battery performance, investigators can build a comprehensive understanding of the drone's flight behavior. This graph is especially useful for identifying anomalies in vertical movement, validating flight claims, and ensuring compliance with altitude restrictions in regulated airspace.

Figure 7.27 displays the "Gimbal Roll Pitch Yaw" tab of the Digital Drone Forensics Software and demonstrates the dynamics of the drone's camera gimbal over time. The graph titled "Gimbal Roll Pitch Yaw Over Time" captures the angular movements of the gimbal–Roll, Pitch, and Yaw–during the flight. This data provides insight into how the gimbal was adjusted to stabilize the camera or respond to external movements like wind or flight maneuvers. The Y-axis of the graph measures the angular values of the gimbal's movements, likely in degrees, ranging approximately from -200 to 200 units. The X-axis represents timestamps (hh:mm:ss:ms), marking specific instances throughout the drone's operation. The gimbal's movements are divided into three components, each color-coded: Roll (orange), Pitch (yellow), and Yaw (green).

From the graph, the Yaw (green line) exhibits variations throughout the flight, with sharp spikes and drops indicating rapid changes in the horizontal rotation of the gimbal. This behavior likely corresponds to directional changes or rotations of the drone, where the gimbal compensates to maintain a stable view. The Pitch (yellow) remains relatively stable around a central value, with minor fluctuations.

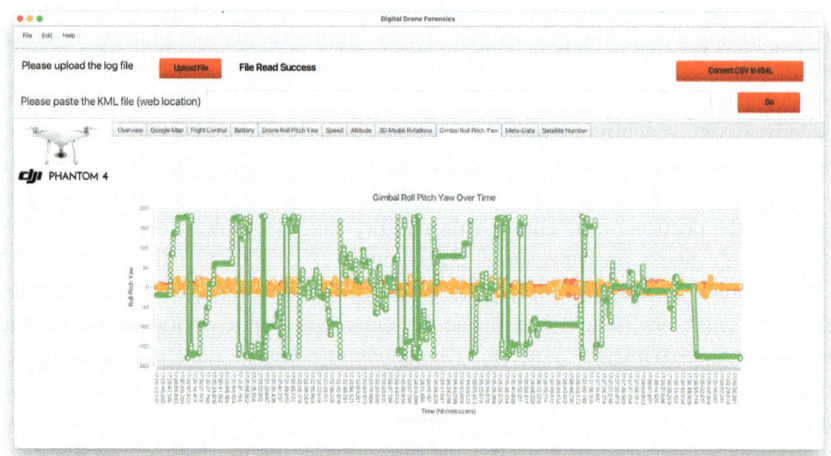

Fig. 7.27 Digital drone forensics software: gimbal roll, pitch, and yaw

This suggests that the camera maintained a forward-facing orientation for much of the flight but adjusted slightly to counteract the drone's tilt or altitude changes. The Roll (orange) is also stable, remaining close to the centerline, indicating minimal side-to-side tilt of the gimbal.

The sections with extreme spikes and variations, especially in the Yaw values, are notable. These likely correspond to significant drone movements, such as turns, accelerations, or altitude adjustments, where the gimbal actively compensated to stabilize the camera's perspective. Toward the end of the flight, there is a pronounced drop in the yaw values, which aligns with a likely descent or landing phase where the gimbal's orientation becomes fixed.

From a forensic perspective, the "Gimbal Roll Pitch Yaw" graph is used to understand the behavior of the drone's camera system during the flight. Investigators can analyze how the gimbal responded to flight maneuvers and environmental factors to ensure that the drone's video or image captures were stabilized. Deviations in gimbal angles could point to turbulence, user interventions, or mechanical issues that affected the camera's orientation. By correlating these data with other telemetry, such as the drone's roll-pitch-yaw dynamics, speed, and altitude, investigators can gain a comprehensive understanding of both the drone's behavior and the gimbal's performance. This analysis is useful for reconstructing flight events, verifying camera stabilization claims, and identifying anomalies during critical phases such as turns, ascents, or landings.

Figure 7.28 represents the "Satellite Number" tab of the Digital Drone Forensics Software and presents a graph titled "Satellites Over Time". The graph tracks the number of satellites connected to the drone during its flight and is plotted against timestamps. Based on this data, we can understand the drone's GPS stability, positioning accuracy, and potential interruptions in satellite connectivity. The Y-axis

7.9 Digital Drone Forensics Software

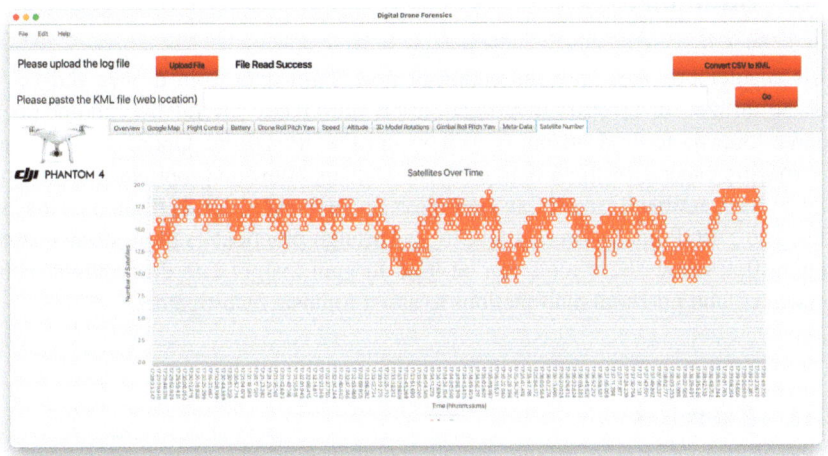

Fig. 7.28 Digital drone forensics software: number of satellite

represents the number of satellites, ranging from 0 to 20, while the X-axis displays the flight timestamps in the format hh:mm:ss:ms. It records satellite connection data throughout the flight duration. The graph is populated with circular orange points that indicate the number of satellites to which the drone was connected at each specific moment.

From the graph, the drone started its operation with a steady increase in satellite connections, stabilizing at approximately 16 to 18 satellites early in the flight. Maintaining this range indicates strong GPS connectivity and stable positioning during this phase. However, there are notable fluctuations in the middle portion of the graph, where the number of satellites drops between 10 and 12. These dips could indicate temporary interruptions or weaker GPS signals, possibly caused by environmental factors such as obstacles, poor weather, or interference. Toward the end of the flight, the graph shows a recovery, with the satellite count returning to around 18 to 19 satellites. It suggests improved GPS stability before the drone's landing phase. The consistency in the satellite numbers during most of the flight indicates that the drone maintained reliable navigation and flight control under strong GPS support.

GPS connectivity directly impacts the drone's ability to follow predefined flight paths, stabilize its position, and return to its home location. Drops in satellite connectivity, as seen in the middle of this graph, could be correlated with erratic behavior or deviations in the flight path. Such anomalies can be correlated to identify the cause of issues during flight, such as signal interference, obstruction of the drone's line of sight to satellites, or hardware malfunctions. By analyzing this graph alongside other telemetry data, such as speed, altitude, and roll-pitch-yaw dynamics, investigators can determine whether changes in flight behavior align with satellite disruptions.

7.10 Summary

This chapter provides an examination of how flight data from drones and UAVs can be utilized in forensic investigations. It introduces several free and/or open-source forensic tools: DatCon, CsvView, and DROP (DRone Open source Parser) for specific aspects of flight data analysis. DatCon converts proprietary drone data logs into accessible formats, CsvView allows visual inspection of flight data in CSV files, and DROP parses drone flight logs. Through practical examples, the chapter demonstrates how these tools can be used to gain insight into drone flight paths, behaviors, and potential malfunctions to assist forensic investigators.

7.11 Exercises

1. What is contained in the drone flight logs as a forensic artifact?
2. Name the three tools discussed in this chapter for analyzing drone flight data.
3. What is the purpose of the DatCon tool?
4. How does CsvView help investigators in analyzing flight data?
5. What is the role of DROP tool in drone forensic investigations?
6. Describe how DatCon facilitates deeper analysis of drone data by converting proprietary logs into accessible formats.
7. Discuss the advantages of using CsvView for visually inspecting flight data over raw log analysis.
8. Evaluate the challenges involved in parsing drone flight logs with tools such as DROP and propose potential solutions.
9. Analyze how flight data analysis can provide insight into drone behavior and its potential involvement in forensic cases.
10. Propose a workflow that combines the use of DatCon, CsvView, and DROP for a complete forensic analysis of drone flight data.

References

1. E. Mantas, C. Patsakis, Who watches the new watchmen? The challenges for drone digital forensics investigations. Array **14**, 100135 (2022)
2. S. Silalahi, T. Ahmad, H. Studiawan, DFLER: drone flight log entity recognizer to support forensic investigation on drone device. Software Impacts **15**, 100457 (2023)
3. R. Kumar, A.K. Agrawal, Drone GPS data analysis for flight path reconstruction: a study on DJI, Parrot & Yuneec make drones. Forensic Sci. Int. Digital Investigation **38**, 301182 (2021)
4. D.A. Hamdi et al., Drone forensics: A case study on DJI phantom 4, in *2019 IEEE/ACS 16th International Conference on Computer Systems and Applications (AICCSA)* (IEEE, 2019), pp. 1–6
5. S. Atkinson, et al., Drone forensics: the impact and challenges, in *Digital Forensic Investigation of Internet of Things (IoT) Devices* (2021), pp. 65–124

References

6. A.L.P.S. Renduchintala, A. Albehadili, A.Y. Javaid, Drone forensics: digital flight log examination framework for micro drones, in *2017 International Conference on Computational Science and Computational Intelligence (CSCI)* (IEEE. 2017), pp. 91–96
7. H. Studiawan, et al., Forensic event reconstruction for drones, in *2021 4th International Seminar on Research of Information Technology and Intelligent Systems (ISRITI)* (2021), pp. 41–45
8. S. Silalahi, T. Ahmad, H. Studiawan, Drone flight logs sequence mining, in *2022 IEEE International Conference on Cybernetics and Computational Intelligence (CyberneticsCom)* (2022), pp. 107–111
9. F.E. Salamh, et al., A comparative uav forensic analysis: static and live digital evidence traceability challenges. Drones **5**(2), 42 (2021)
10. M. Yousef, F. Iqbal, Drone forensics: a case study on a DJI Mavic Air, in *2019 IEEE/ACS 16th International Conference on Computer Systems and Applications (AICCSA)* (2019), pp. 1–3
11. D.R. Clark, et al., DROP(DRone Open source Parser) your drone: forensic analysis of the DJI Phantom III. Digital Investigation **22**, S3–S14 (2017)
12. T. Latzo, et al., Maraudrone's map: an interactive web application for forensic analysis and visualization of DJI drone log data, in *Nordic Conference on Secure IT Systems* (Springer, 2022), pp. 329–345
13. Z. Zhao, Y. Wang, G. Liao, Digital forensic research for analyzing drone and mobile device: focusing on DJI mavic 2 Pro. *Drones* **8**(7), 281 (2024)
14. A. Renduchintala, et al., A comprehensive micro unmanned aerial vehicle (UAV/Drone) forensic framework. Digital Investigation **30**, 52–72 (2019)

Chapter 8
Forensic Investigation of UAV Faults and Anomalies

Abstract In the chapter, we navigate the diagnosis and analysis of malfunctions and security breaches within Unmanned Aerial Vehicle (UAV) systems. This includes methodologies for anomaly and fault detection. This chapter also presents case studies with public datasets and hands-on Python script. Finally, we conclude that the techniques can assist the forensic analysis of faults and anomalies in drones and UAVs.

8.1 Introduction

Increasingly complex operational environments expose UAVs to a wide range of faults and anomalies that can critically impact their safety, reliability, and performance. These occur in sensors, actuators, and motors or are the result of mechanical wear, software errors, environmental factors, and human mistakes. Such faults could cause catastrophic failure if they are not detected and managed early enough.

As such, the detection of faults and anomalies in UAVs is an important area of research. Several methodologies have been proposed, starting from model-based approaches, data-driven techniques, and hybrid systems that incorporate both. For example, autoencoders and variational autoencoders have shown effective anomaly detection by training on normal data and deviation identification based on reconstruction error [1]. Other methods include deep learning algorithms, e.g., long short-term memory (LSTM) networks that work effectively to detect temporal anomalies within UAV data streams. These are considered proper procedures for addressing this problem.

The structure of the chapters is as follows. Chapter 8.2 reviews related work on UAV faults and anomaly detection. Chapter 8.3 focuses on obtaining faults and anomaly datasets. Chapter 8.4 explains the classification of faults and anomalies, covering methodologies, feature selection, and evaluation metrics. Chapter 8.5 demonstrates the application of the model to unseen data. Finally, Chap. 8.6 summarizes the chapter.

8.2 Related Work

The field of fault detection and anomaly detection in unmanned aerial vehicles (UAVs) has seen advancements in recent years. It is driven by the increasing reliance on autonomous systems for mission-critical operations. Existing research has explored various methodologies, including machine learning techniques and hybrid frameworks, to address the challenges of fault detection. These approaches aim to improve UAV safety and operational efficiency by analyzing time series sensor data. They also reduce reliance on predefined models and adapting to various fault scenarios. This section reviews recent contributions in this area. The chapter discusses key methodologies, datasets, and validation strategies in the current state of UAV fault detection and anomaly detection research.

Expanding the resources available for fault detection research, the ALFA dataset has emerged as a benchmark for UAV anomaly detection and fault isolation [2]. This dataset comprises 47 autonomous flights, capturing eight types of control surface faults and providing raw and processed sequences for analysis. Including both normal and post-fault flight data, the ALFA dataset supports diverse research needs. The authors offer tools in Python, MATLAB, and C++ for data analysis and evaluation of detection methods.

Zhang et al. [3] propose a machine learning-based framework for prediction, detection, and classification of failures in mission-critical autonomous UAV flights. By applying long short-term memory (LSTM) networks, the framework analyzes time series data from UAV sensors to predict failures on control surfaces such as engines, rudders, elevators, and ailerons. The research utilizes the ALFA dataset, a comprehensive benchmark of real flight data with various failure scenarios, which is pre-processed into the Modified-ALFA dataset to improve the performance of machine learning models. The framework achieves a prediction accuracy of 93% and a detection accuracy of 100%, with predictions made on average 19 seconds before failure and detections completed in 0.74 seconds on average.

Building on the foundation of model-based fault detection, another study compares three distinct model-based methods for identifying stuck control surface faults in small unmanned aircraft systems (UAS) [4]. The methods include a baseline parity-space approach, a robust linear parameter-varying (LPV) observer-based method, and a multiple model adaptive estimation (MMAE) method. These approaches are evaluated using the Vireo aircraft, a small fixed-wing drone, with flight data to determine their detection performance and computational efficiency. The study demonstrates that each method has distinct strengths and weaknesses depending on the requirements and constraints of the application.

Shifting the focus to real-time anomaly detection, another work proposes a method for autonomous aerial vehicles (AAVs) to ensure safety and reliability during operations [5]. The proposed approach employs the Recursive Least Squares (RLS) algorithm to model and monitor input-output signal relationships in real-time. By identifying unexpected changes in these signals, the system can detect anomalies such as actuator and sensor failures. Unlike many existing methods, this approach is adaptable across various types of aircrafts and fault scenarios due to its independence from predefined aircraft models. Experimental validation using a fixed-wing UAV highlights its effectiveness, and a public dataset with annotated flight logs is provided for comparison.

Focusing on machine learning applications, a study explores the use of support vector machines (SVM) for real-time fault detection in small fixed-wing UAVs [6]. The work addresses challenges such as noisy data, computational constraints, and environmental disturbances, proposing a data-driven approach that avoids reliance on detailed UAV models. With 6 hours of real-flight data, including 2 hours of fault scenarios, the study demonstrates high detection accuracy under specific conditions and provides a public dataset for future research. To further advance real-time anomaly detection, a hierarchical approach has been proposed that combines supervised and unsupervised methods for autonomous aerial vehicles (AAVs) [7]. This framework integrates high-dimensional data reduction techniques with machine learning algorithms to identify anomalies in flight operations. Validation using simulations and real-world flight data demonstrates the effectiveness of this approach in minimizing false positives and detecting various types of fault.

Addressing the challenges of high-dimensional sensor data, another study introduces an anomaly detection framework using the Mahalanobis distance metric [8]. By pre-processing data to identify statistically dependent sensor attributes, the method reduces dimensionality for accurate real-time monitoring. Experiments involving UAVs and laboratory-based UGVs validate the robustness of this approach. For scenarios with limited labeled data and diverse operating conditions, a hybrid fault detection framework combines deep domain adaptation techniques with the Hampel filter [3]. This method employs BiLSTM networks for feature extraction and domain adaptation to align source and target data distributions. Extensive experiments with real flight data validate the accuracy and robustness of the approach and demonstrate its practical application.

Comprehensive reviews have also contributed to the consolidation of knowledge in UAV fault detection, such as a survey categorizing methods into knowledge-based, model-based, and data-driven approaches [9]. Highlighting datasets such as ALFA and Thor, the review discusses challenges and future directions, including hybrid models and the interpretability of detection methods. Innovative algorithms continue to emerge, such as a data-driven navigation sensor fault detection system combining Kalman filters with Adaptive Neuro Fuzzy Inference Systems (ANFIS) [10]. This approach minimizes the reliance on predefined models by detecting various types of faults in real-time through adaptive learning. Field tests and simulations demonstrate the superior performance of the system compared to traditional methods. Unsupervised learning methods have gained traction, as exemplified by a study that employs stacked autoencoders for fault detection in UAVs [11]. By training exclusively on safe flight data, the model detects faults through reconstruction loss and identifies new fault types without labeled datasets.

Finally, Cabahug et al. [12] propose a failure detection system for quadcopter UAVs that uses vibration analysis and k-means clustering as part of a three-step safety framework to improve UAV safety. The study focuses on the first step—

failure detection—employing vibration data captured by an inertial measurement unit sensor to classify health states into three categories: normal, faulty, and failure. The system uses a combination of accelerometer and gyroscope parameters to achieve accurate and real-time failure detection. The k-means clustering algorithm is applied to the vibration data to classify these states, and a visual LED subsystem is used to represent the detection results. Experimental validation in a Parrot ANAFI quadcopter demonstrates the effectiveness of the method. The method can detect failures in less than one second and achieve high accuracy.

8.3 Obtaining Faults and Anomalies Dataset

This section describes a dataset purposefully created for the needs of fault detection and anomaly detection methods with fixed-wing UAVs. This is called the Air Lab Fault and Anomaly (ALFA) Dataset, which contains a large number of real flight data recorded by different fault scenarios that are useful for researchers working in FDI and AD methods [2].

These are the different types of fault that can occur with the control surfaces of a fixed-wing UAV. First, "Engine Full Power Loss" simulates the complete loss of engine power while in flight. Second, "Rudder Stuck" means that the rudder is stuck either fully left or fully right. Third, "Zeroed Elevator Stuck at Zero" means that the elevator being in a neutral position that does not allow the aircraft to pitch properly. It also includes cases in which one or both ailerons are stuck in a neutral position, affecting the roll control of the aircraft. For example, there is a combination in which the rudder and aileron are at zero, and it is a very destructive restriction to airplane control.

The anomalies that are not predicted by the UAV system can crop up from unmodeled faults or even deviations from normal conditions of operation. These raw and processed flight data are contained within this dataset in order to provide a framework for the development and evaluation of such methods for anomaly detection. The processed data contain specific fault scenarios, including ground-truth information, while the raw data are suitable for unsupervised detection methods.

This dataset may be used for benchmarking testing and comparing different fault or anomaly detection methods. It includes some helper tools developed in a number of programming languages (C++, Python, MATLAB) that assist in working with the data and in evaluating new methods. The tools provided can be used to access the data and compute relevant metrics such as detection time and accuracy. There are several types of data files in the ALFA dataset and can be categorized into four main types and explained as follows.

Autonomous Flight Sequences with Failures These are off-line flight sequences, processed to include data from autonomous flights with the ground truth information about the failures. Each file in this section has data from a flight sequence with a maximum of one fault. The data comes in three ways: Comma-Separated Values

8.3 Obtaining Faults and Anomalies Dataset

(CSV) format, MATLAB's MAT format, and original ROS bag files. The dataset comprises:

- 47 flight sequences with faults of 8 different types.
- 10 sequences without any errors (used as common normal reference data).

47 sequences with faults are provided in three formats: CSV, MAT, and ROS bag, thus summing to $47 \times 3 = 141$ files. It also includes 10 sequences without faults given in three different formats: 10 sequences $\times$ 3 formats = 30 files. Overall, this category includes 171 files.

Raw Flight Sequences These are raw flight data files, including flights in all modes: manual, autopilot-supported, and autonomous. In addition, they can include several fault scenarios or none at all. The raw data are provided in ROS bag format only.

TX2 Telemetry Logs This category contains the telemetry data with the fault ground truth information, recorded by the Nvidia Jetson TX2 computer onboard during the tests, or without the fault ground truth information when used for unsupervised detection methods. A new set of telemetry logs appears to be created for each test day. An exact number of telemetry log files is not provided, but one can assume it will equal the number of test days and sequences.

Dataflash Logs of Pixhawk These files are written with the Pixhawk autopilot's recorded raw data during all tests, without any kind of preprocessing. Unlike telemetry logs, they do not contain ground-truth fault information and are intended to be used in unsupervised detection methods. Furthermore, for every day of testing, a dataflash log set is produced; the actual number of files generated is also unspecified, but their count matches the number of tests conducted.

The ALFA dataset contains several types of file and each data has its own set of columns or fields. These files are provided in different formats (CSV, MAT, and ROS bag), and the columns or fields represent various types of flight and fault data. Below is a breakdown of the key columns or fields found in the different types of data file in the ALFA dataset:

1. Processed flight sequences (CSV, MAT, and ROS bag formats)
   These files contain data from autonomous flight sequences and include ground truth information about faults.

   - time: The timestamp of the data point.
   - mavros/global_position/*: Global position information, including latitude, longitude, and altitude.
   - mavros/local_position/*: Local position data in terms of x, y, and z coordinates.
   - mavros/imu/*: Data from the Inertial Measurement Unit (IMU), including accelerations, angular velocities, and orientations.
   - mavros/setpoint_raw/*: Setpoint messages, including desired roll, pitch, yaw, and throttle values.

- `mavros/state`: The current state of the Flight Control Unit (FCU), such as mode and armed status.
- `mavros/wind_estimation`: Estimated wind speed and direction.
- `mavros/vfr_hud`: Data for the Head-Up Display (HUD), such as altitude, airspeed, and climb rate.
- `failure_status/engines`: Indicates whether an engine failure has occurred (Boolean).
- `failure_status/aileron`: Indicates an aileron failure, with values representing which aileron (left, right, or both) is affected.
- `failure_status/rudder`: Indicates a rudder failure, with values representing the position where the rudder is stuck (center, left, or right).
- `failure_status/elevator`: Indicates an elevator failure, with values representing whether the elevator is stuck at zero or all the way down.
- `mavros/nav_info/airspeed`: Airspeed data, including commanded and measured airspeed.
- `mavros/nav_info/roll`: Commanded and measured roll angles.
- `mavros/nav_info/pitch`: Commanded and measured pitch angles.
- `mavros/nav_info/yaw`: Commanded and measured yaw angles.
- `mavros/nav_info/velocity`: Commanded and measured velocity in the x, y, and z directions.
- `mavros/nav_info/errors`: Tracking errors, including airspeed and altitude errors.
- `mavctrl/path_dev`: Path deviation, indicating how much the UAV deviates from its planned path.
- `mavctrl/rpy`: Measured roll, pitch, and yaw values.

2. Raw flight sequences (ROS bag format)
   These files contain unprocessed flight data and do not have predefined columns like the CSV files. The fields in the ROS bag files correspond to topics used in the ROS framework.

   - Topics similar to those in the processed files: These include data on global and local position, IMU, setpoints, FCU state, wind estimation, and VFR HUD.
   - Additional topics: Depending on the specific flight and fault scenarios, there might be additional topics recorded that are not present in the processed data files.

3. Telemetry logs from TX2 (telemetry log format)
   These logs record telemetry data from the NVIDIA Jetson TX2 computer on board the UAV.

   - `flight.tlog`: The primary telemetry log file containing various flight parameters.
   - `flight.tlog.raw`: Raw telemetry data without any preprocessing.
   - `mav.parm`: Parameters used during the flight, including PID values and other settings.
   - Common telemetry fields:

- Position, velocity, and acceleration data: Similar to the fields in the processed files.
- Control surface positions: Actual positions of ailerons, rudder, elevator, and throttle.

4. Dataflash logs from Pixhawk (dataflash log format):
   These logs record data directly from the Pixhawk autopilot.
   - *.bin: The main binary log file containing detailed flight data.
   - *.bin-<number>.mat: MATLAB format files derived from the binary logs for analysis.
   - *.bin.gpx: GPX format file for geographic information, often used for mapping the flight path.
   - *.bin.param: A file containing the parameters set in the Pixhawk during the flight.
   - *.kmz: Google Earth file format for visualizing the flight path.
   - *.log: A plain text log file summarizing the flight data.

The summary of key data fields in each file is as follows.

1. Global Position: Latitude, longitude, and altitude.
2. Local Position: x, y, and z coordinates relative to a local reference frame.
3. IMU Data: Acceleration, angular velocity, orientation (roll, pitch, yaw).
4. Setpoints: Desired control surface positions (roll, pitch, yaw, throttle).
5. Fault Status: Boolean or enumerated values indicating faults in engines, ailerons, rudder, or elevator.
6. Airspeed: Commanded and measured airspeed.
7. Path Deviation: Distance from the planned flight path.
8. Wind Estimation: Speed and direction of the wind affecting the UAV.

8.4 Classification of Faults and Anomalies

In this section, we dive deep into the code to train and test the XGBoost model on the ALFA dataset. Here, every step, from loading the data, preprocessing it, training the model, and then proceeding to the testing part, is elaborated. In order to classify the ALFA dataset into two classes, normal and faults or anomalies, using Python and XGBoost, the following steps are necessary:

1. Install necessary libraries
   First, ensure we have the necessary libraries installed. We can install them using pip:

   ```
   pip install xgboost scikit-learn pandas numpy
   ```

2. Load and preprocess the data
   Load the ALFA dataset and preprocess it to prepare for training. This involves reading the data, handling missing values, and labeling the data as either normal (0) or fault/anomaly (1). Different sensor records were generated with different frequencies. For instance, airspeed data was generated with 20–25 Hz frequency, while the IMU data was generated with 10 Hz frequency. Therefore, the number of records generated during a certain period of time are not same. For this reason, we will only use one feature to be used for training the model. Otherwise, we have to perform data alignment by using either downsampling to reduce the number of records of higher frequency, or interpolation to upsample the number of records of lower frequency data [2].
3. The ALFA dataset has several fault types and anomalies
   We will take engine failure as an experimental case. After downloading the processed dataset, we use the carbonZ_2018-07-30-16-39-00_3_no_failure as the source of the normal data. On the other hand, we use file namely carbonZ_2018-07-30-16-39-00_1_engine_failure as the source of anomaly data. From both folders, we will load the data from a file that has information namely mavros-nav-info-airspeed in it. Figure 8.1 shows the visualization of the airspeed data both during normal and fault flights. From the error data in the third plot, it can be seen that there is no noticeable difference between the normal and fault data. However, from the second plot we can see

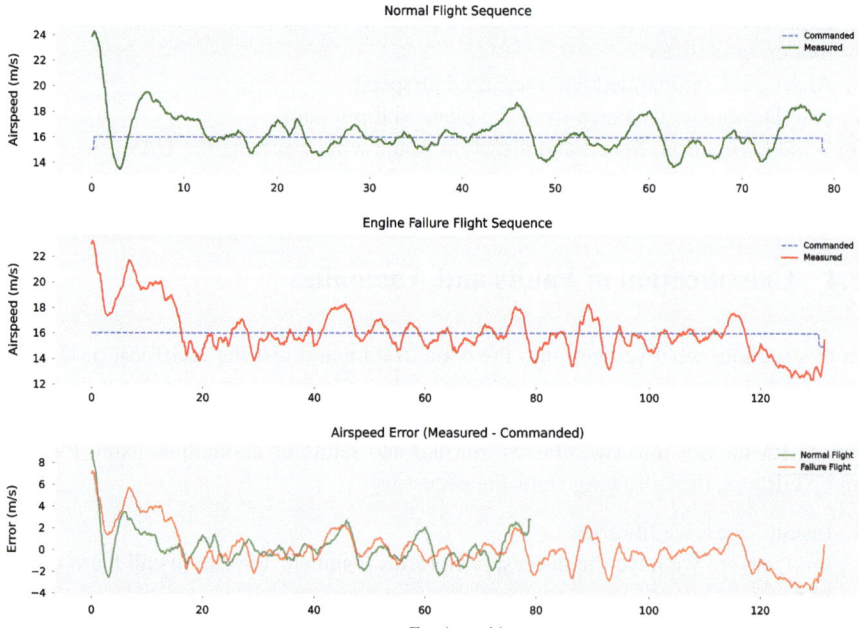

Fig. 8.1 A visualization of the airspeed data from the selected folder

8.4 Classification of Faults and Anomalies

that the deviations between the measured and commanded data during fault flight are bigger than in normal flight. The Python code for preprocessing and training is shown below.

```python
import pandas as pd
import numpy as np
from sklearn.model_selection import train_test_split
from xgboost import XGBClassifier
from sklearn.metrics import accuracy_score,
    ↪classification_report

# Load failure flight data
failure_df = pd.read_csv(failure_flight_path)
failure_df.columns = ['timestamp', 'seq', 'header_stamp',
    ↪'frame_id', 'commanded_airspeed', 'measured_airspeed']

# Load normal flight data
normal_df = pd.read_csv(normal_flight_path)
normal_df.columns = ['timestamp', 'seq', 'header_stamp',
    ↪'frame_id', 'commanded_airspeed', 'measured_airspeed']

# Label the data
failure_df['label'] = 1 # Engine failure
normal_df['label'] = 0 # Normal flight

# Handle missing values, if any
failure_df = failure_df.fillna(X.mean())
normal_df = normal_df.fillna(X.mean())

# Drop irrelevant or non-feature columns
# we use only 'commanded_airspeed' and 'measured_airspeed'
    ↪columns.
failure_df = failure_df.drop(columns=['timestamp', 'seq',
    ↪'header_stamp', 'frame_id']) # Drop non-feature
    ↪columns
normal_df = normal_df.drop(columns=['timestamp', 'seq',
    ↪'header_stamp', 'frame_id'])

# Create sliding windows for our time series data
window_size = 60 # one window contains records from 3 seconds
overlap = 40 # two consecutive window share an overlapping
    ↪records of 2 seconds.

def create_windows(self, df: pd.DataFrame) -> tuple:
    """
```

```python
    Create sliding windows from time series data
    """
    # Features we'll use
    features = [
        'commanded_airspeed',
        'measured_airspeed',
        (df['measured_airspeed'] -
            ↪df['commanded_airspeed']).values # Airspeed
            ↪error
    ]

    windows = []
    labels = []
    stride = window_size - overlap

    for i in range(0, len(df) - window_size + 1, stride):
        # Create window with all features
        window = np.column_stack([feature[i:i + window_size]
                            if isinstance(feature,
                                ↪np.ndarray)
                            else df[feature].values[i:i +
                                ↪window_size]
                            for feature in features])

        # Use the label from the last timestamp in the window
        label = df['label'].iloc[i + window_size - 1]

        windows.append(window)
        labels.append(label)

    return np.array(windows), np.array(labels)


# Create windows for both flights
X_failure, y_failure = create_windows(failure_df)
X_normal, y_normal = create_windows(normal_df)

# Combine the data
X = np.concatenate([X_failure, X_normal])
y = np.concatenate([y_failure, y_normal])

# Reshape X to 2D for scaling (flatten the windows)
X_reshaped = X.reshape(X.shape[0], -1)
```

```
# Scale the features
X_scaled = self.scaler.fit_transform(X_reshaped)

# Split the data into training and testing sets
X_train, X_test, y_train, y_test =
    ↪train_test_split(X_scaled, y, test_size=0.2,
    ↪random_state=42)

# Convert to DMatrix for XGBoost (optional, but often used
    ↪for performance reasons)
import xgboost as xgb
dtrain = xgb.DMatrix(X_train, label=y_train)
dtest = xgb.DMatrix(X_test, label=y_test)
```

The `pd.read_csv()` function is used to read a CSV file into a pandas DataFrame. Pandas is a powerful data manipulation library in Python that allows we to work with tabular data efficiently. The string `failure_flight_path` and `normal_flight_path` should be replaced with the actual path to our CSV file containing the ALFA dataset. This file should include the features (input variables) and possibly the labels (target variable) indicating whether each data point is normal or represents a fault or an anomaly.

Labeling the data is an important step in the process, as it involves defining what constitutes a "fault/anomaly" (class 1) and what is "normal" (class 0). Since we have the information from the folder name, we label the data from `no_failure` folder as normal (class 0), and the data loaded from `engine_failures` folder as anomaly (class 1). As a result, a new label column is created in the DataFrame, which will be used as the target variable for classification.

The `fillna(X.mean())` function is used to replace any missing values (NaNs) in the dataset with the mean value of the respective columns. Handling missing data is essential because many machine learning algorithms, including XGBoost, cannot process missing values directly. As a result, the DataFrame X is now prepared for training, with all missing values addressed.

The line `df.drop(columns=[...])` creates a new DataFrame X that contains all the features (input variables) used for training, while excluding the timestamp, seq, header_stamp, and frame_id columns. These columns contain exactly the same information on both failure and normal flight data. Thus, we regard them to be not relevant for training. Therefore, we have only two columns remaining used for training, include commanded_airspeed and measured_airspeed. From these two columns, we can create a new column to represent the deviation between the commanded and the measured airspeed, as shown in Fig. 8.1. Then, from these three columns, we will construct the sliding window as the features during the training. Note that the data is recorded with a frequency of 20Hz, meaning that there are 20 records every second. The label is taken from the last element of each window. On this occasion, we will try using 60 and 50 as the window and overlap size, respectively.

After constructing the sliding windows, we combine the failure and normal data and scale the value. Scaling the value of input features is a crucial before training a neural network model, especially when different features use different scales. By scaling all the features' values, we avoid an exploding value due to the multiplication nature during the training.

The `train_test_split()` function from scikit-learn is used to divide the data into training and testing sets. Here, `X` and `y` represent the features and labels, respectively. The `test_size=0.2` parameter specifies that 20% of the data should be reserved for testing, with the remaining 80% used for training. The `random_state=42` parameter ensures that the split is reproducible; the number 42 is arbitrary, but using the same number allows for consistent splits each time the code is run. As a result, we now have four variables: `X_train` and `y_train` for training the model, and `X_test` and `y_test` for testing the model.

4. Train the XGBoost classifier

   Next, we train the XGBoost classifier on the training data:

```
# Initialize the XGBoost classifier
xgb_model = XGBClassifier(use_label_encoder=False,
    ↪eval_metric='logloss')

# Fit the model
xgb_model.fit(X_train, y_train)

# Predict on the test set
y_pred = xgb_model.predict(X_test)
```

   The `XGBClassifier()` function initializes an XGBoost classifier model. The `use_label_encoder=False` argument is included to avoid a warning related to label encoding, which is not necessary in this context. The next parameter is `eval_metric='logloss'` which sets the evaluation metric for the training process, with `logloss` being a commonly used metric for binary classification problems.

   The `fit(X_train, y_train)` function is used to train the model using the training data, where `X_train` contains the input features and `y_train` contains the corresponding output labels. During this training process, the model learns to associate the input features with the output labels. As a result, the `xgb_model` is now trained and ready to make predictions on unseen data.

   The `predict(X_test)` function is used to generate predictions for the test set using the trained model. It applies the learned model to the input features in `X_test` to predict the corresponding output labels. As a result, `y_pred` is an array of predicted labels, where each label is either 0 for normal or 1 for fault/anomaly, corresponding to each instance in the test set.

5. Evaluate the model

   After training the model, evaluate its performance using accuracy, classification report, and confusion matrix:

8.4 Classification of Faults and Anomalies

```
# Calculate accuracy
accuracy = accuracy_score(y_test, y_pred)
print(f'Accuracy: {accuracy:.2f}')

# Classification report
print('Classification Report:')
print(classification_report(y_test, y_pred))

# Confusion Matrix (optional)
from sklearn.metrics import confusion_matrix
cm = confusion_matrix(y_test, y_pred)
print('Confusion Matrix:')
print(cm)

# Plot the confusion matrix
import matplotlib.pyplot as plt
import seaborn as sns

plt.figure(figsize=(8, 6))
sns.heatmap(cm, annot=True, fmt='d', cmap='Blues')
plt.title('Confusion Matrix')
plt.ylabel('True Label')
plt.xlabel('Predicted Label')
plt.show()
```

The accuracy_score(y_test, y_pred) function calculates the accuracy of the model by determining the ratio of correctly predicted instances to the total number of instances in the test set. The statement print(f'Accuracy: accuracy:.2f') then outputs this accuracy, rounded to two decimal places.

The classification_report(y_test, y_pred) function generates a comprehensive report that includes precision, recall, F1-score, and support for each class (normal and fault/anomaly). The confusion_matrix(y_test, y_pred) function computes the confusion matrix, displaying the counts of true positives, false positives, true negatives, and false negatives, which is a valuable tool for understanding the performance of the classification model. The purpose of these evaluations is threefold: Accuracy provides an overall measure of the model's performance; the Classification Report offers insights into how well the model is predicting each class, which is particularly important in imbalanced datasets; and the Confusion Matrix helps to visualize and understand the types of errors the model is making, as shown in Fig. 8.2.

6. Interpretation

Accuracy provides a basic measure of how well the model is performing. The classification report offers additional insights by detailing precision, recall, and F1-score for each class, including normal and fault/anomaly. Meanwhile, the

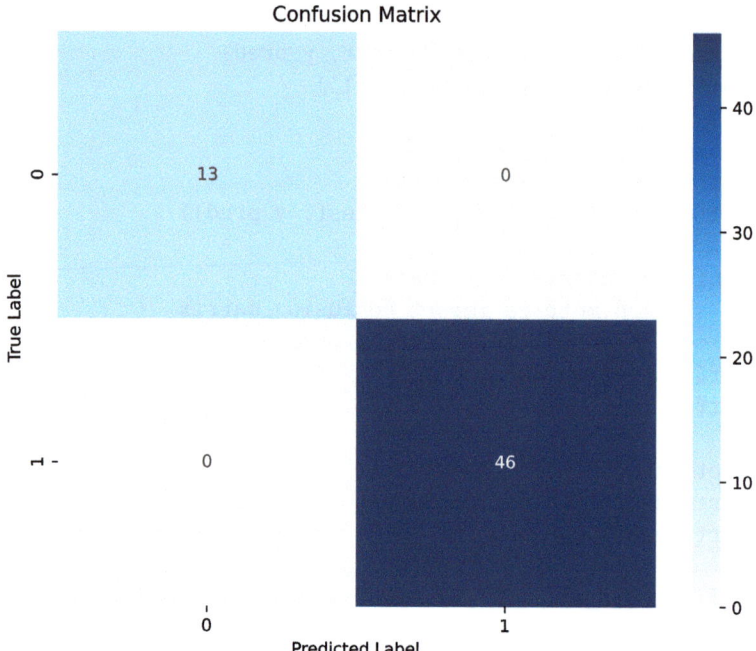

Fig. 8.2 Confusion matrix of the prediction on test data

confusion matrix presents a breakdown of the true positives, false positives, true negatives, and false negatives, giving a clearer view of the model's performance. Dataset and model have certain very essential considerations in working. One such consideration is the selection of features, as there might be some important features in determining a fault. The feature selection or engineering can further be executed to enhance the performance of the model. We should consider using SMOTE, undersampling, or other techniques, or else tune the hyperparameter `scale_pos_weight` in XGBoost to deal with it. Finally, run hyperparameter tuning with grid search or random search to really fine-tune the performance of the XGBoost model.

To save the trained XGBoost model in a file, we can use the `joblib` or `pickle` library in Python. Here is how we can do it:

1. Import the necessary library
   We can use `joblib` for saving the model, which is more efficient for large models. `joblib` is a library that is particularly efficient for saving and loading large Python objects, such as machine learning models. It is optimized for performance and often faster than using Python's built-in pickle module, especially with large models or large datasets.

   ```
   import joblib
   ```

2. Save the trained model to a file
   After training our XGBoost model, save it to a file with the following code:

   ```
   # Assuming xgb_model is our trained model
   joblib.dump(xgb_model, 'xgboost_alfa_model.pkl')
   ```

   The joblib.dump function is used to serialize an object, such as a trained XGBoost model, and save it to a file. In this context, xgb_model refers to the trained XGBoost model that we want to save. By this stage, the model has already been trained using our training data. The file will be saved with the name 'xgboost_alfa_model.pkl', where the .pkl extension indicates that the file is saved using the pickle or joblib library, though we can choose any filename we prefer. Saving the model allows we to persist the trained model on disk so that we can load it later and make predictions on new data without having to retrain the model.

8.5 Run the model to unseen data

When we want to test the model on another dataset, we can load it back using:

```
# Load the saved model
loaded_model = joblib.load('xgboost_alfa_model.pkl')

# Use the loaded model to make predictions on new data
# Assuming new_data is a DataFrame containing our new data
y_pred_new = loaded_model.predict(new_data)

# Continue with our evaluation on the new data
```

The joblib.load function is used to read the serialized object, such as a saved model, from a file and load it back into memory. The loaded_model variable then contains the XGBoost model that was previously saved, allowing us to use it just like any model we have trained directly. Loading the model is particularly useful because it enables us to reuse the trained model without the need for retraining, which can be time consuming. This is valuable in scenarios where training the model takes a long time or when we want to deploy the model in a production environment where retraining is not feasible.

The loaded_model.predict(new_data) method is used to make predictions on new data using the loaded model. The model will classify each instance in the new dataset as "normal" (class 0) or "fault/anomaly" (class 1). The new_data refers to the dataset on which we want to make predictions. For this purpose, we can take the data from carbonZ_2018-07-18-16-37-39_1_no_failure and carbonZ_2018-07-18-15-53-31_1_engine_failure folders for the normal and anomaly, respectively. They should be preprocessed in the same way as the training data, which means that any feature engineering, scaling, or handling of missing values applied during training should also be applied to these new data. The `y_pred_new` variable will contain an array or a list of predictions, where each element corresponds to the predicted class for each instance in the new dataset.

> **Important**

As a side note, ensure that the new data (`new_data`) we are testing follows the same structure and preprocessing as the data used to train.

8.6 Summary

This chapter describes procedures related to the processes that are followed in the diagnosis and analysis of malfunctions or failures of UAV systems. This chapter presents a theoretical and implementation-oriented insight into methodologies developed for the detection of anomalies and faults. In addition, case studies are based on public datasets, accompanied by hands-on examples utilizing Python scripts to demonstrate the application of those techniques. These methods can assist in the forensic analysis of faults and anomalies in drones/UAVs.

8.7 Exercises

1. Define the main objective of forensic investigation in UAV systems.
2. What are two examples of malfunctions that might occur in UAV systems?
3. List one methodology used for anomaly detection in UAV systems.
4. What is the significance of using public datasets in this chapter's case studies?
5. What kinds of anomalies can occur in UAVs or drones?
6. Discuss how UAV forensic methodologies differ from traditional forensic methodologies. Provide examples.
7. Analyze the challenges in detecting security breaches within UAV systems and suggest potential solutions.
8. Explain the role of machine learning in anomaly detection for UAV systems. Provide examples of algorithms that might be useful.
9. Evaluate the benefits and limitations of using case studies to validate forensic techniques for UAV faults and anomalies.
10. What is real-time anomaly detection, and why is it important for UAV operations?

References

1. R. Dhakal, et al., UAV fault and anomaly detection using autoencoders, in *2023 IEEE/AIAA 42nd Digital Avionics Systems Conference (DASC)* (2023). https://doi.org/10.1109/DASC58513.2023.10311126
2. A. Keipour, M. Mousaei, S. Scherer, ALFA: a dataset for uav fault and anomaly detection. Int. J. Robot. Res. **40**(2–3), 515–520 (2021)
3. Y. Zhang, et al., An intelligent fault detection framework for FW-UAV based on hybrid deep domain adaptation networks and the hampel filter. Int. J. Intell. Syst. **2023**(1), 6608967 (2023)
4. R. Venkataraman, et al., Comparison of fault detection and isolation methods for a small unmanned aircraft. Control Eng. Practice **84**, 365–376 (2019)
5. A. Keipour, M. Mousaei, S. Scherer, Automatic real-time anomaly detection for autonomous aerial vehicles, in *2019 International Conference on Robotics and Automation (ICRA)* (2019), pp. 5679–5685
6. M. Bronz, et al., Real-time fault detection on small fixed-wing UAVs using machine learning, in *2020 AIAA/IEEE 39th Digital Avionics Systems Conference (DASC)* (2020), pp. 1–10
7. R. Lin, E. Khalastchi, G.A. Kaminka, Detecting anomalies in unmanned vehicles using the mahalanobis distance, in *2010 IEEE International Conference on Robotics and Automation* (2010), pp. 3038–3044
8. J. Bu, et al., Integrated method for the UAV navigation sensor anomaly detection. IET Radar Sonar Navigation **11**(5), 847–853 (2017)
9. L. Yang, et al., A survey of unmanned aerial vehicle flight data anomaly detection: technologies, applications, and future directions. Sci. China Technol. Sci. **66**(4), 901–919 (2023)
10. R. Sun, et al., A novel online data-driven algorithm for detecting UAV navigation sensor faults. Sensors **17**(10), 2243 (2017)
11. K.H. Park, E. Park, H.K. Kim, Unsupervised fault detection on unmanned aerial vehicles: encoding and thresholding approach. Sensors **21**(6), 2208 (2021)
12. J. Cabahug, H. Eslamiat, Failure detection in quadcopter UAVs using K-means clustering. Sensors **22**(16), 6037 (2022)

Chapter 9
Forensic Analysis of Drone and UAV Telemetry Logs

Abstract This chapter explores the forensic analysis of telemetry logs from drones or unmanned aerial vehicles (UAVs). With the growing use of drones in civilian, commercial, and potentially malicious sectors, the need for drone forensic analysis has become increasingly important. The chapter details methods for analyzing telemetry logs, particularly focusing on logs generated by drones using the ArduPilot platform. Using GRYPHON tools, which automate the extraction and examination of telemetry and Dataflash logs, forensic investigators can reconstruct flight paths, verify command executions, and identify anomalies such as unexpected altitude changes or hardware failures. This approach provides valuable information on drone incidents and aids in accident analysis.

9.1 Introduction

Drones and unmanned aerial vehicles (UAVs) have become ubiquitous in modern society, serving purposes ranging from civilian and commercial applications to potentially malicious activities. This rapid expansion has increased the need for effective forensic analysis of telemetry logs, which serve as critical records of drone operations. Telemetry logs, particularly those generated by platforms such as ArduPilot, offer a wealth of information, including flight paths, command executions, and performance metrics. These records allow investigators to trace events, recognize irregularities, and pinpoint problems such as unanticipated altitude variations or equipment failures.

An open source tool called GRYPHON is introduced in the paper by Mantas and Patsakis [1]. It deals specifically with the forensic analysis of flight data logs, especially those generated by drones utilizing the Ardupilot platform. GRYPHON has the capability to extract and analyze forensic evidence from two types of logs: dataflash logs, which would be stored in the onboard memory of the drone, and telemetry logs generated by the ground control station or drone controller. The approach suggested here is a six-step, step-by-step procedure for analysis of Dataflash logs: verification of the integrity of the UAV's firmware, trajectory flight path investigation, verification of execution of commands, error identification, sensor data analysis, and timeline construction.

GRYPHON extracts the required data from the dataflash logs and maps GPS coordinates to find unexpected altitude variations or verify whether given commands have been executed. The authors tested GRYPHON on datasets provided by VTO Labs and provided a clear understanding into important events that might occur during a drone flight, such as crashes due to battery failure. It also provides functionalities that replay the data from telemetry logs for a more detailed look at drone behavior.

This chapter explores methodologies for analyzing drone telemetry logs using GRYPHON tools to automate extraction and examination processes. Using these tools, forensic investigators can uncover findings to understand drone incidents and enhance their investigative toolkit. Finally, this chapter is organized as follows. Section 9.2 reviews related work on drone flight logs research. Section 9.3 introduces the ArduPilot platform that is discussed in the chapter and describes its telemetry data format, including its structure and the information it encapsulates for analysis. Section 9.4 provides a practical guide to the installation and usage of GRYPHON tools. Section 9.5 presents the results of the telemetry parsing. Section 9.6 explores additional types of telemetry logs beyond the primary dataset. Finally, Sect. 9.7 summarizes the chapter.

9.2 Related Work

Recent advances in drone technology have impacted digital forensics. They need new methodologies and tools for thorough investigations, especially for telemetry logs. Various studies have emerged focusing on forensic analysis of DJI drones and other UAVs. It shows the challenges posed by proprietary data structures and encrypted file systems. These works contribute valuable frameworks and tools, such as the DRone Open source Parser (DROP) [2] and the Drone Data Parser (DRDP) [3], for the extraction and analysis of flight data and user-related metadata. Moreover, recent innovations such as Maraudrone's Map offer more intuitive visualization to assist forensic investigators by supporting newer log formats and interactive analysis of sensor data and GPS tracks [4]. Collectively, these efforts demonstrate the need for specialized forensic solutions as drone use continues to grow in civilian and unauthorized activities.

The first related effort in this area is a forensic analysis of flight logs of the DJI Phantom III drone [2]. It focuses on examining its proprietary data structures and providing methodologies for evidence extraction. Furthermore, it identifies the encrypted and encoded DAT files in the drone's internal storage and the TXT files from the mobile device used to control the drone. These files contain data such as GPS coordinates, flight details, and user-related metadata. To address these challenges, the study introduces the DRone Open source Parser (DROP), a Python-based tool developed specifically to decode these proprietary formats. Through comprehensive testing, the authors demonstrate the forensic robustness of their methods and discuss their findings, including how to best acquire drone data without compromising its integrity.

Renduchintala et al. [5] introduce a comprehensive forensic framework for UAVs or drones. This framework addresses the increasing misuse of drones in illegal activities and focuses on both hardware and digital forensics. The research proposes methodologies for investigating the physical components of drones and extracting digital evidence from flight logs. A standalone application, developed using JavaFX, is presented as a digital forensic tool to analyze log data and visualize flight paths in 2D and 3D. The study utilizes DJI Phantom 4 and Yuneec Typhoon H drones for case studies. Other research focuses on how malicious entities exploit these drones and their associated security vulnerabilities [6]. It offers a comprehensive analysis of drone roles in civilian, military, and terrorist contexts. The key aspects discussed include drone architecture, communication types, vulnerabilities, and the security risks they pose. The paper presents an overview of countermeasures for both drone protection and malicious usage detection.

Furthermore, Zhao et al. [3] explore the challenges and methodologies of digital forensics for drones specifically on the DJI Mavic 2 Pro. It addresses the need for extracting data from internal memory and encrypted file systems of drones in criminal investigations. To address these challenges, the authors develop the Drone Data Parser (DRDP), a tool designed to analyze proprietary file types such as DAT, TXT, and default files stored on drones and mobile devices. The study presents case studies to validate the tool's performance and it also demonstrates its ability to decrypt and analyze flight logs, GPS data, and user-related information.

To analyze and visualize DJI drone flight logs, Maraudrone's Map, an interactive and web-based forensic tool, is designed specifically for those who use the DATv3 log format [4]. The study extends the features of the existing DRone Open source Parser (DROP) to support newer log formats and provide a user-friendly platform for investigating incidents involving DJI drones. By integrating visualization, anomaly detection, and interactive playback functions, the tool assists digital forensic investigators in analyzing sensor data, GPS tracks, and anomalies related to drone flights. The authors demonstrate the effectiveness of the tool through evaluations using real-world flight data.

9.3 ArduPilot and Telemetry Data Format

The Ardupilot platform is an open source autopilot system that is used in a variety of UAVs, including drones, helicopters, and fixed-wing aircraft. It is a versatile platform that can be used for a variety of applications, including aerial photography, mapping, and search and rescue operations. The platform is based on the Arduino hardware and software platform, which makes it easy to modify and extend. The Ardupilot platform is widely used in the UAV community and has a large and active user base.

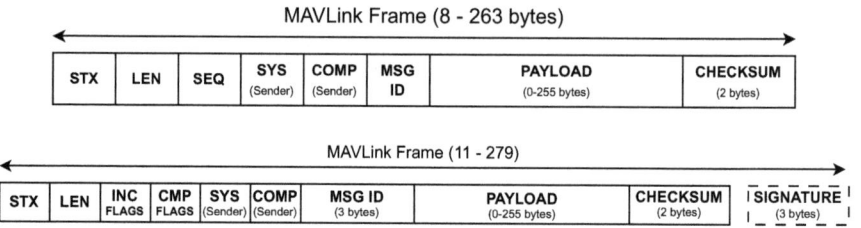

Fig. 9.1 MAVLink frame structure [1]

The MAVLink frame is the unit of communication in the MAVLink protocol, and it is widely used by drones running ArduPilot and other autopilot systems. MAVLink frames encapsulate data or commands exchanged between drones, ground control stations, and other components. The protocol ensures reliable and efficient communication, with features to detect errors, maintain synchronization, and improve security.

As shown in Fig. 9.1, each MAVLink frame begins with the Start-of-Frame (STX) indicator, which signals the start of a data packet. This is followed by the Payload Length (LEN) field, specifying the size of the data payload, ranging from 0 to 255 bytes. The Sequence Number (SEQ) field facilitates that each frame can be uniquely identified and helps to detect lost or out-of-order packets in the communication stream. To identify the sender, the frame includes a System ID (SYS) for the overall system (e.g., a specific drone) and a Component ID (COMP) for the specific component within the system (e.g., a GPS module or flight controller). The Message ID (MSG ID) defines the type of message that is transmitted, such as heartbeat signals, telemetry data, or control commands. In MAVLink version 2, this field is extended to 3 bytes to accommodate additional message types. The actual data are contained within the Payload section, which holds up to 255 bytes of information, depending on the message type. To ensure the integrity of the data, the frame includes a 2-byte Checksum to detect errors during transmission.

MAVLink version 2 introduces additional features for enhanced functionality and security. It includes Flags fields, such as Incompatibility Flags and Compatibility Flags, to indicate protocol features. It also offers an optional Signature field for authentication, adding a layer of security to prevent tampering with the communication. These upgrades make the MAVLink v2 frame slightly larger than its predecessor but more robust for modern applications. The MAVLink protocol is used for the operation of ArduPilot drones to perform tasks such as transmitting telemetry data, sending commands, and receiving system updates. For instance, heartbeat messages maintain a live connection, telemetry data provide real-time feedback into the drone's status, and command messages provide precise control of the drone's operations.

9.4 Installation and Usage of GRYPHON

This section provides a detailed guide on installing and running GRYPHON, a forensic analysis tool for UAV telemetry logs.

1. Clone the repository
   First, clone the GRYPHON repository from GitHub and navigate to the project directory:

   ```
   git clone https://github.com/emantas/GRYPHON_dft.git
   cd GRYPHON_dft
   ```

2. Create and activate a Conda environment
   Set up an isolated environment using Conda to ensure all dependencies are installed separately from other projects. Here, we specify Python 3.12 for compatibility:

   ```
   conda create --name gryphon_env python=3.12
   conda activate gryphon_env
   ```

3. Install dependencies using `requirements.txt`
   Several system libraries and Python packages are required for GRYPHON. First, install `libgtk-3-dev` to support graphics and visualization, followed by `pip3` and other essential Python packages:

   ```
   apt install libgtk-3-dev python3-pip
   pip3 install pymavlink mavproxy opencv-python wxPython
       ↪GitPython termcolor
   ```

4. Run GRYPHON
   Once installation is complete, analyze a UAV dataflash log file by running the following command, replacing `LOGFILE.bin` with the path to the log file:

   ```
   python gryphon.py LOGFILE.bin
   ```

The above steps conclude the setup for GRYPHON and we can run it as presented in Fig. 9.2. The explanation of the parsing results is provided in the next section. We should now be able to load and analyze UAV log files to assist in forensic investigations. To obtain the ArduPilot raw artifacts, follow the following steps.

1. Access the CFReDS portal
   Go to the CFReDS portal, which hosts a variety of publicly available datasets curated by NIST for digital forensic research and analysis. CFReDS, the Computer Forensic Reference Data Sets portal, provides essential resources for forensic investigators, including drone datasets.[1]

---

[1] https://cfreds.nist.gov/all/SteveWatson%2FVTOInc./DroneDataSet.

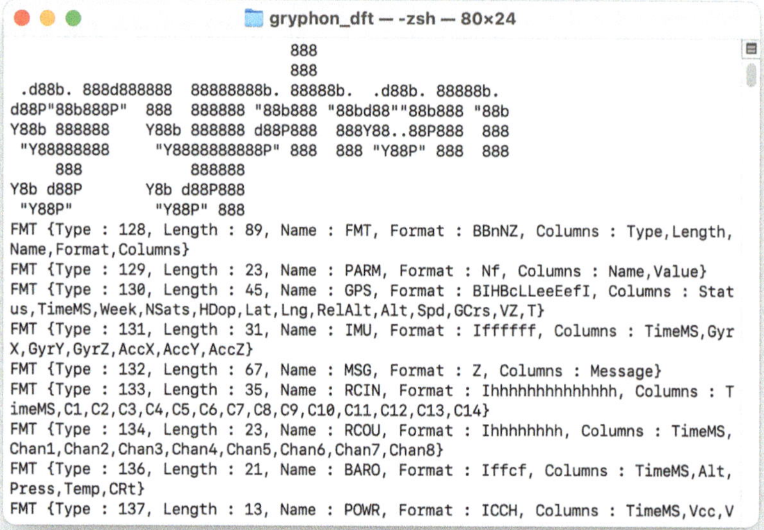

```
    ● ● ●                  gryphon_dft — -zsh — 80×24
                                 888
                                 888
     .d88b. 888d888888   88888888b. 88888b.   .d88b. 88888b.
    d88P"88b888P"   888   888888 "88b888 "88bd88""88b888 "88b
    Y88b 888888     Y88b 888888 d88P888   888Y88..88P888  888
     "Y88888888      "Y8888888888P" 888  888 "Y88P" 888   888
          888             888888
    Y8b d88P         Y8b d88P888
     "Y88P"           "Y88P" 888
    FMT {Type : 128, Length : 89, Name : FMT, Format : BBnNZ, Columns : Type,Length,
    Name,Format,Columns}
    FMT {Type : 129, Length : 23, Name : PARM, Format : Nf, Columns : Name,Value}
    FMT {Type : 130, Length : 45, Name : GPS, Format : BIHBcLLeeEefI, Columns : Stat
    us,TimeMS,Week,NSats,HDop,Lat,Lng,RelAlt,Alt,Spd,GCrs,VZ,T}
    FMT {Type : 131, Length : 31, Name : IMU, Format : Iffffff, Columns : TimeMS,Gyr
    X,GyrY,GyrZ,AccX,AccY,AccZ}
    FMT {Type : 132, Length : 67, Name : MSG, Format : Z, Columns : Message}
    FMT {Type : 133, Length : 35, Name : RCIN, Format : Ihhhhhhhhhhhhhh, Columns : T
    imeMS,C1,C2,C3,C4,C5,C6,C7,C8,C9,C10,C11,C12,C13,C14}
    FMT {Type : 134, Length : 23, Name : RCOU, Format : Ihhhhhhhh, Columns : TimeMS,
    Chan1,Chan2,Chan3,Chan4,Chan5,Chan6,Chan7,Chan8}
    FMT {Type : 136, Length : 21, Name : BARO, Format : Iffcf, Columns : TimeMS,Alt,
    Press,Temp,CRt}
    FMT {Type : 137, Length : 13, Name : POWR, Format : ICCH, Columns : TimeMS,Vcc,V
```

Fig. 9.2 A screenshot for the command: `python gryphon.py 1-12-31-1969-17-00.bin`

2. Redirect to Google Drive
   Upon clicking the link, we will be automatically redirected to a Google Drive folder that contains the specific dataset needed for this experiment. This repository provides access to various data files related to drone forensic analysis.
3. Navigate to the ArduPilot_Drone directory
   Inside the Google Drive, locate and select the directory labeled `ArduPilot_Drone`. This directory contains multiple subdirectories, each holding distinct sets of mission data and logs collected from ArduPilot-equipped drones.
4. Select a specific directory
   Within `ArduPilot_Drone`, choose a directory to explore. For example, select `df058_ArduPilot_Drone`. Inside this folder, we will find additional directories organized by date. Navigate further into the folder labeled `2018_June`, which contains drone mission data logged in June 2018.
5. Locate and download mission planner logs
   Within the `2018_June` folder, open the `Mission_Planner_Logs` directory. This directory includes log files and other critical data recorded during various drone missions. Download the file `Mission_Planner_Logs.zip` to access the complete set of mission planner logs. These logs contain detailed records of the flight data of the drone.

9.5 Telemetry Parsing Results

The telemetry logs are part of the ArduPilot drone log files, which include various data types.[2] The parsing results of GRYPHON have their own format and associated columns, as shown below.

```
FMT {Type : 128, Length : 89, Name : FMT, Format : BBnNZ,
    ↪Columns : Type,Length,Name,Format,Columns}
FMT {Type : 129, Length : 23, Name : PARM, Format : Nf,
    ↪Columns : Name,Value}
FMT {Type : 130, Length : 45, Name : GPS, Format :
    ↪BIHBcLLeeEefI, Columns :
    ↪Status,TimeMS,Week,NSats,HDop,Lat,Lng,RelAlt,Alt,
    ↪Spd,GCrs,VZ,T}
FMT {Type : 131, Length : 31, Name : IMU, Format : Ifffffff,
    ↪Columns : TimeMS,GyrX,GyrY,GyrZ,AccX,AccY,AccZ}
FMT {Type : 132, Length : 67, Name : MSG, Format : Z, Columns
    ↪: Message}
```

Here is a breakdown of what each of these log entries means:

1. FMT (Format Definition) Logs
   The FMT logs describes the format of other log entries. Each FMT log specifies the structure of subsequent entries, including details such as the type, length, name, data format, and columns. These entries act as a "header" or "blueprint" for proper interpretation of the following log data. For example, an FMT entry might be recorded as:

   ```
   Type: 128, Length: 89, Name: FMT, Format: BBnNZ, Columns:
       ↪Type,Length,Name,Format,Columns
   ```

   This entry itself is an FMT entry that defines the format for other logs.

2. PARM (Parameter) Log
   The PARM (Parameter) log records parameters and their corresponding values. Its format is Nf and it includes columns for the parameter name and value. For example, a log entry captures the name of a parameter along with its associated value.

3. GPS (Global Positioning System) Log
   The GPS log captures GPS-related data using the format BIHBcLLeeEefI and includes columns such as status, timestamp (TimeMS), week, satellite count (NSats), HDop, latitude, longitude, relative altitude (RelAlt), altitude, speed, ground course (GCrs), vertical speed (VZ), and temperature (T). For example, it records details such as GPS status, time, satellite count, position, altitude, and movement metrics.

---

[2] https://ardupilot.org/ardupilot/index.html.

4. IMU (Inertial Measurement Unit) Log
   The IMU log records sensor data from the gyroscope and accelerometer using the format `Iffffff`. It includes columns for timestamp (`TimeMS`), gyroscopic readings (`GyrX, GyrY, GyrZ`), and accelerometer readings (`AccX, AccY, AccZ`). For example, it captures the time along with gyroscopic and accelerometer measurements.
5. MSG (Message) Log
   The message log records textual messages using the format `Z`. It includes a single column for the message, which can contain important system messages or status updates.
6. RCIN (Radio Control Input) Log
   The RCIN log records input from the radio control (RC) channels using the format `Ihhhhhhhhhhhhhhh`. It includes columns for timestamp (`TimeMS`) and inputs for channels `C1` through `C14`. For example, it logs the RC inputs for each channel at a specific time.
7. RCOU (Radio Control Output) Log
   The RCOU log records the output signals sent to the drone's actuators, such as motors, using the format `Ihhhhhhhh`. It includes columns for timestamp (`TimeMS`) and output values for RC channels `Chan1` through `Chan8`. For example, it logs the output values sent to the actuators for each channel.
8. BARO (Barometer) Log
   The BARO log records barometric altitude data using the format `Iffcf`. It includes columns for timestamp (`TimeMS`), altitude (`Alt`), pressure (`Press`), temperature (`Temp`), and climb rate (`CRt`). For example, it logs altitude, pressure, temperature, and climb rate from the barometer sensor.
9. POWR (Power) Log
   The POWR log records power-related data using the format `ICCH`. It includes columns for timestamp (`TimeMS`), main power voltage (`Vcc`), servo voltage (`VServo`), and power management flags. For example, it logs the main power voltage, servo voltage, and any relevant power flags.
10. CMD (Command) Log
    The CMD log records commands issued to the drone using the `IHHHfffffff` format. It includes columns for timestamp (`TimeMS`), total commands (`CTot`), command number (`CNum`), command ID (`CId`), parameters (`Prm1` to `Prm4`), and optional location data (latitude, longitude, and altitude). For example, it logs the command type, associated parameters, and location information.
11. RAD (Radio) Log
    The RAD log records radio telemetry data using the format `IBBBBBHH`. It includes columns for timestamp (`TimeMS`), signal strength (`RSSI, RemRSSI`), transmission buffer (`TxBuf`), noise levels (`Noise, RemNoise`), received errors (`RxErrors`), and fixed packets (*Fixed*). For example, it logs signal strength, noise levels, and error data for the telemetry link.
12. CAM (Camera) Log
    The CAM log records camera trigger events along with associated GPS data using the format `IHLLeeccC`. It includes columns for GPS time (`GPSTime`),

9.5 Telemetry Parsing Results

GPS week (`GPSWeek`), latitude (`Lat`), longitude (`Lng`), altitude (`Alt`), relative altitude (`RelAlt`), and orientation data (`Roll`, `Pitch`, `Yaw`). For example, it logs when a camera is triggered along with the drone's position and orientation at that moment.

13. ATUN (Autotune) Log

    The ATUN log records data from the autotuning process for stabilization parameters using the format `BBfffff`. It includes columns for axis (`Axis`), tuning step (`TuneStep`), minimum and maximum rates (`RateMin`, `RateMax`), and gain values (`RPGain`, `RDGain`, `SPGain`). For example, it logs the tuning process for specific axes and their corresponding gains.

14. ATDE (Autotune Debug) Log

    The ATDE log records debugging information for the autotune process using the format `cf`. It includes columns for angle (`Angle`) and rate (`Rate`). For example, it logs the angle and rate during the tuning process.

15. CURR (Current) Log

    The CURR log records current consumption data using the format `IhIhhhf`. It includes columns for timestamp (`TimeMS`), throttle output (`ThrOut`), throttle input (`ThrInt`), voltage (`Volt`), current (`Curr`), main power voltage (`Vcc`), and total current consumption (`CurrTot`). For example, it logs throttle activity, voltage, current, and cumulative current consumption.

16. OF (Optical Flow) Log

    The OF log records data from the optical flow sensor using the format `hhBccee`. It includes columns for detected movement (`Dx`, `Dy`), sensor quality (`SQual`), position (`X`, `Y`), and orientation (`Roll`, `Pitch`). For example, it logs movement detected by the sensor along with orientation data.

17. NTUN (Navigation Tuning) Log

    The NTUN log records data related to the navigation tuning process using the format `If`. It logs information used for tuning navigation algorithms.

The format string `hhBccee` specifies the data types and the structure of the data in the corresponding log entry.[3] Here is a breakdown of each character in the format string:

- h: Represents a 16-bit signed integer (short). In this context, it typically means a small integer, such as a coordinate or a measurement.
- B: Represents an 8-bit unsigned integer (byte). This is usually used for small numeric values, such as quality indicators or flags.
- c: Represents a 16-bit signed integer, typically a compact or char, but it can also be interpreted as a short integer in this context.
- e: Represents a 32-bit signed integer (float). It typically indicates a floating-point number, such as a coordinate or an angle.

---

[3] https://github.com/ArduPilot/ardupilot/blob/master/libraries/AP_Logger/README.md.

So, for hhBccee, the breakdown is as follows:

- hh: Two 16-bit signed integers.
- B: One 8-bit unsigned integer.
- cc: Two 16-bit signed integers.
- ee: Two 32-bit floating-point numbers.

Each of these corresponds to a column in the telemetry data and determines how the raw bytes in the log file should be interpreted for each specific field. The columns associated with this format, according to the FMT entry, are values such as coordinates (Dx, Dy), quality (SQual), or orientation data (X, Y, Roll, Pitch).

9.6 Other Telemetry Logs

The telemetry logs are entries from an ArduPilot drone's parameter settings. Each entry corresponds to a specific system parameter and give details about its name and value at a particular time. These parameters control various aspects of the drone's operation and behavior. From GRYPHON parsing results, we have the following other results as an example.

```
PARM {Name : SYSID_SW_MREV, Value : 120.0}
PARM {Name : SYSID_SW_TYPE, Value : 10.0}
PARM {Name : SYSID_THISMAV, Value : 1.0}
PARM {Name : SYSID_MYGCS, Value : 255.0}
PARM {Name : SERIAL0_BAUD, Value : 115.0}
PARM {Name : SERIAL1_BAUD, Value : 57.0}
PARM {Name : TELEM_DELAY, Value : 0.0}
PARM {Name : RTL_ALT, Value : 1500.0}
PARM {Name : RNGFND_GAIN, Value : 0.800000011920929}
PARM {Name : FS_BATT_ENABLE, Value : 0.0}
```

Here is a breakdown of each parameter:

1. SYSID_SW_MREV (Software Mission Revision ID)
   This parameter identifies the software mission revision. It is used for version tracking and compatibility checks within the ArduPilot system.
2. SYSID_SW_TYPE (Software Type ID)
   This parameter indicates the type of ArduPilot software running on the drone, such as ArduCopter, ArduPlane, etc. Each type has a unique ID.
3. SYSID_THISMAV (MAVLink System ID)
   This is the MAVLink system ID for the drone. It uniquely identifies the drone in the MAVLink network and distinguish it from other vehicles.
4. SYSID_MYGCS (Ground Control Station ID)
   This parameter sets the ID of the Ground Control Station (GCS) that the drone is configured to communicate with. The value 255 typically represents a broadcast or default setting to enable communication with any GCS.

9.6 Other Telemetry Logs 241

5. SERIAL0_BAUD (Baud Rate for Serial Port 0)
   This parameter sets the baud rate for the first serial port (SERIAL0). It is usually used for the primary telemetry or console link. The value 115 represents 115,200 baud, which is a common communication speed.
6. SERIAL1_BAUD (Baud Rate for Serial Port 1)
   This parameter sets the baud rate for the second serial port (SERIAL1). The value 57 represents 57,600 baud, another common speed used for telemetry or other peripherals.
7. TELEM_DELAY (Telemetry Startup Delay)
   This parameter defines the delay before telemetry starts after the drone is powered on. A value of 0.0 means no delay, and telemetry starts immediately.
8. RTL_ALT (Return-to-Launch Altitude)
   This parameter sets the altitude in centimeters that the drone will climb or descend to before returning to the launch point when the Return-to-Launch (RTL) mode is activated. 1500.0 cm equals 15 meters.
9. RNGFND_GAIN (Rangefinder Gain)
   This parameter sets the gain for the rangefinder, which is used to adjust how the sensor data is interpreted for altitude or obstacle detection. The value 0.8 adjusts the sensitivity or scaling of the rangefinder readings.
10. FS_BATT_ENABLE (Battery Failsafe Enable)
    This parameter determines whether the battery failsafe is enabled. A value of 0.0 means the battery failsafe is disabled.
11. FS_BATT_VOLTAGE (Battery Failsafe Voltage)
    This is the voltage threshold at which the battery failsafe will trigger. If the battery voltage drops below 10.5 volts, the failsafe actions (such as RTL or land) may be initiated, depending on configuration.
12. FS_BATT_MAH (Battery Failsafe Capacity)
    This parameter sets the battery capacity threshold for the failsafe. If the remaining battery capacity drops below this value, the failsafe is triggered. A value of 0.0 usually means this check is disabled.
13. FS_GPS_ENABLE (GPS Failsafe Enable)
    This parameter enables or disables the GPS failsafe. A value of 0.0 means the GPS failsafe is turned off, so the drone will not automatically respond to GPS issues such as loss of signal.
14. FS_GCS_ENABLE (Ground Control Station Failsafe Enable)
    This parameter controls whether the failsafe will trigger if the Ground Control Station (GCS) connection is lost. A value of 0.0 means this failsafe is disabled.
15. GPS_HDOP_GOOD (Good HDOP Threshold for GPS)
    This parameter sets the threshold for Horizontal Dilution of Precision (HDOP) that is considered good or acceptable. Lower HDOP values indicate better GPS accuracy. The value 230.0 corresponds to 2.3, which is typically a reasonable threshold for acceptable GPS accuracy.
16. MAG_ENABLE (Magnetometer Enable)
    This parameter enables or disables the magnetometer (compass) sensor. A value of 1.0 means the magnetometer is enabled, and the drone will use it for heading information.

17. FLOW_ENABLE (Optical Flow Sensor Enable)
    This parameter enables or disables the optical flow sensor. A value of 0.0 means the optical flow sensor is disabled, and its data will not be used for position estimation.
18. SUPER_SIMPLE (Super Simple Mode Enable)
    This parameter controls the activation of Super Simple Mode, which simplifies the control of the drone's direction relative to its takeoff position. A value of 0.0 means this mode is disabled.
19. RTL_ALT_FINAL (Final Return-to-Launch Altitude)
    This parameter sets the final altitude the drone will descend to during the Return-to-Launch (RTL) procedure. A value of 0.0 usually means the drone will land at its current altitude after reaching the launch point.

9.7 Summary

In this chapter, the focus is on the forensic examination of drone telemetry logs, specifically those generated by the ArduPilot platform. The introduction highlights the need for forensic capabilities due to the increasing use of drones in various sectors. The chapter introduces GRYPHON, an open source tool designed to facilitate the forensic analysis of two key types of drone logs, specifically dataflash logs stored onboard the drone and telemetry logs transmitted to ground control stations (GCS). Through datasets from real-world operations, the chapter demonstrates how telemetry logs can reveal details about drone operations.

9.8 Exercises

1. What is the main purpose of the GRYPHON tool?
2. What platform is the focus of the telemetry log analysis in this chapter?
3. What are the two types of logs examined in the ArduPilot platform?
4. How many steps are included in the procedure for analyzing dataflash logs?
5. What does MAVLink stand for in the context of drone communication?
6. Explain the significance of the different components in a MAVLink frame structure and how do the Flags and Signature fields in MAVLink version 2 enhance communication security?
7. Compare and contrast the forensic analysis approaches of GRYPHON with other tools such as DROP and Maraudrone's Map from Chap. 7. What unique capabilities does each tool offer?
8. Analyze the various log types presented in the chapter (FMT, GPS, IMU, PARM, etc.). How do these different log types collectively contribute to a comprehensive forensic investigation of a drone's operations?

9. Critically evaluate the six-step procedure for dataflash log analysis based on the literature review discussed in this chapter. How could this procedure be adapted or challenged when dealing with logs from different drone platforms or in unique investigative scenarios?
10. Discuss the potential ethical and legal implications of advanced drone telemetry forensics. How might the detailed logging and analysis capabilities described in this chapter impact privacy, surveillance, and the use of drone technology in civilian contexts?

References

1. E. Mantas, C. Patsakis, GRYPHON: drone forensics in dataflash and telemetry logs, in *Advances in Information and Computer Security: 14th International Workshop on Security, IWSEC 2019, Tokyo, Japan, August 28–30, 2019, Proceedings 14* (Springer, 2019), pp. 377–390
2. D.R. Clark, et al., DROP (DRone Open source Parser) your drone: forensic analysis of the DJI Phantom III. Digital Investigation **22**, S3–S14 (2017)
3. Z. Zhao, Y. Wang, G. Liao, Digital forensic research for analyzing drone and mobile device: focusing on DJI mavic 2 pro. Drones **8**(7), 281 (2024)
4. T. Latzo, et al., Maraudrone's map: an interactive web application for forensic analysis and visualization of DJI drone log data, in *Nordic Conference on Secure IT Systems* (Springer. 2022), pp. 329–345
5. A. Renduchintala, et al., A comprehensive micro unmanned aerial vehicle (UAV/Drone) forensic framework. Digital Investigation **30**, 52–72 (2019)
6. J.-P. Yaacoub, et al., Security analysis of drones systems: attacks, limitations, and recommendations. Internet Things **11**, 100218 (2020)

Chapter 10
Forensic Timeline Analysis of Drones and UAVs

Abstract This chapter presents an approach for conducting forensic timeline investigations on drone images, including data from internal memory and images from iOS and Android controllers. It explores techniques for extracting and constructing forensic timelines from these sources, with a particular focus on utilizing log2timeline plaso as the primary forensic timeline tool. The chapter offers a hands-on tutorial for applying plaso to drone images and provides practical and step-by-step guidance. Finally, it demonstrates how to analyze the constructed timeline to gain a deeper understanding of the security incident under investigation.

10.1 Introduction

A forensic timeline is a tool used in the field of digital forensics to help investigators understand the sequence and timing of events that occurred on a digital device, such as a computer, smartphone, or even a drone. The creation of a forensic timeline is important to the investigation process because it provides a chronological view of all actions taken on a device, helping to identify what occurred, when it occurred, and often who was involved. In addition, this timeline is used in digital forensic investigations to help investigators understand the sequence of events, determine what happened during a specific period, and identify suspicious activities. In this chapter, we will use log2timeline plaso,[1] or plaso in short, as a forensic tool [1].

Timelines are constructed from a variety of data sources within a device. These sources include system logs, browser histories, file creation and modification dates, metadata from photos and documents, and application logs. Data from external devices, such as network logs and security camera records, can also be integrated to provide a more comprehensive timeline. Tools such as Plaso are commonly used to automate the extraction and organization of timeline data. These tools parse data from various sources and standardize timestamps. It can provide investigators with a unified view of events that can be filtered and searched to isolate relevant

---

[1] https://github.com/log2timeline/plaso

information. Forensic timelines are used to track the sequence of unauthorized actions, such as unauthorized access, malware installation, and data exfiltration. This will help investigators reconstruct the actions of the perpetrator.

This chapter is organized into several sections to guide the reader through the process of forensic timeline analysis for drone images. Section 10.2 provides an overview of related work that is relevant to forensic timeline investigations. Section 10.3 introduces the process of creating forensic timelines using log2timeline plaso. Section 10.4 describes the drone dataset used for the analysis of forensic timelines. Section 10.5 focuses on building a forensic timeline specifically from an iOS drone controller, while Sect. 10.6 explores the same process for an Android drone controller. Section 10.7 explains the analysis of the constructed timelines. Finally, Sect. 10.8 summarizes the key points discussed in the chapter.

10.2 Related Work

Forensic timeline analysis is one of the aspects of digital forensics that involves the chronological reconstruction of digital events. Researchers have extensively explored various methodologies and tools to improve timeline analysis. Previous work in this area includes the development of tools, analysis of event reconstruction, and drone incident analysis based on timeline. This section reviews contributions in drone forensic timeline analysis.

Previous work highlights that the main importance of forensic timeline analysis in drone investigations is the prime usefulness it offers in event reconstruction and linking to crime [2]. A thorough examination of the timestamps and flight data helps yield a detailed timeline that enables the investigators to recreate the sequence of events. The importance of this examination is that it connects an incident with the operational aspect of the drone. Another paper notes that forensic timeline analysis can explain drone operations or incidents [3]. When flight information is displayed in a time series format, the different events can be matched to help investigators understand how events unfolded during the flight of a drone. Such analysis forms the basis for incident reconstruction and anomaly detection and can also serve as a snapshot of the drone during its flight. The authors state that demonstrating when specific digital artifacts can be found and creating a timeline is one of the most important analyses in all digital forensic investigations involving drones, as it provides clarity to investigators about the events as they occurred over time.

Forensic timeline analysis assists drone investigations to create a chronology of activities related to drone flights [4]. It could shed light on the movement of suspects, for example, when they started stalking a victim. Since most of the process involves the removal of data, investigators can visualize a timeline using this information that removes some context from the evidence and what it means in criminal cases. In the context of drone investigations, forensic timeline analysis helps investigators determine a chronological order of events related to drone activities. This means that investigators can analyze GPS data to reconstruct flight paths in order to create a timeline connecting drone movements with specific events of interest [5]. This method contributes to a more comprehensive understanding of drone use and may support the detection of behavioral patterns relevant to crime approaches.

Timeline analysis is a field of forensics that often focuses on computers, but the authors demonstrate its applicability to drone investigations as well. This suggests that reconstructing events using a timeline analysis may provide useful information regarding the sequence of actions taken by the drone. This analysis can assist investigators in correlating different pieces of evidence and give them a better view of what is being investigated [6]. Hence, the authors suggest that timeline analysis be part of the proposed forensic model to improve the quality of drone forensic investigations. An indication that reverse engineering of the drone operation timeline can provide valuable information on the emergence of such events [7]. Establishing a clear indication of events helps facilitate the way forward for legal proceedings and also the general comprehension of drone activities. The authors of the study therefore propose timeline analysis as an addition to their proposed forensic model to further enhance the thoroughness of the investigations.

Another previous study suggests that forensic timeline analysis is a valuable approach to understanding drone-related incidents by tracing the sequence of actions within log files [8]. This timeline reconstruction process provides information on the operational history of drones, enabling investigators to detect potentially unlawful activities. The directed graph model helps to highlight important actions, while sentiment analysis helps identify suspicious events that warrant further investigation. Furthermore, the paper by Alotaibi et al. [9] discuss that forensic timeline analysis is important for understanding drone-related incidents. This analysis helps investigators reconstruct event sequences, such as flight paths, recorded data, and any unusual activities, which can provide a comprehensive overview of a drone's actions before, during, or after an incident. Establishing a clear timeline helps to pinpoint relevant evidence and helps to build a logical narrative.

In the area of application of NLP to drone forensic timeline, Silalahi et al. [10] highlight that forensic timeline analysis is used for reconstructing event sequences and identifying critical moments during a drone flight using NLP. Drone Flight Log Entity Recognizer (DFLER) contributes to this analysis by providing a structured timeline that marks key events detected in flight logs and provides an easier review of actions and potential anomalies in drone operations. By focusing on human-readable messages, DFLER allows investigators to interpret incidents with greater clarity and precision. The argument in this paper is to address a gap in drone forensic tools that previously relied mainly on raw sensor data without contextual message insights. The authors believe that the NER model can simplify the process of creating forensic timelines by highlighting key events and entities directly within the logs [11]. This approach helps investigators focus on critical information quickly and identify relevant actions and potential issues in incidents involving drones.

Another study presents DroneTimeline, a specialized tool developed for forensic timeline analysis of drones [12]. Traditional forensic tools largely focus on data from digital devices such as computers and smartphones and neglect drones and their unique requirements. The authors address this gap by introducing a system that constructs comprehensive timelines from drone data, including micro-, macro-, and super-timelines. DroneTimeline uses data from various sources—internal and external storage and control devices—to build these timelines. It allows investigators to track the activity sequence of a drone from launch to landing or incident. Subsequently, it compiles them into a unified "super timeline". This timeline simplifies event tracking by merging and visualizing data for effective analysis.

The anomalies can be detected in drone forensic timelines using the Sigma rules [13]. To help forensic investigators identify unusual behaviors on drones, the paper proposes using the log2timeline Plaso tool to create a forensic timeline of drone activities and captures details such as timestamps and event sources. Sigma rules, which are rule-based formats for log analysis, enable the detection of anomalous events by applying pre-defined keywords to the forensic timeline. These rules highlight anomalous activities, such as battery errors or weak signals. The paper also explores the advantages of rule-based anomaly detection and notes its cost effectiveness and ease of implementation. However, it acknowledges limitations in handling extreme or unknown anomalies.

Finally, another work discusses the analysis of the forensic timeline in drones as a means of understanding event sequences that lead to incidents [14]. By applying sentiment analysis, the authors argue that it becomes easier to detect specific entries within the logs that may signal an anomaly. The proposed approach can assist investigators in quickly pinpointing issues. This technique brings a new perspective to drone forensics by enabling a structured, sentiment-based analysis, potentially assisting in how drone operator data are evaluated in cases of failure or security events. This chapter focuses on drone forensic timeline analysis, while Chap. 11 describes application of NLP for drone log forensics.

10.3 Creating Forensic Timeline with log2timeline Plaso

Log2timeline is a suite of forensic tools specifically designed to assist in the creation of detailed timelines during digital forensic investigations. Log2timeline has several repositories, including: Plaso, dftimewolf, l2tdevtools, etc. One of the tools within this suite is Plaso, which is important in generating comprehensive timelines from various digital artifacts and logs found on electronic devices. Plaso is capable of extracting time-related information from a wide array of file types and standardizing these data into a uniform format. This makes it easier for forensic investigators to analyze the timeline of events, such as file accesses, system log entries, and internet browsing histories. The unified timeline created by Plaso can be applied to understand the context and sequence of activities that took place on a device.

To generate a timeline, Plaso reads from a source (such as a hard drive or a drone forensic image), processes the contained data, and outputs a timeline that can be viewed and analyzed by investigators. This output is highly customizable and can be filtered based on various parameters, such as date ranges or specific types of activity. Using this approach, forensic professionals can focus on the most relevant data in their investigations. Before using Plaso, we need to install Docker Desktop by following the steps in the next section.

10.3.1 Installation of Docker Desktop

To install Docker Desktop on the forensic workstations, follow these detailed steps to confirm that it is correctly set up and ready for use. Docker Desktop provides a user-friendly interface for managing Docker containers and images.

1. Download the installer

   - Visit the official Docker webpage at https://www.docker.com/products/docker-desktop.
   - Navigate to the Docker Desktop section and select the version of Docker Desktop that corresponds to our operating system (Windows, macOS, or Linux).
   - Click the download link for the appropriate installer. The file will typically be quite large, so it may take a few minutes depending on our internet connection speed.

2. Run the installer

   - Once the download is complete, locate the installer file in our downloads folder or wherever we have saved it.
   - Double-click on the installer file to begin the installation process and the configuration window will appear (Fig. 10.1). On Windows, the file may have a `.exe` extension, while on macOS, it might be a `.dmg` file.

3. Installation process

   - Follow the on-screen instructions to proceed with the installation.
   - During installation, we might be asked to authorize the installer or enter our system password, especially on macOS or Linux, to allow the installation to make changes.
   - Wait for the installation to complete. This process may take several minutes as Docker sets up all necessary components as shown in Fig. 10.2.

4. Verify the installation

   - After installation is successful (Fig. 10.3), we can verify that Docker Desktop is installed correctly by opening the application.
   - On Windows, we might find Docker Desktop in the Start menu and open it (Fig. 10.4). On macOS, we can look for it in the Applications folder.
   - Once opened, Docker Desktop should show a welcome screen or a tutorial if it is our first time installing Docker on this machine.

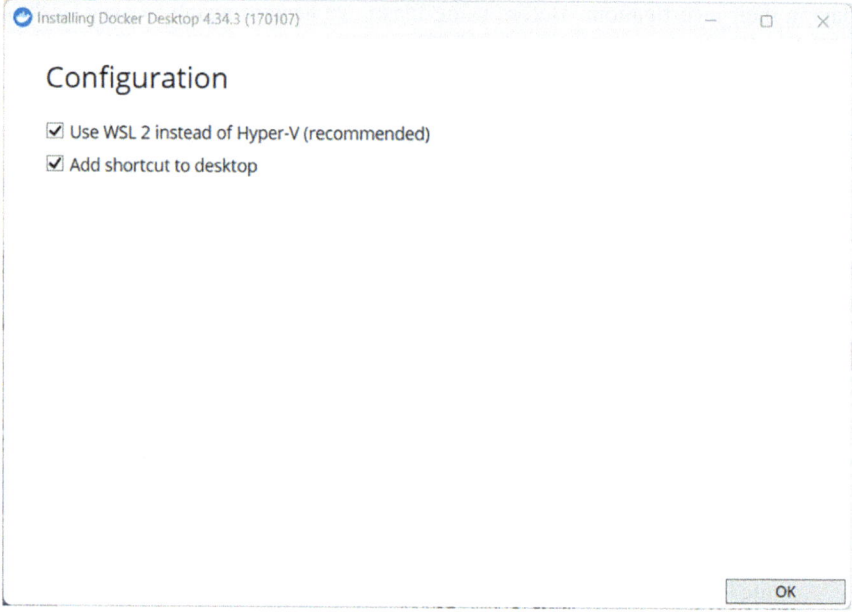

Fig. 10.1 Docker configuration before installation

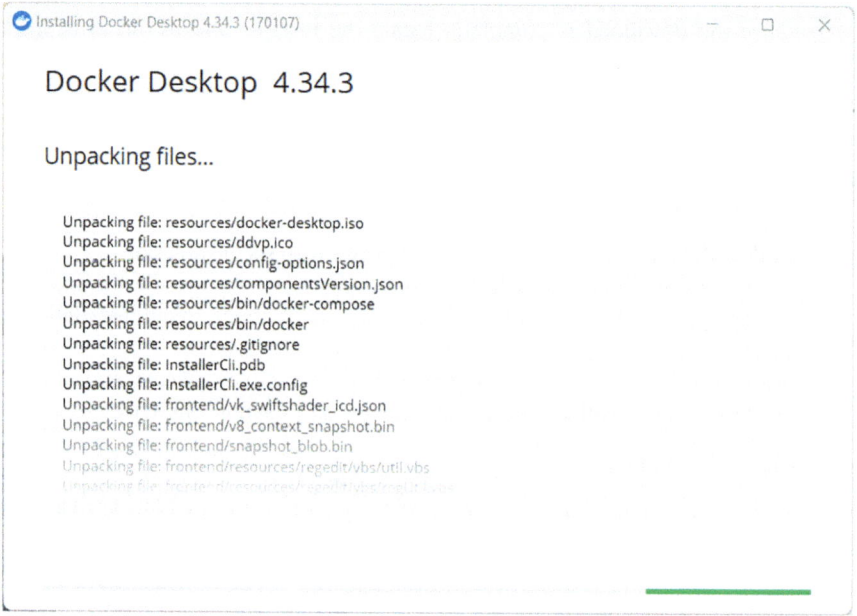

Fig. 10.2 Docker installation progress

10.3 Creating Forensic Timeline with log2timeline Plaso

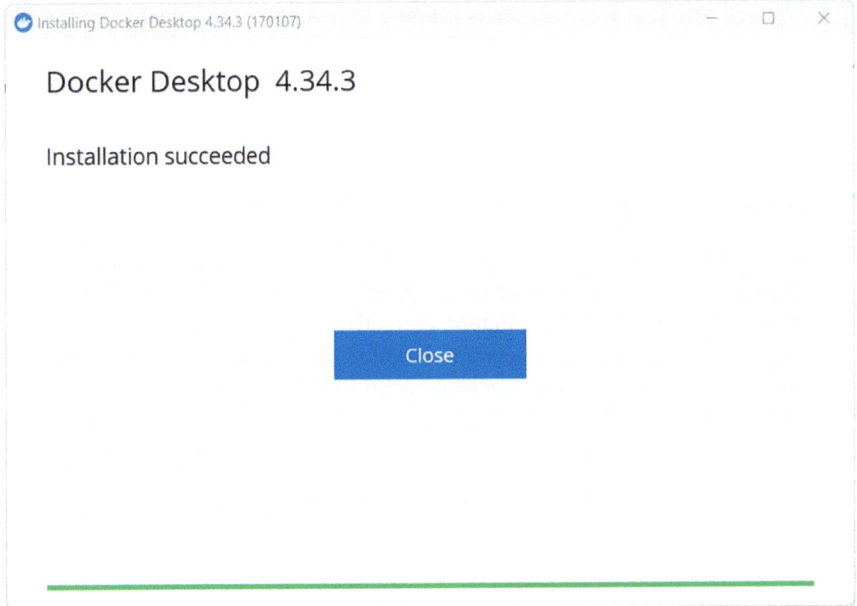

Fig. 10.3 Docker installation successful

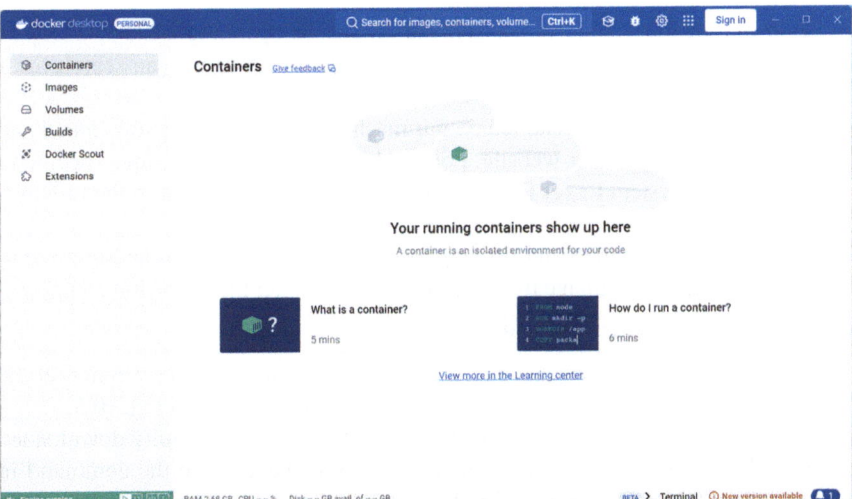

Fig. 10.4 Docker installation successful: Docker Desktop Personal

- We can also check if Docker is running by opening a terminal or command prompt and typing:

  `docker --version`

  This command should display the version of Docker installed, indicating that it is ready to use.

5. Post-installation setup
   - Depending on the operating system and Docker requirements, we may need to log out or reboot our machine to complete the installation, especially to verify that the Docker daemon is running correctly.
   - Once restarted, we may also want to adjust Docker settings such as memory and CPU allocation, which can be done through the settings/preferences menu in Docker Desktop.

10.3.2 Installation of Plaso

To install Plaso using Docker, we can follow these straightforward instructions to set up it on our system. Docker provides a containerized environment that simplifies the installation process and guarantees that Plaso runs smoothly regardless of the underlying operating system. Here is how we can get started:

1. Begin by opening the Terminal application on our operating system. This can typically be found in the utilities or applications folder, or we can search for it using the OS search feature.
2. If Docker is not already installed on our machine, we will need to download and install it from the Docker website. Follow the installation instructions specific to our operating system to get Docker set up properly as explained in the previous section.
3. Once Docker is installed and running, use the command line in the Terminal to pull the official Plaso image from the Docker Hub. Enter the following command:

   `docker pull log2timeline/plaso`

   This command communicates with Docker Hub to fetch the latest version of the Plaso container image and install it on our machine as shown in Fig. 10.5.
4. After pulling the image, we can verify that it has been successfully downloaded to our Docker environment by listing all Docker images. Use the command in the terminal: `docker images`. Look for the `log2timeline/plaso` entry in the list to confirm that it is installed. We can also check the entry in Docker Desktop as presented in Fig. 10.6.

At this stage, we will have Plaso installed in a Docker container and ready to use in investigations. Using this set-up means that Plaso runs in a controlled environment. It also minimizes issues related to dependencies or conflicts with other software on our system.

10.4 Drone Dataset for Forensic Timeline Analysis

Fig. 10.5 Downloading plaso using `docker pull` command

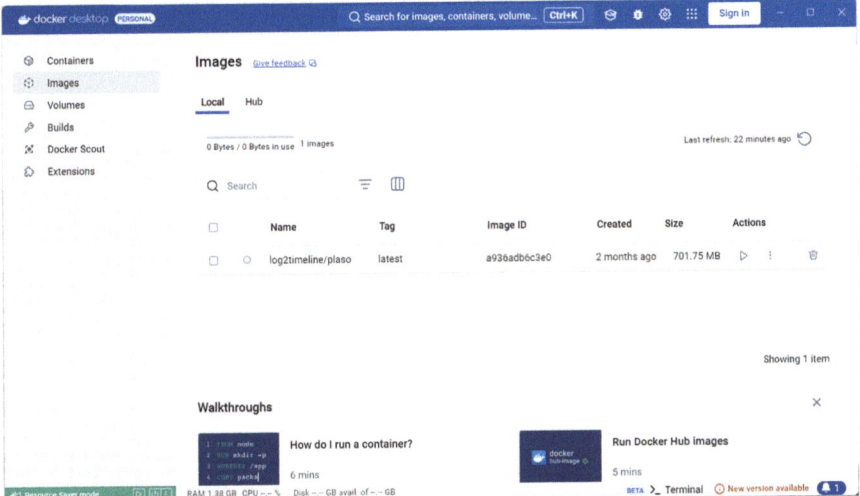

Fig. 10.6 log2timeline/plaso Docker image shown in Docker Desktop

10.4 Drone Dataset for Forensic Timeline Analysis

We are utilizing a drone dataset provided by the VTO Lab and it is accessible through the CFReDS portal hosted by NIST. This dataset can be found at CFReDS portal,[2] where it is available for download as part of the comprehensive forensic resources provided by CFReDS. This section will use DJI Mavic Air forensic images. To access the data, start by opening the CFReDS portal and navigate to the dataset link, which leads to a Google Drive folder.[3] Within Google Drive, look for the directory titled `DJI_Mavic_Air`. Once there, locate the subfolder named

---

[2] https://cfreds.nist.gov/all/SteveWatson%2FVTOInc./DroneDataSet.

[3] https://drive.google.com/drive/folders/1-UrxFGpCo54bVujwFmmqNbsZEV28dSNz?usp=sharing.

df048_DJI_Mavic_Air. Inside this subfolder, we will find another folder labeled 2018_April, which contains the specific forensic images needed for analysis.

Within the 2018_April folder, three forensic images are available for download. The first, internal_memory_intact_physical_acquisition, contains a complete physical acquisition of the drone's internal memory. This image captures the raw data from the internal memory and preserves its state for forensic examination. The second image, mobile_android_logical, is a logical acquisition of data from an Android device associated with the drone, including user data and system files accessible by logical acquisition methods. The third image, mobile_iOS_backup, is a backup of data from an iOS device associated with the drone. The backup provides access to the app data, the settings, and the user information specific to iOS.

10.5 Building a Timeline from a iOS Drone Controller

To create a comprehensive timeline from an iPhone-based drone controller, we need to utilize the log2timeline command, which will generate a Plaso storage file containing a structured timeline of events from the extracted data. This process involves mounting the local directory and the extracted data directory so that log2timeline can access and process files from the iPhone drone controller effectively.

1. Set up paths

   In this step, we need to replace the placeholders /local-directory and /ios-extracted-directory-here with the actual paths on our computer. The /local-directory path is where we want to save the resulting .plaso file, while /ios-extracted-directory-here should point to the directory where our extracted iOS files from the drone controller are stored.

2. Run the Docker command for log2timeline

   Use Docker to execute log2timeline within a Plaso container and map the specified paths so that they are accessible within the container. This log2timeline command will scan and process the extracted iOS directory, build a timeline, and save it as a .plaso file in the specified local directory. The running process is shown in Fig. 10.7.

   ```
   docker run -v /local-directory/:/data log2timeline/plaso
       ↪log2timeline --storage-file /data/timeline.plaso
       ↪/data/ios-extracted-directory-here
   ```

   Explanation of the command:

   - docker run -v /local-directory/:/data: This mounts our chosen local directory to /data within the Docker container.
   - log2timeline/plaso: This specifies the Docker image for Plaso.

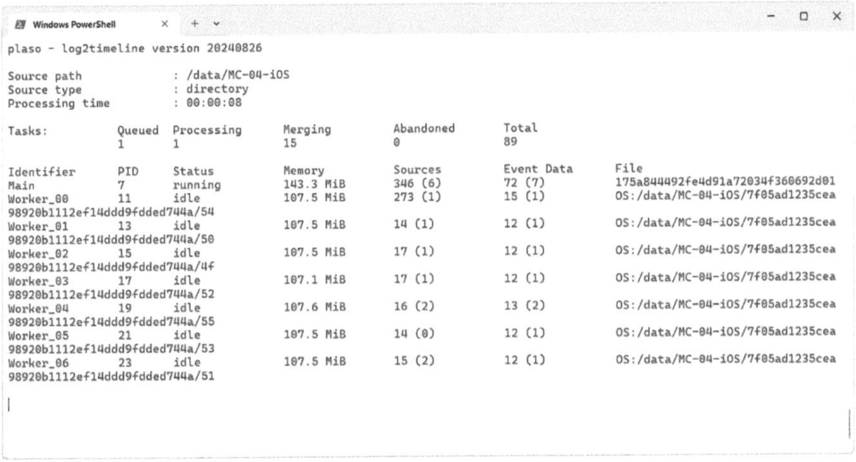

10.5 Building a Timeline from a iOS Drone Controller

Fig. 10.7 Building timeline from iOS drone controller

- `log2timeline -storage-file`: Runs the log2timeline command. The `-storage-file` option specifies the output file where the timeline data will be stored. This file is created in Plaso's own format (.plaso), which is a structured storage file that can hold a detailed and timestamped log of events extracted from the analyzed data sources.
- /data/timeline.plaso /data/ios-extracted-directory-here: Saving the output to `/data/timeline.plaso` and scanning the directory specified by `/data/ios -extracted-directory-here`.

After running this command, we should have a .plaso file in our local directory containing the event timeline. We can then proceed to analyze the timeline using other forensic tools compatible with Plaso storage files or convert the .plaso file to .csv.

3. Run the Docker command for `psort`

   To convert a Plaso storage file (.plaso) into a CSV file, we can use the `psort` command within the Plaso Docker container. This process will extract and structure the timeline data from the Plaso file into a readable and easily analyzable CSV format. Use Docker to execute `psort` within a Plaso container. This command will process the .plaso file and output a .csv file to the specified directory as presented in Fig. 10.8.

   ```
   docker run -v /local-directory/:/data log2timeline/plaso
     ↪psort -w /data/timeline.csv /data/timeline.plaso
   ```

   Explanation of the command:

   - `docker run -v /local-directory/:/data`: This mounts our local directory to `/data` inside the Docker container so that `psort` can access the .plaso file and save the output in the same location.

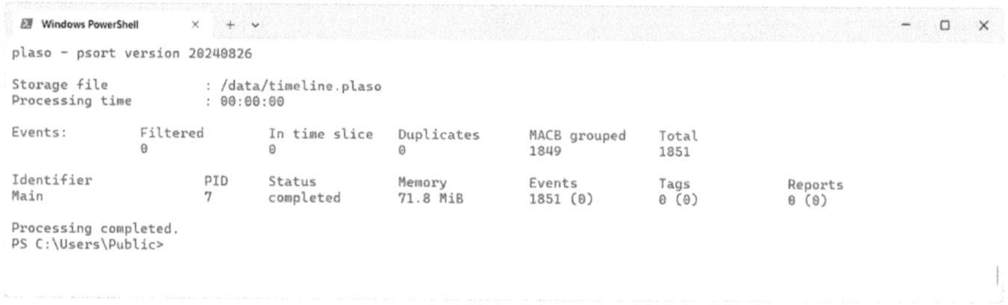

```
plaso - psort version 20240826
Storage file         : /data/timeline.plaso
Processing time      : 00:00:00

Events:     Filtered       In time slice    Duplicates     MACB grouped   Total
            0              0                0              1849           1851

Identifier          PID    Status           Memory         Events         Tags         Reports
Main                7      completed        71.8 MiB       1851 (0)       0 (0)        0 (0)

Processing completed.
PS C:\Users\Public>
```

Fig. 10.8 Running psort to convert .plaso to a CSV file

- `log2timeline/plaso`: This specifies the Docker image for Plaso.
- `psort -w`: This runs the `psort` command. It is part of the Plaso suite and allows forensic analysts to transform the data from Plaso's format to more accessible formats such as CSV for further analysis and reporting. The -w flag stands for "write" and is followed by the output file path. This file will store the extracted events in the specified format (e.g., .csv).
- `/data/timeline.csv /data/timeline.plaso`: Reading .plaso data from /data/timeline.plaso and writing the output to /data/timeline.csv.

Once the command completes, a `timeline.csv` file will be created in our local directory. This file will contain the events organized in CSV format. Investigators then can perform easy sorting, filtering, and analysis in tools such as Excel or other spreadsheet processing software.

The CSV file generated by log2timeline from a Plaso storage file contains several important columns that describe each event captured during the timeline creation process. Each column provides specific details about the event to be analyzed and the timeline data interpreted by investigators. Here is a breakdown of each column in the CSV file.

Datetime This column records the exact date and time of the event, formatted as %Y-%m-%d %H:%M:%S.%f%z. For instance, 2023-12-26 00:34:24.708138+00:00. In this datetime format, '%Y' represents the four-digit year, such as '2023', while '%m' denotes the two-digit month, with '12' representing December. The '%d' specifier corresponds to the two-digit day of the month, like '26'. For the time, '%H' indicates the two-digit hour in 24-hour format (e.g., '00'), '%M' stands for the two-digit minute (e.g., '34'), and '%S' represents the two-digit second (e.g., '24'). The '.%f' component provides microsecond precision, allowing up to six digits, such as '708138'. Finally, '%z' specifies the UTC offset in '±HHMM' format, with '+0000' indicating UTC time. It provides the specific timestamp of when the event occurred or was logged. This column is useful for arranging events in chronological order or for filtering events based on a specific time range.

The inclusion of microseconds is used for a detailed analysis of events that might occur in rapid succession, so that even tiny time differences between events are recorded. The time zone offset informs the exact global time at which the event took place, which is important when events are logged from devices in different time zones.

Timestamp_desc This column explains the nature of the timestamp (e.g., Creation Time, Last Access Time, Modification Time). It tells us what the datetime value represents in the context of the event. This field helps identify each event in terms of whether it represents a creation, access, modification, or other types of action.

Source This short code or abbreviation represents the source from which the event was extracted, such as LOG, FILE, or WEBHIST (for web history). It gives a quick idea of the origin of the event and enables users to focus on specific sources relevant to the investigation.

Source_long This is a more detailed description of the source and provides additional context. For example, it might specify Windows Event Log for an event with the source value LOG. This field provides a better view of the source and helps to interpret the nature and context of each event.

Message A human-readable summary or description of the event. This column usually contains detailed information about what happened during the event. This is one of the most informative columns, often used to understand the actual content or action associated with each event (e.g., "User logged into the system").

Parser This column specifies the name of the parser or plugin used to extract and interpret event data, such as winreg, sqlite, or chrome_history. This field helps identify which parser was responsible for interpreting the event data, which can be useful for understanding potential limitations or assumptions associated with the data.

Display_name This column shows the path or location of the file from which the event data was extracted. For example, it may point to a specific file or registry path like C:\Windows\System32\config\SAM. This field provides information on the structure of the file system and helps investigators locate the source of the data for verification or further analysis.

Tag This column allows for the tagging of events based on user-defined criteria or analysis needs. Tags can be applied to highlight certain events, such as those that are suspicious, critical, or related to a specific case aspect. It is useful for grouping, filtering, or flagging specific events during the investigation process. We can identify key events easier using tags when working with large datasets.

10.6 Building a Timeline from a Android Drone Controller

In this section, we use Docker to execute log2timeline within a Plaso container and map the specified paths so that they are accessible within the container. log2timeline then can scan and process the extracted Android directory, building

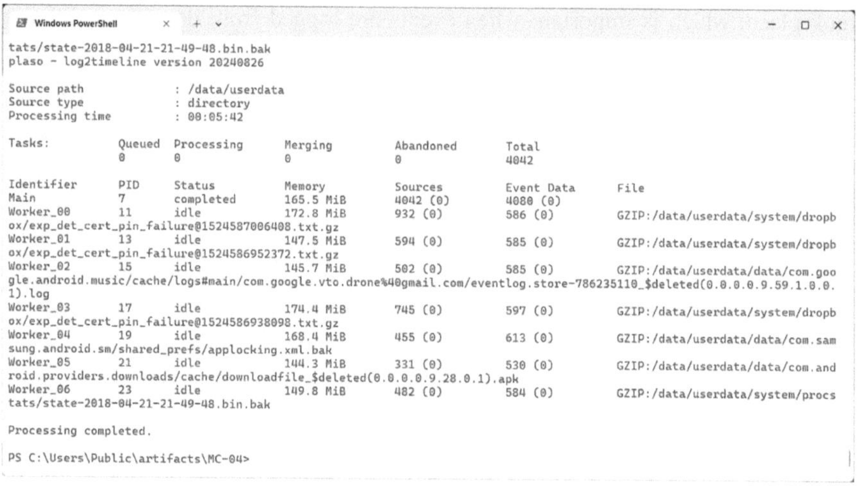

Fig. 10.9 Running plaso on Android drone controller

a timeline and saving it as a .plaso file in the specified local directory. Note that this command is very similar to the one used for an iOS drone controller.

```
docker run -v /local-directory/:/data log2timeline/plaso
    log2timeline --storage-file /data/timeline.plaso
    /data/android-extracted-directory-here
```

In this command, replace `android-extracted-directory-here` with the path to the extracted Android directory within `/local-directory` on the computer. Adjust the path as needed to match the location of our data. The result of running this command for the Android drone controller is shown in Fig. 10.9. Next, the command for `psort` is the same as that used for the iOS drone controller.

Tips

The `log2timeline` and `psort` commands can also be used in the same way to analyze the internal memory of a drone (Fig. 10.10).

10.7 Timeline Analysis

Forensic timeline analysis is a method used in digital forensics to systematically organize and examine event data chronologically. It supports the investigator in answering investigative questions about what actions took place on a digital device and when they occurred [15]. This approach provides a structured sequence of events. It is often derived from timestamps found in file systems, web activity

10.7 Timeline Analysis

```
Windows PowerShell                                                          — □ ×
plaso - log2timeline version 20240826
Source path          : /data/df048_Mavic_Air_Internal_Memory.E01
Source type          : storage media image
Processing time      : 00:01:07

Tasks:       Queued   Processing   Merging   Abandoned   Total
             0        0            0         0           26

Identifier   PID   Status      Memory       Sources   Event Data   File
Main         7     completed   153.1 MiB    26 (0)    25 (0)
Worker_00    11    idle        193.7 MiB    11 (0)    5  (0)       TSK:/DCIM/100MEDIA/DJI_0003.MP4
Worker_01    13    idle        183.8 MiB    3  (0)    5  (0)       TSK:/DCIM/100MEDIA/DJI_0002.MP4
Worker_02    15    idle        187.0 MiB    0  (0)    3  (0)       TSK:/DCIM/100MEDIA/DJI_0005.MP4
Worker_03    17    idle        117.8 MiB    10 (0)    3  (0)       TSK:/MISC/GIS/dji.gis
Worker_04    19    idle        186.6 MiB    1  (0)    2  (0)       TSK:/DCIM/100MEDIA/DJI_0004.MP4
Worker_05    21    idle        123.8 MiB    0  (0)    6  (0)       TSK:/MISC/THM/100/DJI_0005.THM
Worker_06    23    idle        177.6 MiB    0  (0)    1  (0)       TSK:/DCIM/100MEDIA/DJI_0001.MP4

Processing completed.

PS C:\Users\Public>
```

Fig. 10.10 Running plaso on the drone internal memory

logs, and other digital artifacts. By creating a timeline, investigators can reconstruct incidents, identify critical moments, and uncover potentially malicious or suspicious activity. In this section, we use log2timeline results to assist in this process in building an organized and analyzable timeline. Finally, it helps visualize the sequence and context of digital events for forensic investigations.

Based on Lin [15], in forensic timeline analysis, the process of "zooming" into specific periods around a key event, such as file deletion, is important for identifying the relevant context and sequence of actions. However, we first need to search for events of interest. Timeline analysis involves organizing collected event data to isolate moments of interest. There are several types of "zooming" as follows.

Temporal Zooming Investigators can zoom into the timeline to different levels of detail. For example, they may focus on minute-by-minute or even second-by-second events surrounding the file deletion. This level of granularity helps identify the exact actions that occurred immediately before and after deletion. This type of zooming is supported by the "datetime" column from log2timeline CSV results.

Event Type Zooming This approach allows investigators to focus specifically on particular types of events, such as file access, modification, or user logins, which are directly relevant to the investigation. Filtering for these event types allows investigators to narrow down the data set to only those actions that could have influenced the deletion event. Event types can be easily obtained using the columns "source" and "source_long" from the CSV timeline file.

Description Level Zooming This type of zooming allows investigators to dig into the metadata or specific contents of an event. This information is provided in the "message" column from the log2timeline CSV file. For instance, they can examine the user account details, application logs, or the exact path of the deleted file, which provides more context about the deletion.

After zooming in, investigators often apply filters to reduce "noise" from unrelated events. By filtering irrelevant actions, such as background processes or system updates, they focus on human-driven activities like user logins, file manipulations, or access patterns. This focused view helps detect whether unauthorized access or suspicious user actions coincide with deletion.

If deletion aligns with specific user actions, such as a log-in by an unusual account or an unexpected command executed, investigators can highlight these moments in the timeline. This helps create a clear, evidence-based narrative showing whether the deletion might have been intentional or accidental, and if an unauthorized party might have been involved.

10.8 Summary

This chapter of the book provides a structured approach to conducting forensic timeline investigations on drone-captured images. We describe the process of constructing a forensic timeline, a crucial tool in digital forensics that reveals the chronological sequence of events on a device. The chapter focuses on using the log2timeline Plaso forensic tool to build timelines from drone data and provides a hands-on tutorial for applying this tool specifically to drone images. Through the timeline, investigators can uncover details about security incidents, identify events, their timing, and potential sources of anomalies. As a side note, we want to develop a Plaso plugin which can read the content of drone flight logs so these information can be integrated to the resulted forensic timeline.

10.9 Exercises

1. What is the main purpose of forensic timeline analysis in the context of drone investigations?
2. What is Plaso, and how is it used in forensic timeline analysis?
3. What are the key steps involved in installing Docker and Plaso?
4. What is the purpose of the `log2timeline` command?
5. What is the purpose of the `psort` command?
6. How does DroneTimeline tool address the limitations of traditional forensic tools in analyzing drone data?
7. Discuss the challenges and potential solutions for detecting anomalies in drone forensic timelines using Sigma rules.
8. Explain the significance of the datetime, source, and message columns in the CSV output generated by psort.

9. How can forensic timeline analysis be used to investigate a suspected drone-related incident, such as unauthorized flight?
10. Discuss the potential for developing a custom Plaso plugin specifically designed for analyzing flight logs from a particular drone model.

References

1. J. Metz, K. Gudjonsson, D. White, et al., *log2timeline Plaso: Super timeline all the things* (2024). https://github.com/log2timeline/plaso.2024
2. D.-Y. Kao, et al., Drone forensic investigation: DJI spark drone as a case study. Proc. Comput. Sci. **159**, 1890–1899 (2019)
3. E. Mantas, C. Patsakis, GRYPHON: drone forensics in dataflash and telemetry logs, in *Advances in Information and Computer Security: 14th International Workshop on Security, IWSEC 2019, Tokyo, Japan, August 28–30, 2019, Proceedings 14* (Springer. 2019), pp. 377–390
4. S. Atkinson, et al., Drone forensics: the impact and challenges, in *Digital Forensic Investigation of Internet of Things (IoT) Devices* (2021), pp. 65–124
5. R. Kumar, A.K. Agrawal, Drone GPS data analysis for flight path reconstruction: a study on DJI, Parrot & Yuneec make drones. Forensic Sci. Int. Digital Investigation **38**, 301182 (2021)
6. A. Al-Dhaqm, et al., Research challenges and opportunities in drone forensics models. Electronics **10**(13), 1519 (2021)
7. F.M. Alotaibi, et al., A comprehensive collection and analysis model for the drone forensics field. Sensors **22**(17), 6486 (2022)
8. H. Studiawan, et al., Forensic event reconstruction for drones, in *2021 4th International Seminar on Research of Information Technology and Intelligent Systems (ISRITI)* (2021), pp. 41–45
9. F. Alotaibi, A. Al-Dhaqm, Y.D. Al-Otaibi, A conceptual digital forensic investigation model applicable to the drone forensics field. Eng. Technol. Appl. Sci. Res. **13**(5), 11608–11615 (2023)
10. S. Silalahi, T. Ahmad, H. Studiawan, DFLER: drone flight log entity recognizer to support forensic investigation on drone device. Software Impacts **15**, 100457 (2023)
11. S. Silalahi, T. Ahmad, H. Studiawan, Transformer-based named entity recognition on drone flight logs to support forensic investigation. IEEE Access **11**, 3257–3274 (2023)
12. H. Studiawan, et al., DroneTimeline: forensic timeline analysis for drones. SoftwareX **20**, 101255 (2022)
13. H. Studiawan, et al., Anomaly detection on drone forensic timeline with sigma rules, in *2023 International Conference on Emerging Smart Computing and Informatics (ESCI)* (IEEE, 2023), pp. 1–5
14. S. Silalahi, T. Ahmad, H. Studiawan, Transformer-based sentiment analysis for anomaly detection on drone forensic timeline, in *2023 11th International Symposium on Digital Forensics and Security (ISDFS)* (IEEE, 2023), pp. 1–6
15. X. Lin, Timeline analysis, in *Introductory Computer Forensics: A Hands-on Practical Approach* (Springer International Publishing, Cham, 2018), pp. 257–269. ISBN: 978-3-030-00581-8. https://doi.org/10.1007/978-3-030-00581-8_12

Chapter 11
Bringing Natural Language Processing to Drone and UAV Forensics

Abstract This chapter discusses the incorporation of natural language processing (NLP) into drone forensics to efficiently analyze human-readable drone log files. These artifacts are a rich source of evidence in unmanned aerial vehicles (UAVs). By applying NLP techniques, such as named-entity recognition and sentiment analysis, it can automate the extraction and interpretation of data from these logs. Therefore, the method improves the speed and accuracy of forensic analysis. This integration of NLP into drone forensics not only optimizes the analysis process but also opens new possibilities to improve drone security and forensics.

11.1 Introduction

The increase in drone use in commercial, recreational, and surveillance activities has expanded the scope of digital forensics. It introduces different challenges and opportunities for investigation. Drone forensic artifacts stand out due to the specialized technologies and varied data they generate. Flight logs, for example, are comprehensive records that detail each drone's flight path, altitude, speed, and status throughout its operation. These data are needed to piece together the drone's movements and activities. Moreover, telemetry data, which include real-time information such as positional coordinates, battery levels, and sensor readings, is important to understanding the drone's operational environment and conditions. Control command logs also play an important role, as they document every command sent from the controller to the drone, indicating the user's actions and intentions. Lastly, media files such as photos and videos taken by the drone are invaluable, as they can be analyzed for metadata and content. They also offer contextual information and evidence for forensic investigations.

One of the most accessible types of drone forensic artifacts is readable log file messages. Unlike binary or encoded data, these human-readable logs offer a direct window into the drone's operations and the user's interactions with the device. Logs can include information on system messages, warnings, errors, and operational data.

Using natural language processing (NLP) techniques, investigators can analyze these logs as they are in human-readable format [1]. NLP allows for automated parsing, understanding, and categorization of text-based data. Through techniques such as sentiment analysis, keyword extraction, and topic modeling, NLP can uncover patterns, anomalies, and specific details within logs that might be indicative of particular behaviors or incidents [2]. For instance, NLP analysis might reveal frequent error messages before a drone's malfunction or suggest potential tampering or hardware issues. Similarly, patterns in command logs could indicate specific user behaviors or operational profiles.

This chapter is structured to guide the reader through the application of NLP techniques in drone log forensics. Section 11.2 reviews related work on NLP in digital forensics. Section 11.3 introduces drone flight log messages, explaining their structure, types, and importance in forensic analysis. Section 11.4 discusses the use of named entity recognition (NER) to extract key entities, such as locations and timestamps, from these logs. Section 11.5 presents DFLER (Drone Flight Log Entity Recognizer), its architecture, methodology, and performance in identifying relevant entities. Section 11.6 explores the role of sentiment analysis in the interpretation of qualitative insights from drone logs. Finally, Sect. 11.7 summarizes the key points discussed in this chapter.

11.2 Related Work

In recent years, the field of application of NLP to drone forensics has seen improvements and innovations. Several studies have focused on addressing drone flight logs analysis with NLP. This section reviews the most relevant works and their contributions towards solving these challenges.

Amato et al. [3] discuss the integration of semantic methodologies and natural language processing (NLP) techniques for the analysis of digital evidence. It describes the challenges faced in correlating and interpreting digital data, especially in cases involving large volumes of information. The authors propose a structured approach that applies these technologies to improve the extraction and representation of relevant concepts from digital evidence.

Another work presents a platform based on machine learning that aims to improve digital investigations of online communications [4]. It addresses the growing need for effective tools to analyze digital interactions, particularly in the context of criminal investigations and civil litigations. The authors discuss the importance of integrating feature selection and natural language processing to improve the identification of relevant evidence from communication data. Through experiments carried out on a real-world dataset, the proposed platform demonstrates good performance compared to existing methods.

The detection of anomalies in drone flight logs is investigated, specifically focusing on classifying these anomalies based on their severity levels [5]. The authors analyze human-readable log messages generated by drones during their operation. The research builds on existing methodologies for log-based anomaly detection and extends them to include a severity classification system. The proposed

11.2 Related Work

framework categorizes anomalies into four distinct classes: normal, low, medium, and high. This classification not only aids in identifying unusual events but also allows investigators to prioritize their efforts on more severe anomalies.

Another previous study identified and annotated named entities (for example, process IDs, file extensions, and hash values) in forensic timeline message fields using a rule-based approach implemented with the spaCy library [6]. Forensic timelines, generated using tools such as log2timeline Plaso, often contain unstructured textual data that require detailed analysis for digital forensic investigations. The proposed system processes timelines, applies predefined entity recognition rules, and highlights identified entities. Subsequently, it stores the results in an SQLite database for further analysis. Experimental results, conducted on Linux and Windows disk images, demonstrate the method's effectiveness with 150 entity rules and 800 file extension rules. While rule-based NER offers advantages such as ease of customization and interpretation without training data, it faces challenges in maintaining domain-specific rules and adapting to evolving datasets. The study concludes with suggestions for future improvements, such as expanding the coverage of rules through crowdsourcing and exploring machine learning techniques based on the generated rules.

Furthermore, another study presents the DroNER dataset, which is specifically designed for named entity recognition (NER) in the context of drone forensics [7]. It addresses the lack of publicly available NER datasets for drone-related data by compiling a comprehensive collection of flight log messages from various DJI drone models. The authors detail the process of data collection, annotation, and formatting the unique challenges posed by the diverse nature of drone log messages. The dataset aims to support drone forensic research and improve the accuracy of NER systems in this domain.

In addition, Rodrigues et al. [8] explore the use of natural language processing (NLP) in digital forensics to overcome the challenges associated with the processing of large and diverse datasets. It introduces a pipeline that employs Transformer models for named entity recognition (NER) and relation extraction (RE) to construct knowledge graphs. These graphs help automate the analysis process and reduce human effort. The pipeline also incorporates hyperparameter tuning to improve performance and is designed to support multilingual applications, with a specific focus on English and Portuguese.

Another research provides a review of the role and application of natural language processing (NLP) in digital forensics and cybersecurity [9]. It highlights how NLP has become a tool for addressing the challenges posed by the increasing volume and complexity of cyber threats. It provides an overview of existing NLP-based systems, their applications in various forensic tasks, and their integration with technologies such as machine learning and deep learning. The paper also discusses challenges such as data diversity, evolving cyberattacks, and system limitations. It offers perspectives into future directions for improving NLP-based cybersecurity solutions.

Moreover, Silalahi et al. [10] focuses on applying named entity recognition (NER) in drone forensics as a relatively new domain within digital forensics.

It proposes using advanced NLP techniques to extract information from drone flight logs to assist forensic investigations. The study introduces a model named Drone Named Entity Recognition (DroNER) based on pre-trained language models, specifically BERT and DistilBERT, fine-tuned for drone forensic data. The research demonstrates the feasibility of using NER to identify important entities in drone logs and highlights its effectiveness with high F1 scores.

In the context of anomaly detection, previous work introduces a method called Drone Log seVerity (DroLoVe), designed for multiclass anomaly detection in drone flight logs, and focuses on the severity levels of anomalies [11]. It addresses the challenges of analyzing large and diverse data sets generated during drone operations and provides a severity-oriented approach to help investigators prioritize critical anomalies. Using a multitask label representation and a severity-oriented decoding procedure, the method offers improved performance in identifying and classifying anomalies. The results demonstrate that DroLoVe can improve detection accuracy and confidence compared to baseline models.

Another study in drone anomaly detection proposes the use of sentiment analysis for anomaly detection within drone flight logs [12]. By focusing on human-readable log messages, the method identifies anomalies based on sentiment, categorizing negative sentiments as potentially anomalous events. This research employs pre-trained Transformer-based language models, fine-tuned to detect these anomalies effectively. The proposed method is applied to create a forensic timeline to help investigators analyze drone incidents. The study demonstrates promising results, with the fine-tuned models achieving high accuracy in detecting anomalies.

Finally, the same authors explore the use of named entity recognition (NER) based on Transformer models in the domain of drone forensics [12]. The contextual and consistent NER discussed in the paper are illustrated in Figs. 11.1 and 11.2, respectively. It aims to help forensic investigators extract meaningful information from drone flight logs, which are often unstructured and difficult to analyze. The study develops a novel NER dataset specifically for drone forensics, defines six relevant entity types, and proposes an enhanced model that integrates cosine similarity into the Transformer's attention mechanism. The proposed approach

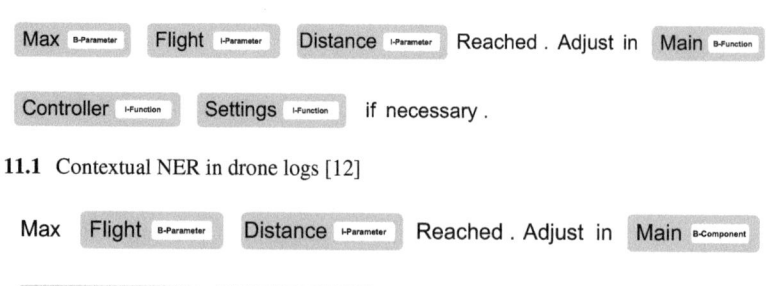

Fig. 11.1 Contextual NER in drone logs [12]

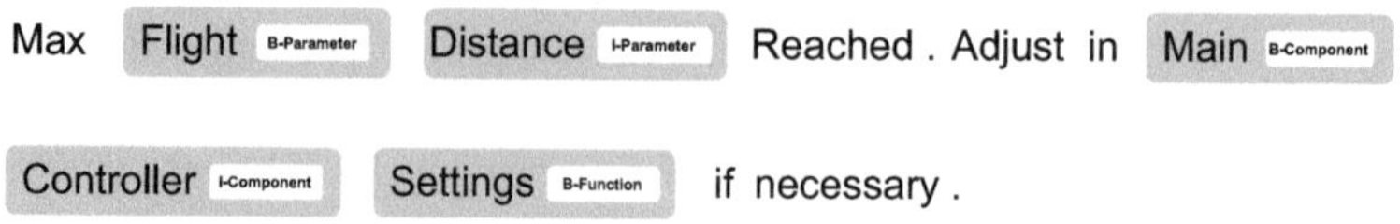

Fig. 11.2 Consistent NER in drone logs [12]

demonstrates significant performance improvements in the recognition of entities in flight logs. It achieves high F1 scores and demonstrates its potential as a tool for forensic investigations.

11.3 Drone Flight Log Messages

During a flight, the drone records various types of log, including sensors, component status, flight modes, surrounding environmental conditions, and many more. In addition to storing this information in tabular data with hundreds of columns, drone also writes human-readable messages to describe the occurring events. Where can we find these messages? Silalahi et al. [7, 12] explained the procedure of extracting human-readable messages from flight log files acquired from Android and iOS-based phone controllers. Table 11.1 shows a sample of flight log messages from various DJI-made models. The flight log message covers all events from different parts of drone devices, including hardware components: sensors, motors, gimbal, battery, controller, and camera; software features: obstacle avoidance, flight mode, auto-RTH, and GPS; configuration parameters: home point, maximum flight distance, altitude, and RTH altitude. Therefore, analyzing the flight log messages is the same as analyzing the whole drone at once. In contrast, analyzing the flight log tabular data requires more effort since each column records a certain type of flight data. For instance, to analyze the flight path taken by drone during flight, GPS records (altitude, latitude, and longitude) are the target columns.

Table 11.1 Samples of drone flight log messages

Message
...
Abnormal compass function or GPS signal detected. Aircraft switched to ATTI mode.
Battery Alert. Battery installation error. Please check the batteries are inserted correctly.
Camera sensor error. Hardware malfunction: Contact DJI Support to arrange for repairs.
Critical low battery. Aircraft landing automatically.
CrystalSky is too hot. Cool down the monitor to prevent overheating.
Forward vision sensor calibration error. Auto calibration in progress.
Gimbal motor overloaded. Check whether gimbal can rotate freely.
High wind velocity. Fly with caution and land in a safe place ASAP.
Insufficient SD card space. Change card or delete files.
Magnetic Field Interference. Exit P-GPS Mode. Yaw Error.
Mobile device CPU fully loaded. Related performance will be affected.
Max Flight Altitude Reached. Adjust in Main Controller Settings if necessary.
Remote controller battery level low. Recharge promptly.
...

Reading and interpreting human-readable logs is relatively easier and more direct than analyzing hundreds of rows of sensors value stored in tabular data. However, reading flight log messages and analyzing them manually is labor intensive and prone to inconsistency. On the other hand, automating the information extraction process can help the investigator to pinpoint event of interest faster and more efficiently, not to replace the human's expert role completely. Rather, to eliminate unrelated instances of logs, focusing more on a segment of events that are likely to be related to the incident. As shown in Table 11.1, if the investigator suspects that the incident is related to a hardware issue, the flight messages that do not describe hardware-related events can be filtered out.

Several attempts have been made on utilizing flight log messages to perform information extraction as part of the evidence analysis phase. The following subsections demonstrate two approaches on how to make use of drone flight log messages for investigative purposes.

11.4 Named Entity Recognition for Drone Log Forensics

Several drone logs are designed to be easily understood by people. They record various data points and narratives related to drone operations, such as flight paths, operational statuses, and system alerts. The human-readable format of these logs presents an opportunity to apply analytical techniques, notably named-entity recognition (NER) techniques, from the domain of natural language processing (NLP).

Named entity recognition (NER) is an NLP technique that identifies and categorizes key information in text into predefined categories or entities. In the general domain, these entities can include names of people, organizations, locations, expressions of times, quantities, monetary values, percentages, and more. Applying NER to drone logs can unveil detailed insights into the drone's operations, enhance data organization, and improve the efficiency of analyzing large volumes of text data. Consider a sample of drone log entry that reads: "Aircraft is close to the Home Point. Initiating Return to Home will now trigger Auto Landing." By applying NER to this text, we can extract valuable entities that provide structured information about the drone's flight:

- Aircraft (identified as a drone component or hardware)
- Home Point (identified as a configuration parameter)
- Return to Home (identified as an action performed by the drone)
- Auto Landing (identified as a flight or operational mode)

This processed information, structured through NER, simplifies data management and analysis. For instance, by extracting and analyzing locations, investigators can quickly map out the drone's operational areas. Similarly, by recognizing numerical values such as altitude and wind speed, it becomes easier to correlate drone performance or incidents with specific environmental conditions. NER's application extends beyond individual log analysis. Aggregating and structuring data across numerous logs can help in trend analysis, operational planning, and improving drone safety protocols. For example, if a series of logs indicates frequent wind speed warnings at certain locations or times, operators can adjust flight schedules or paths accordingly.

11.4.1 DroNER Dataset

The DroNER dataset was developed using drone flight log messages, which were sourced from drone image datasets made publicly available by VTO Labs as part of their Drone Forensic Program [13]. The construction of the dataset involved several steps: extracting, decrypting, parsing, cleaning, applying a uniqueness filter, annotating, dividing, and analyzing the data. It is presented in CoNLL format and utilizes the IOB2 tagging system to identify six types of entities. The sample of DroNER dataset is presented in Table 11.2.

A total of 1850 log messages were collected from 12 different models of DJI drones. These messages were then categorized according to the drone model, 1412 designated for training purposes, and 438 designated for testing. On average, log messages contain 6.5 words, and training and testing sets average 6.6 and 8.8 words per message, respectively. We can download the DroNER dataset at this link: https://data.mendeley.com/datasets/fwcjyc754h/1 [7].

11.4.2 Running NER on Drone Logs

The operation of named entity recognition (NER) in drone logs involves the process of automatically identifying and categorizing entities such as actions, issues, and other terms within the data collected from drone operations. This could include extracting information such as flight paths, timestamps, GPS coordinates, equipment identifiers, or mission-specific terms from logs generated by drone onboard systems. Applying NER to drone logs can assist in better data organization, enable quick retrieval of critical information, and assist in tasks such as forensic analysis, mission planning, or compliance monitoring by highlighting key entities that are relevant to specific operational scenarios.

Although many studies focus on drone frameworks, tools, and case studies, fewer explore specific data artifacts such as telemetry, dataflash, and flight logs. The research by Silalahi et al. [12] propose using log message data from drone forensic images, apply a deep learning-based NLP technique named entity recognition (NER) using the Transformer, and use cosine similarity to replace the dot product in self-attention. A novel NER architecture is evaluated using a dataset constructed from DJI drone logs. To avoid technical complexity, we will use a library called

Table 11.2 Sample of DroNER dataset annotated using consistent tagging procedure on df061 data

Id	Drone_model	Controller	Sentence_id	Words	Labels
10373	DJI_Phantom_4_Pro_V2	mobile_android_physical	1629	Backward	O
10374	DJI_Phantom_4_Pro_V2	mobile_android_physical	1629	Obstacle	B-FUNCTION
10375	DJI_Phantom_4_Pro_V2	mobile_android_physical	1629	Sensing	I-FUNCTION
10376	DJI_Phantom_4_Pro_V2	mobile_android_physical	1629	Is	O
10377	DJI_Phantom_4_Pro_V2	mobile_android_physical	1629	Not	O
10378	DJI_Phantom_4_Pro_V2	mobile_android_physical	1629	Functioning	O
10379	DJI_Phantom_4_Pro_V2	mobile_android_physical	1629	.	O
10380	DJI_Phantom_4_Pro_V2	mobile_android_physical	1630	Ambient	O
10381	DJI_Phantom_4_Pro_V2	mobile_android_physical	1630	Light	O
10382	DJI_Phantom_4_Pro_V2	mobile_android_physical	1630	Is	O
10383	DJI_Phantom_4_Pro_V2	mobile_android_physical	1630	Too	O
10384	DJI_Phantom_4_Pro_V2	mobile_android_physical	1630	Weak	O
10385	DJI_Phantom_4_Pro_V2	mobile_android_physical	1630	.	O

simpletransformers[1] that can be used to fine-tune various Transformer-based models for various downstream NLP tasks, including text classification, named entity recognition, sequence-to-sequence modeling, machine translation, question answering, and many more.

To train a NER model on a drone log dataset, we can fine-tune pre-trained language models such as BERT with the following steps.

1. Clone the GitHub repository

   First, we need to clone the GitHub repository to our local machine. Open the terminal or command prompt and execute the following command:

   ```
   git clone https://github.com/swardiantara/DroNER.git
   cd DroNER
   ```

   This command line will download the repository's contents into a directory named DroNER on our computer.

2. Create a Python virtual environment

   It is good practice to use a virtual environment to manage dependencies for our project. We can create a virtual environment using Conda by running the following command:

   ```
   conda env create -f environment.yml
   ```

   This command will create a new Conda environment named DroNER.

3. Activate the virtual environment

   After creating the virtual environment, we need to activate it. Use the following command:

   ```
   conda activate DroNER
   ```

   The terminal prompt should now indicate that we are working within the DroNER environment.

4. Train the model

   Training a supervised learning-based model require a certain amount of dataset that represent the task's problem. Since the size of the dataset is relatively small in size, we better use pre-trained models to exploit the knowledge inherited from the pre-trained dataset, which is a large corpus from general domain text. There are several pre-trained models we can use on Huggingface.[2] To start the training, issue the following command:

   ```
   python fine_tune_ner.py
   ```

   Note that the above script is designed to fine-tune several pre-trained models at once. The script can be customized to train one model for an initial experiment. The evaluation on the test set is stored within a folder named ner-results/model-name. For example, after training a bert-base-cased

---

[1] https://simpletransformers.ai/.

[2] https://huggingface.co/models.

model, the evaluation score and other files for prediction and error analysis are stored in `ner-results/bert-base-cased` directory.
5. Test the model
   Once the model training is finished, we can test the model on raw message by using below scripts.

```
from transformers import pipeline
model = pipeline('ner', model='path/to/model')

model.predict("Unknown Error, Cannot Takeoff. Contact DJI
    ↪ support.")
```

Load the fine-tuned model by changing the `path/to/model` with the actual directory where the model is stored. By default, the script stores the model in the `outputs/model-name` folder. For instance, to load the bert-base-cased model, the directory is `outputs/bert-base-cased`. The above command will result an output a list of dictionary looks like the following:

```
[{'entity': 'B-ISSUE',
  'score': np.float32(0.9898274),
  'index': 1,
  'word': 'Unknown',
  'start': 0,
  'end': 7},
 {'entity': 'I-ISSUE',
  'score': np.float32(0.9994),
  'index': 2,
  'word': 'E',
  'start': 8,
  'end': 9},
 {'entity': 'I-ISSUE',
  'score': np.float32(0.9951474),
  'index': 3,
  'word': '##rro',
  'start': 9,
  'end': 12},
 {'entity': 'I-ISSUE',
  'score': np.float32(0.9777035),
  'index': 4,
  'word': '##r',
  'start': 12,
  'end': 13},
 {'entity': 'B-ACTION',
  'score': np.float32(0.9980229),
  'index': 8,
  'word': 'Take',
```

11.4 Named Entity Recognition for Drone Log Forensics

```
  'start': 21,
  'end': 25},
 {'entity': 'I-ACTION',
  'score': np.float32(0.9903217),
  'index': 9,
  'word': '##off',
  'start': 25,
  'end': 28}]
```

The above input is a result from a bert-base-cased model. Note that some words are split into several subwords (the ones that have ## prefix) as the consequences of Wordpiece tokenization used by BERT. From the above output, we can see that the model successfully recognizes two entity spans: (Unknown Error, Issue) and (Takeoff, Action). The following details what each element in the above dictionary means.

- entity: The predicted tag in IOB2 tagging scheme.
- score: The probability score of the predicted tag.
- index: The integer index of the token in the sequence after tokenization.
- word: The literal tokenized word.
- start: The integer index of the starting character of the entity.
- end: The integer index of the ending character of the entity (exclusive).

Huggingface is a hub to host Transformer-based models on various NLP downstream tasks. This can help increase the transparency, visibility, and discoverability of our research results. After successfully fine-tuning a model in the previous step, the model can be pushed to Huggingface hub by following these steps:

1. Create a model on Huggingface. Go to the profile → New Model (Fig. 11.3).
   After creating the model, we will push the fine-tuned model to this repository. There are several ways to push a model to a repo on Huggingface. In this case, we will use the `huggingface-cli` interface. Install the library by issuing a command:

   ```
   pip install huggingface_hub
   ```

   Then, to log into Huggingface via the CLI, use the following command:

   ```
   huggingface-cli login
   ```

   It will ask the user to enter an authentication token. To get this token, go to the profile setting in Huggingface, then go to the Access Token menu. It will show us the current access token of the login-cli, as depicted in Fig. 11.4. To get the token, click the 3 dots on the right, and Invalidate and refresh. Copy the token, and paste it into the command line.

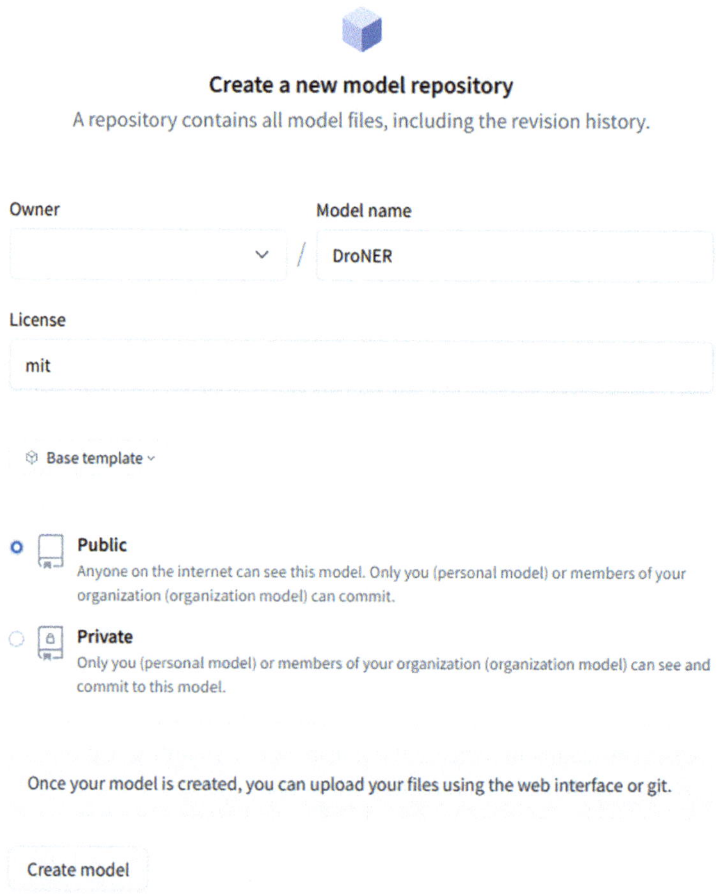

Create a new model repository

A repository contains all model files, including the revision history.

Owner / Model name: DroNER

License: mit

Base template

○ **Public**
Anyone on the internet can see this model. Only you (personal model) or members of your organization (organization model) can commit.

○ **Private**
Only you (personal model) or members of your organization (organization model) can see and commit to this model.

Once your model is created, you can upload your files using the web interface or git.

Create model

Fig. 11.3 Create a new model on Huggingface

Now, we have successfully logged into our Huggingface account via CLI. We can push the fine-tuned model by using below scripts:

```
from transformers import AutoTokenizer,
    BertForTokenClassification

# Load your model
model =
    BertForTokenClassification.from_pretrained(model_
    path)
tokenizer = AutoTokenizer.from_pretrained(model_path)

# Save it in Hugging Face's format
model.save_pretrained(model_path)
```

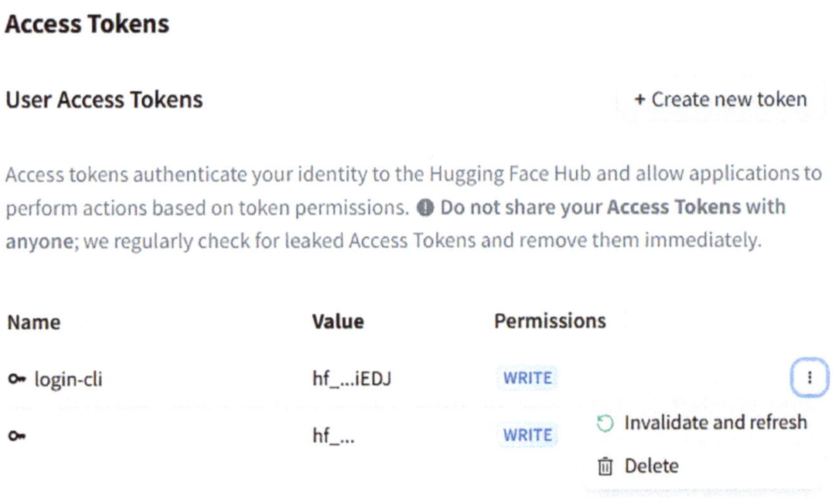

Access Tokens

User Access Tokens   + Create new token

Access tokens authenticate your identity to the Hugging Face Hub and allow applications to perform actions based on token permissions. ⓘ Do not share your **Access Tokens** with anyone; we regularly check for leaked Access Tokens and remove them immediately.

Name	Value	Permissions	
🔑 login-cli	hf_...iEDJ	WRITE	⋮
🔑	hf_...	WRITE	↻ Invalidate and refresh 🗑 Delete

Fig. 11.4 Invalidate and refresh Huggingface access token for authentication via CLI

```
tokenizer.save_pretrained(model_path)

# Push model to Hugging Face
model.push_to_hub("your-username/DroNER")
tokenizer.push_to_hub("your-username/DroNER")
```

Note that the variable `model_path` is the actual path where the fine-tuned model is stored. Since we fine-tune the model by using `simpletransformers` library, we need to save the model once again by using `transformers` module after loading it to ensure that the saved model is compatible to the Huggingface's format and dependencies. Finally, the model is ready to push to the Huggingface hub.

11.5 DFLER: Drone Flight Log Entity Recognizer

DFLER (Drone Flight Log Entity Recognizer) is a tool to perform named entity recognition (NER) on drone flight logs and aims to assist forensic investigations [14]. It uses a fine-tuned BERT model to extract entities from decrypted DJI flight logs, generates a forensic timeline, and PDF report. The tool is compatible with smartphone-based drone controllers such as Android and iOS logs. It requires several dependencies and a pre-trained model hosted on HuggingFace. To install

DFLER, follow these steps:

1. Clone the repository

   ```
   git clone https://github.com/swardiantara/dfler.git
   cd dfler
   ```

2. Create a virtual environment and activate
   To isolate the dependencies used by this tool, create a separate virtual environment and activate it by issuing the following commands:

   ```
   conda env create -f environment.yml
   conda activate DFLER
   ```

3. Ensure `wkhtmltopdf` is installed on our system
   `wkhtmltopdf` is a tool that we can use to convert an HTML file into a PDF file. The tool DFLER depends on `wkhtmltopdf` to generate the forensic report and save it into a PDF file. To install `wkhtmltopdf`, follow the instruction in the official website.[3] After the installation, make sure that the executable path is correctly configured in the `config.json` file.

4. Download the pre-trained model from HuggingFace by issuing the following commands:

   ```
   git lfs install
   git clone https://huggingface.co/swardiantara/drone-term-
       ↪extractor
   ```

   From the cloned repository, copy all the files into the /DFLER/model directory. Now, all the prerequisites have been fulfilled. To run the tool, use below command:

   ```
   python dfler.py
   ```

   After running the command, it will display a CLI-based menu to choose, as shown in Fig. 11.5. DFLER has four core features:

   - Evidence checking
     Before performing evidence analysis, the tool runs an evidence checking process to make sure that the evidence are ready to analyze. It expects flight log files acquired either from Android- or iOS-based controllers. The flight log files must be placed into `flight_logs/android` or `flight_logs/ios` folders. This steps produce several files containing the date, time, and message information from each evidence file.
   - Forensic timeline construction
     The tool will construct a forensic timeline containing timestamp and log message information extracted from the provided log files.
   - Drone entity recognition

---

[3] https://wkhtmltopdf.org/.

11.5 DFLER: Drone Flight Log Entity Recognizer

```
========================================================================
==============    Drone Flight Log Entity Recognizer   ==============
========================================================================

Action to perform:

        1. Evidence Checking
        2. Forensic Timeline Construction
        3. Drone Entity Recognition
        4. Forensic Report Generation
        0. Exit

Enter option:
```

Fig. 11.5 The user interface of DFLER

Drone Forensic Report

This report is generated on: 11/27/2024 18:40:48

Computer Name	adminuser-HURACAN-G21CN-G21CN
Report Type	Entity Recognition
Number of log files	55

Source evidence

- DJIFlightRecord_2018-06-15_(11-17-44).csv
- 19-06-2018-11VKF5500202NZ
- DJIFlightRecord_2018-06-19_(14-57-20).csv
- 15-06-2018
- DJIFlightRecord_2018-06-19_(15-34-56).csv

Fig. 11.6 The generated forensic report by DFLER after executing on the provided sample evidence

This is the main feature, where the tool will perform entity recognition on the collected log messages.

- Forensic report generation

  After performing the entity recognition, this step will generate a simple forensic report in a PDF file, as depicted in Fig. 11.6. The last page of the report shows the constructed forensic timeline along with highlighted entities on the message column, as shown in Fig. 11.7 with the color code explained in Fig. 11.8.

Upon successfully executing all the functionalities, the output is stored in /outputs/yyyymmdd_HHMMSS directory. During the evidence checking, the parsing results are stored under the /outputs/parsed directory. Meanwhile, the constructed forensic timeline from all provided evidence can be seen in the /outputs/forensic_timeline.csv. Finally, the forensic report is featured

Highlighted Forensic Timeline

Timestamp	Message
2018-06-15 11:07:46	Aircraft is in Attitude mode, so that it will not hover. Please fly with caution.
2018-06-15 11:08:00	Aircraft is in Attitude mode, so that it will not hover. Please fly with caution.
2018-06-15 11:08:02	Aircraft is in Attitude mode, so that it will not hover. Please fly with caution.
2018-06-15 11:08:15	Aircraft is in Attitude mode, so that it will not hover. Please fly with caution.
2018-06-15 11:08:19	Cannot Takeoff in Travel Mode. Exit Travel Mode.
2018-06-15 11:08:38	Cannot Takeoff in Travel Mode. Exit Travel Mode.
2018-06-15 11:08:48	Cannot Takeoff in Travel Mode. Exit Travel Mode.

Fig. 11.7 The constructed forensic timeline with highlights on successfully recognized entities

Highlights Color Code

Color Code	Entity Type	Description
■	Issue	Words/phrases that indicate some issues happen to the drone.
■	Parameter	Words/phrases that represent some parameters of configuration in a drone.
■	Action	Words/phrases that indicate some actions taken by the drone.
■	Component	Words/phrases that reflect physical components of a drone.
■	Function	Words/phrases that denote some functionalities or features of a drone equipped with.
■	State	Words/phrases that notify a state/mode of a drone operates in during a flight.

Fig. 11.8 Color code for the highlights in the forensic timeline

with forensic timeline where mentioned entities are highlighted. This report aims to assist the investigator pinpointing critical information easier and faster.

11.6 Sentiment Analysis for Drone Log Forensics

This accessibility of human-readable drone logs opens up new challenges for analysis, particularly through the view of natural language processing (NLP) techniques. Among these techniques, we choose sentiment analysis for its ability to extract the emotional tone behind text-based data. Human-readable drone logs are essentially text files that record various aspects of drone operations, including system messages, errors, and informational alerts. These logs are designed to be easily interpreted by humans. Therefore, they become a valuable resource for understanding the drone's behavior and identifying potential issues.

The application of sentiment analysis to these logs involves processing the text to detect sentiments or emotions conveyed within the messages. This could range from detecting stress or urgency in warning messages to recognizing neutral or positive sentiments in standard operational logs. Sentiment analysis employs NLP algorithms to categorize text into sentiment classes, typically positive, negative, or

11.6 Sentiment Analysis for Drone Log Forensics

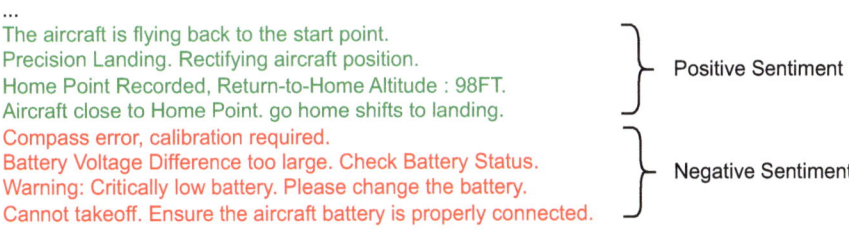

Fig. 11.9 Sentiment analysis in drone logs

neutral. Imagine a scenario where a drone log contains various messages related to its flight operations, as shown in Fig. 11.9. Several messages may include warnings about low battery levels, while others may report successful completion of a flight path or mission. By applying sentiment analysis, we can categorize these messages based on their emotional tone.

For instance, a message such as "Warning: Battery level critical, immediate landing recommended", would likely be classified as conveying a negative sentiment due to the urgency and potential risk involved. In contrast, a message stating, "Flight completed successfully with all objectives met", would be identified as positive, reflecting a successful operation without issues. The sentiment analysis process can be further refined to detect nuances in the logs, such as the urgency level of warnings or confidence in successful messages. This sentiment analysis can assist forensic investigators in preemptively identifying potential issues or confirming the reliability of the drone's operations.

11.6.1 Dataset for Sentiment Analysis for Drone Logs

We obtained the sentiment dataset for our drone log analysis from the GitHub repository.[4] This dataset provides the necessary input for our sentiment analysis tasks. This dataset enables us to evaluate and interpret sentiment within drone-related logs. In addition, the GitHub repository not only hosts the dataset but also serves as the source code for the tool we employ to perform sentiment analysis. This tool, designed for sentiment evaluation in drone logs, is integrated with the dataset. The sample dataset from `train_new.csv` under the `dataset` folder for this task is provided in Table 11.3, where label 1 represents anomalies (negative sentiment) and 0 means normal messages (positive sentiment).

---

[4] https://github.com/swardiantara/drone-sentiment/tree/main/dataset.

Table 11.3 Sample of drone sentiment dataset

Text	Labels
Taking off.	0
Camera settings adjusted to ActiveTrack	0
Home point recorded. RTH altitude: 30 m.	0
The aircraft is flying back to the start point	0
Precision landing. Rectifying aircraft position.	0
Home point recorded, return-to-home altitude: 98 FT.	0
Aircraft close to home point. go home shifts to landing.	0
Compass error, calibration required.	1
Battery voltage difference too large. Check battery status.	1
Warning: critically low battery. Please change the battery.	1
Cannot takeoff. Ensure the aircraft battery is properly connected.	1

11.6.2 Running Sentiment Analysis on Drone Logs

Sentiment analysis on drone logs is as simple as binary text classification, where the input is the raw flight log message and the output is either 0 or 1 for Normal and Anomaly class, respectively. In this part, we will fine-tune pre-trained models on our drone logs to perform binary classification as an approach to detect anomalous events. We will use `simpletransformers` library to make the training code simpler and get the model ready for our analysis as demonstrated in [12].

1. Clone the GitHub repository
   First, we need to clone the GitHub repository to our local machine. To do so, run the following command:

   ```
   git clone https://github.com/swardiantara/drone-sentiment.git
   ```

   This command line will download the repository's contents into a directory named `drone-sentiment` on the forensic workstation.

2. Create a Python virtual environment
   To manage dependencies separately from our system Python installation, we should create a virtual environment using Conda. Open our terminal or command prompt and execute the following command:

   ```
   conda env create -f environment.yml
   ```

   This command will create a new Conda environment named `droner-sentiment`. Also, all the dependencies listed in `environment.yml` file required by the tool have been installed.

3. Activate the virtual environment
   After creating the virtual environment, we need to activate it. Use the following command:

   ```
   conda activate droner-sentiment
   ```

our terminal prompt should now indicate that we are working within the `droner-sentiment` environment.

4. Run the program:
   Finally, we can run the program to fine-tune a $BERT_{base-cased}$ model for sentiment analysis. Use the following command:

   ```
   python finetune.py --model_type bert --model_name_or_path
     ↪ bert-base-cased
   ```

   This command will start the fine-tuning process using the $BERT_{base-cased}$ model. If we want to train a different variants of BERT or even different pre-trained model such as RoBERTa, we can change the argument for the above command. For instance, below are the command for fine-tuning $RoBERTa_{base}$:

   ```
   python finetune.py --model_type roberta --model_name_or_path
     ↪ roberta-base
   ```

   The model is fine-tuned using 10 epochs and learning rate of $2e-5$. We can customize these hyperparameters by modifying the source code. After the fine-tuning is finished, the resulting model is stored in the `/outputs/model-path` folder. If we perform a fine-tune on the `bert-base-cased`, then the model is stored in the following directory: `/outputs/bert-base-cased`.

5. Result and analysis
   The trained model is evaluated on the test set `test_new.csv` under the dataset folder. The model's performance and evaluation score are available in the `eval_results.txt`. This file contains information such as number of true positives, true negatives, false positives, false negatives, accuracy, F1 score, and many more, as shown in Fig. 11.10. We can use the information to compare the performance of different models to choose the best-performing one. For error analysis, we can analyze the model's prediction stored in `prediction_bert-base-cased.xlsx` for the $BERT_{base-cased}$ model, as shown in Table 11.4.

Fig. 11.10 Model performance and evaluation scores from `eval_results.txt` file

```
accuracy = 0.9230769230769231
auprc = 0.9837251327922056
auroc = 0.9686688311688312
eval_loss = 0.3740938398987055
f1_score = 0.9140057166190282
fn = 4
fp = 5
mcc = 0.8281766860861415
tn = 35
tp = 73
```

Table 11.4 Model's prediction for error analysis in `prediction_bert-base-cased.xlsx` file

Text	Labels	Preds
...	...	...
API automatic takeoff	0	0
Tap fly flight ended landing gear lowered	0	0
Compass interference. Temp max altitude: nnn	1	1
No GPS signal. Unable to hover. Fly with caution	1	1
Critical low battery voltage	1	1
Aircraft will automatically descend in nnn	1	0
Obstacle avoidance disabled. Fly with caution	1	0
GPS position mismatch.	1	0
Motor idle. Check whether propellers are installed	0	1
Compass redundancy switch	0	1
Propulsion output has been limited to ensure battery health.	0	1
Drag a box around or tap a target on screen. Then tap go.	0	1
...	...	...

6. Push model to Huggingface

   After choosing the best model, we can make the model publicly available by pushing it to the Huggingface hub. This can be done by following the exact same procedure explained in the step 6 of Sect. 11.4.2.

11.7 Summary

The chapter offers an overview of how natural language processing (NLP) techniques can improve drone forensic analysis by efficiently parsing through human-readable log files from unmanned aerial vehicles (UAVs). These logs, rich in evidence value, are made accessible for analysis using techniques such as named-entity recognition (NER) and sentiment analysis to improve the speed and accuracy of forensic investigations.

Focusing on human-readable log file messages, the chapter details how NLP can automate the extraction and interpretation of data, making it possible to uncover patterns, anomalies, and potential security threats without extensive manual effort. Techniques such as sentiment analysis and NER are specifically discussed for their ability to analyze textual data for emotional tones or specific entities, respectively. They provide insights that could indicate hardware issues, user behaviors, or operational profiles.

One of the sections is dedicated to the DroNER dataset, a resource developed from drone flight log messages and designed to facilitate NER research. This dataset, comprising 1850 log messages from various DJI drone models, is structured to train and test NER models. Therefore, it represents the practical application of

NLP in drone forensics. This chapter also touches upon the potential of sentiment analysis in forensic contexts. By analyzing the tone of log messages, investigators can identify urgent issues or confirm operational success, thus assisting in preemptive problem identification. Finally, this chapter discusses the integral role of NLP in the advancement of drone and UAV forensics. It shows how techniques such as NER and sentiment analysis can provide deeper insight into drone operations.

11.8 Exercises

1. What is the primary goal of incorporating NLP into drone forensics?
2. What are some common types of drone forensic artifacts?
3. What are some NLP techniques that can be applied to drone log analysis?
4. What is the purpose of the DroNER dataset?
5. What is the main function of the DFLER tool?
6. How can NLP techniques help identify potential tampering or malicious activity in drone operations based on log analysis?
7. Discuss the challenges of applying NLP to drone log analysis, such as data variability, noise, and the evolving nature of drone technology.
8. How can the accuracy and interpretability of NLP-based drone forensics be improved?
9. How can NLP be integrated with other forensic techniques, such as image analysis and geolocation data, to enhance drone investigations?
10. Discuss the ethical and legal implications of using NLP in drone forensics, such as privacy concerns, bias in algorithms, and the potential for misuse.

References

1. H. Studiawan, F. Sohel, C. Payne, Sentiment analysis in a forensic timeline with deep learning. IEEE Access **8**, 60664–60675 (2020)
2. H. Studiawan, F. Sohel, C. Payne, Anomaly detection in operating system logs with deep learning-based sentiment analysis. IEEE Trans. Dependable Secure Comput. **18**(5), 2136–2148 (2021)
3. F. Amato, et al., Analyse digital forensic evidences through a semantic based methodology and NLP techniques. Future Gener. Comput. Syst. **98**, 297–307 (2019)
4. D. Sun, et al., NLP-based digital forensic investigation platform for online communications. Comput. Secur. **104**, 102210 (2021)
5. S. Silalahi, T. Ahmad, H. Studiawan, Drone flight log anomaly severity classification via sentence embedding, in *2023 International Conference on Artificial Intelligence, Blockchain, Cloud Computing, and Data Analytics (ICoABCD)* (2023), pp. 100–105
6. H. Studiawan, M.F. Hasan, B.A. Pratomo, Rule-based entity recognition for forensic timeline, in *2023 Conference on Information Communications Technology and Society (ICTAS)* (2023), pp. 1–6
7. S. Silalahi, T. Ahmad, H. Studiawan. *DroNER: Dataset for Drone Named Entity Recognition* (2022). https://doi.org/10.17632/fwcjyc754h.1

8. F.B. Rodrigues, et al., Natural language processing applied to forensics information extraction with transformers and graph visualization. IEEE Trans. Comput. Soc. Syst. **11**(4), 4727–4743 (2022)
9. D.O. Ukwen, M. Karabatak, Review of NLP-based systems in digital forensics and cybersecurity, in *2021 9th International Symposium on Digital Forensics and Security (ISDFS)* (2021), pp. 1–9
10. S. Silalahi, T. Ahmad, H. Studiawan, Named entity recognition for drone forensic using BERT and DistilBERT, in *2022 International Conference on Data Science and Its Applications (ICoDSA)* (2022), pp. 53–58
11. S. Silalahi, et al., Severity-oriented multiclass drone flight logs anomaly detection. IEEE Access **12**, 64252–64266 (2024)
12. S. Silalahi, T. Ahmad, H. Studiawan, Transformer-based named entity recognition on drone flight logs to support forensic investigation. IEEE Access **11**, 3257–3274 (2023)
13. S. Silalahi, T. Ahmad, H. Studiawan, DroNER: dataset for drone named entity recognition. Data Brief **48**, 109179 (2023)
14. S. Silalahi, T. Ahmad, H. Studiawan, DFLER: drone flight log entity recognizer to support forensic investigation on drone device. Software Impacts **15**, 100457 (2023)

If you have any concerns about our products,
you can contact us on
ProductSafety@springernature.com

In case Publisher is established outside the EU,
the EU authorized representative is:
**Springer Nature Customer Service Center GmbH
Europaplatz 3, 69115 Heidelberg, Germany**

Printed by Libri Plureos GmbH
in Hamburg, Germany